Britain and the World

Britain and the World is a series of books on 'British world' history. The editors invite book proposals from historians of all ranks on the ways in which Britain has interacted with other societies from the sixteenth century to the present. The series is sponsored by the Britain and the World society.

Britain and the World is made up of people from around the world who share a common interest in Britain, its history, and its impact on the wider world. The society serves to link the various intellectual communities around the world that study Britain and its international influence from the seventeenth century to the present. It explores the impact of Britain on the world through this book series, an annual conference, and the *Britain and the World* peer-reviewed journal.

Martin Farr (martin.farr@newcastle.ac.uk) is General Series Editor for the Britain and the World book series. Michelle D. Brock (brockm@wlu.edu) is Series Editor for titles focusing on the pre-1800 period and Eric G. E. Zuelow (ezuelow@une.edu) is Series Editor for titles covering the post-1800 period.

More information about this series at
http://www.palgrave.com/gp/series/14795

William King

Nerve Agents in Postwar Britain

Deterrence, Publicity and Disarmament, 1945–1976

palgrave
macmillan

William King
German Historical Institute London
London, UK

Britain and the World
ISBN 978-3-030-70473-5 ISBN 978-3-030-70474-2 (eBook)
https://doi.org/10.1007/978-3-030-70474-2

This Palgrave Macmillan imprint is published by the registered company Springer Nature Switzerland AG.
The registered company address is: Gewerbestrasse 11, 6330 Cham, Switzerland

For my parents,
Lynn and Andy.

Acknowledgements

I am greatly indebted to all those who have supported, guided and advised me throughout the book-writing process and my recent PhD; this would not have been possible without them.

I have been very fortunate to benefit from, and participate in, an excellent PhD community in the International History Department at the LSE. I would like to especially thank Bastiaan Bouwman, Alexandre Dab, Grace Carrington, Errikos Diamantopoulos, Caroline Green, Sinan Ekim, Corina Mavrodin, Alex Mayhew, Hamish McDougall, Artemis Photiadou and Max Skjönsberg. I would particularly like to thank Eline van Ommen for all her help and friendship over the years (as well as pushing me to appreciate the broader field of international history and grass-roots movements!) and Tommaso Milani for his guidance, friendship and support.

It was a brilliant experience researching and teaching at the LSE, and I am incredibly grateful to the Department of International History, where much of this research and writing took place. Teaching has always been a particular highlight, and I am thankful to all the students whom I have had the genuine pleasure of teaching over the years, at the LSE, King's College London and now at Sciences Po. I would also like to thank Tanya Harmer, Paul Keenan and Piers Ludlow for all their advice and support over the years.

Vital administrative and archival support has greatly facilitated and aided my research. At the LSE, I would particularly like to thank Nayna Bhatti, Susana Carvalho, Demetra Frini and Jacquie Minter. For archival research, the amazing Sandy Ropper at the Harvard-Sussex Program was of immense help. I would also like to thank Mary Curry for her guidance and assistance when I was researching at the National Security Archives, as well as the archivists at the University of East Anglia, the Liddell Hart Centre for Military Archives and the United States National Archives II. I would also like to thank William Sigmund for his time and assistance—it has been fascinating to hear more about the instrumental life of a key anti-CBW campaigner, Elizabeth Sigmund. Such a study of British policy would also not have been possible without the hugely impressive UK National Archives at Kew, with its abundance of easily accessible material and efficient and friendly staff.

My interest in Cold War history and security was shaped and formulated during my undergraduate degree at Lancaster University. I will always appreciate Thomas Mills, Basil Germond, Mark Garnett, and Mark Lacy for being excellent teachers and for introducing me to the field. I am also grateful to Huw Dylan, Mike Goodman, Joe Maiolo and Peter Martland, who throughout my postgraduate studies at KCL and Cambridge encouraged and guided me, and who helped deepen and develop my understanding of history.

I would also like to thank Jeanne Guillemin and Matthew Meselson for their help and guidance. Jeanne was accommodating and supportive, and writing my first journal article with her was a formative and unforgettable experience. Roham Alvandi, Antony Best, Susan Martin and Vladislav Zubok also provided insightful observations and comments throughout this research, for which I am very grateful. I would particularly like to thank Nicholas Sims, who has always been extremely generous with his time and feedback. His comments are always highly valued, pushing me to consider further avenues and explore key interpretations, and it is always a pleasure to discuss CBW history with him. I would also like to thank the anonymous reviewers and the series editors for their helpful suggestions in further refining this book, and especially Hassan Elbahtimy for his valued comments and advice. At Palgrave Macmillan, and in the production of this book, I am grateful to Molly Beck, Vinodh Kumar V, and Ruby Panigrahi. In the past year of difficulties and delays they have been very understanding and patient.

The basis of this book is founded upon my PhD research whilst at the LSE. My two examiners, Brian Balmer and Edward Spiers, provided insightful and detailed comments on the original project. I would like to thank Edward Spiers for his in-depth questioning of the thesis and his very useful comments. Brian has been a very welcome and constant source of support and help, his guidance has been crucial to my development and progression, and I cannot thank him enough. During my time at the LSE, I had the pleasure of learning from and studying under both Matthew Jones and David Stevenson. David, my PhD adviser, has helped me formulate my ideas and push my analysis of history to the next level. His comments, advice and guidance have been greatly appreciated. Matthew, my PhD lead supervisor, has been integral to the thesis, and his insights, kindness, and meticulous feedback have been of immense value. His support and backing throughout this project have been instrumental. Any mistakes are my own.

I would like to thank everyone who has supported me throughout the writing of this book. Kanpunnarin Amphunan has always been supportive, and brightened up my days! I would like to thank my family. I am incredibly thankful to my parents, Lynn and Andy, for their unwavering support over the years, and to my brothers, Michael and James, for their support and friendship.

Contents

1	Introduction	1
2	British Defence Policy and the Nerve Agents in Postwar Britain, 1945–1950	13
3	A Step Too Far: The Nerve Agents and the Global Strategy Paper, 1951–1953	55
4	Drift, Decline, and a Stubborn Prime Minister, 1954–1957	95
5	Another Reverse Course: From Dependence to Acquisition, 1958–1963	133
6	A 'Secret Deterrent' and a 'Campaign of Criticism', 1964–1969	173
7	From the CS Debacle to the Rise of CW Disarmament, 1970–1976	213

8 Conclusion: From Discovery to Disarmament 251

Bibliography 265

Index 281

Abbreviations

BW	Biological Warfare
BWC	Biological Weapons Convention
CBW	Chemical and Biological Warfare
CDAB	Chemical Defence Advisory Board
CIA	Central Intelligence Agency
CND	Campaign for Nuclear Disarmament
CoS	Chiefs of Staff Committee
CW	Chemical Warfare
CWC	Chemical Weapons Convention
CWSC	Chemical Warfare Sub-Committee
DRPC/DRC	Defence Research Policy Sub-Committee/Defence Research Committee
FO/FCO	Foreign Office/Foreign and Commonwealth Office
GSP (1952)	The 1952 Defence Policy and Global Strategy Paper
ICI	Imperial Chemical Industries
JIC	Joint Intelligence (Sub-)Committee
MI6	Secret Intelligence Service
MoD	Ministry of Defence
MP	Member of Parliament (UK)
NATO	The North Atlantic Treaty Organization
NIE	National Intelligence Estimate (United States)
RAF	Royal Air Force
SIGINT	Signals intelligence
SIPRI	Stockholm International Peace Research Institute
UN	United Nations
WMD(s)	Weapon(s) of Mass Destruction

ABBREVIATIONS

CHAPTER 1

Introduction

In August 2020, Alexei Navalny, a Russian opposition leader and anti-corruption activist, suddenly became ill and lost consciousness on a flight to Moscow.[1] In a precarious condition, and with crucial support from the charity the 'Cinema for Peace Foundation', Navalny was flown from Omsk to Berlin for treatment.[2] Given his poor condition and allegations that he had in fact been exposed to a lethal, yet unknown, substance, fears for his health were growing. On 2 September, after investigating the causes of Navalny's sudden illness, Steffen Seibert, spokesperson for the German Federal Government, confirmed that they had 'unequivocal proof of a chemical nerve agent of the Novichok group'.[3] The internationally recognised Organisation for the Prohibition of Chemical Weapons supported this finding.[4] The ramifications are still unfolding, including diplomatic fallout and potential sanctions. Unfortunately, though, an attempted assassination with a nerve agent is not unique nor a lone incident.

[1] Harding, Luke and Roth, Andrew. 2020. A cup of tea, then screams of agony: how Alexei Navalny was left fighting for his life. *The Guardian*, August 20.

[2] Johnson, Ian P. 2020. The German NGO behind Alexei Navalny's rescue. *Deutsche Welle (DW)*, August 23.

[3] Seibert, Steffen. 2020. Der Sprecher der Bundesregierung, Erklärung der Bundesregierung im Fall Nawalny, Pressemitteilung 306, September 2.

[4] OPCW News. 2020. OPCW Issues Report on Technical Assistance Requested by Germany. October 6.

W. King, *Nerve Agents in Postwar Britain*, Britain and the World,
https://doi.org/10.1007/978-3-030-70474-2_1

The first two decades of the twenty-first century continue to see the use of chemical weapons receiving front-page newspaper coverage. Whether Salisbury or Syria, the unfortunate legacy of chemical weapons remains with us to this day.[5] Organophosphorus compounds may also appear more familiar than initially expected. Mention of sarin, VX, or Novichok could trigger recollections of recent reports of assassinations, or attempted ones, the stockpiling of these chemical warfare (CW) agents, and civilians falling victim to this gruesome form of warfare.[6] While recent accusations have been levelled against countries such as North Korea, Russia, and Syria, in the Cold War, these were weapons for mass-stockpiling and potential use by both sides of the divide.

Chemical weapons have a long history, but the mass-stockpiling of chemical weapons is a development particularly associated with the twentieth century. Chemical weapons, alongside nuclear and biological weapons and other scientific developments such as in missile technology, were a core feature of the arms races. Sarin, a lethal nerve agent, was discovered in Nazi Germany and acquired by the Allied powers during the closing stages of the Second World War, the V-agents were uncovered in Britain, and the Novichok agents were developed in the Soviet Union. All were produced in the Cold War. During the Cold War, we see controversial debates over CW deterrence, first-use, secrecy and publicity, as well as extremely controversial research with human experimentation. Scientists in Britain, the United States, France, China and the Soviet Union count among the group of countries that raced to acquire new capabilities. This rush for new developments, and the legacy of it, is reflected in recent news headlines. Historic developments and discoveries to wage war remain with us today. It is the British role, and place in this history, that this book will explore.

[5] Costanzi, Stefano & Koblentz, Gregory D. 2019 Controlling Novichoks after Salisbury: revising the Chemical Weapons Convention schedules, *The Nonproliferation Review*, 26:5–6, 599–601; Tezcür, Güneş Murat & Horschig, Doreen. 2020. A conditional norm: chemical warfare from colonialism to contemporary civil wars. *Third World Quarterly*, 42:2, 366–371.

[6] For examples, see: Shaheen, Kareem. 2016. Almost 1500 killed in chemical weapons attacks in Syria. *The Guardian*, March 11; *BBC News*. 2017. Kim Jong-nam: VX dose was "high and lethal"', February 26; Evans, Martin. 2018. Sergei Skripal: The "spy with the Louis Vuitton bag" allegedly poisoned during quiet retirement in Salisbury. *The Telegraph*, March 5; Varriale, Cristina. 2018. North Korea's Other Weapons of Mass Destruction. *Arms Control Today*, 48:7, 6–10; Ekzayez, Abdulkarim et al. 2019. Chemical Weapons and public health: assessing impact and responses. *Journal of Public Health*, 42:3, e340.

Britain, in particular, has a long history with chemical and biological weapons, stretching back far beyond the recent but crucial nerve agent development. For example, accounts have revealed plans, put forward by Lord Cochrane in the 1800s, to use 'stink-pots' or 'stink-vessels' releasing sulphur.[7] It was the First World War, though, which would witness chemical weapons taking on a previously unprecedented level of importance in policy and planning. Responding to German use, and following a significant mobilisation effort, British forces deployed chemical weapons in 1915.[8] During the conflict, the British Government also authorised the first-use of chemical weapons against the Ottoman Empire in 1917.[9] Throughout the interwar period, and following ratification of the 1925 Geneva Protocol, British capabilities remained to produce sulphur mustard. And, with the outbreak of the Second World War, CW preparations were again intensified and tensions resurfaced regarding first-use, deterrence, and the role and place of chemical weapons in warfare. By the end of the war, despite chemical weapons not being used in the military conflict in Europe, Britain had stockpiled over 40,000 tons of CW agents.[10] The history of Britain and chemical weapons is complex and controversial.

It should be noted here that not all chemical weapons are nerve agents. Other variations include incapacitating agents such as BZ, riot-control agents including CS and CN, asphyxiants such as chlorine and blister agents such as sulphur mustard. But considerations of nerve agent acquisition often took place in relative isolation (with the exception of sulphur mustard in the early postwar period) from these other forms of warfare, following a different path and holding distinct experiences and lessons.[11]

[7] Thomas, Donald. 2002. *Cochrane: Britannia's Sea Wolf.* London: Cassell Military Paperbacks, see: Chapter 6. Announce Lord Cochrane's Degradation.

[8] 'Protocol for the Prohibition of the Use in War of Asphyxiating, Poisonous or Other Gases, and of Bacteriological Methods of Warfare', Geneva, 17 June 1925.

[9] Price, Richard. 2007. *The Chemical Weapons Taboo.* London: Cornell University Press, 54–55.

[10] TNA [The National Archives, UK], PREM 3/89, Norman Brook to Churchill, 6 June 1945, Table C. For an overview of Porton Down and British CW history in this period, see: Carter, Gradon B. 2000. *Chemical and Biological Defence at Porton Down 1916–2000.* London: The Stationery Office, 1–66.

[11] Valuable accounts in these fields include, but are not limited to: Balmer, Brian. 2012. *Secrecy and Science: A Historical Sociology of Biological and Chemical Warfare.* Farnham, Surrey: Ashgate; Spelling, Alex. 2016. "Driven to Tears": Britain, CS Tear Gas, and the Geneva Protocol, 1969–1975. *Diplomacy & Statecraft*, 27:4, 701–725; Mankoo, Alex and

One of the most potent and lethal forms of chemical weapons are the nerve agents. In the Cold War, they were valued for their ability to kill, and quickly. Often odourless and colourless, nerve agents are capable, in small doses, of killing within 1–15 minutes of exposure. They can enter the body through inhalation, or absorption through the skin or the eyes.[12] As their name suggests, they affect the nervous system by disrupting the enzyme acetylcholinesterase, which inhibits the breakdown of the neurotransmitter acetylcholine.[13] This leads to a loss of control of muscles, including respiratory, and causes death by asphyxiation.[14] Their very effectiveness and lethality give them an insidious and even terrifying quality.[15]

By studying the history of the nerve agents, we can further understand the drivers and risks of weapons proliferation. The number of states possessing chemical weapons rapidly increased from 1914 to the 1970s.[16] A key period of chemical weapons proliferation, and development, was the Cold War. Through exploring this controversial history, we can determine the perceived justification of acquisition, the place and role of deterrence, the pressures of publicity and public scrutiny, and the struggles and motivations involved in achieving successful disarmament.

This book is not solely a history of weapons or of British defence policy, but one that explores public protest movements, the press, politics, and

Rappert, Brian. 2018. *Chemical Bodies: The Techno-Politics of Control.* London: Rowman & Littlefield International.

[12] Sidell, Frederick and Borak, Jonathan. 1992. Chemical Warfare Agents: II. Nerve Agents. *Annals of Emergency Medicine*, 21:7, 865–871; Szinicz, L. 2005. History of chemical and biological warfare agents. *Toxicology*, 214: 173–174; Langford, R. Everett. 2004. *Introduction to Weapons of Mass Destruction: Radiological, Chemical and Biological.* Hoboken, NJ: John Wiley & Sons, 227; Sidell, Frederick et al. 2008. Nerve Agents. In *Medical Aspects of Chemical Warfare*, ed. Martha Lenhart. Falls Church, VA: Office of the Surgeon General, 155–161, 167. It should also be noted that there are many variations of nerve agents, with the section here primarily referring to tabun, sarin and soman.

[13] McLeish, Caitríona and Balmer, Brian. 2012. Development of the V-Series Nerve Agents. In *Innovation, Dual Use, and Security: Managing the Risks of Emerging Biological and Chemical Technologies*, ed. Jonathan Tucker. Cambridge, MA: The MIT Press, 273–274.

[14] Sidell and Borak. Chemical Warfare Agents, 865–871.

[15] For further details on the nerve agents, see: Robinson, Julian Perry. 1971. *The Rise of CB Weapons.* Stockholm: Almqvist & Wiksell, 84–85; Spiers, Edward M. 1989. *Chemical Weaponry: A Continuing Challenge.* London: Macmillan, 5–7. Also, see Chapter 2 and the section on discovery.

[16] Kovačević, Damir and Afrimadona & Martin Claar. 2019. Gas, power, and norms: competing logics for the declining use of chemical warfare. *The Nonproliferation Review*, 26:3–4, 251.

secrecy, which culminate in international disarmament negotiations. It explores how Britain faced the challenges of the Cold War, the nature of international cooperation and competition, as well as domestic politics and grass-roots activism. While the scope could have been broadened out to include the entirety of CW policy, this would have diluted the analysis and detracted from the unique nature of the British nerve agent experience. Emphasis has thus been placed on the nerve agents as the bedrock of CW deterrence, the focal point of fears regarding the Soviet CW threat, the centre of the CW arms race and the topic of sustained discussion as to their actual use as a weapon of war. This is not to say significant and related areas will be neglected, as at times, as with Biological Warfare (BW), other aspects of CW policy profoundly impact British nerve agent policy during the postwar period.

In assessing the British experience with nerve agents in the Cold War, the primary focus will be on senior actors and agencies in the political and defence establishments. Rather than focusing on specific accounts of human experiments or the experiences of scientists working in the British programmes, this project has focused on the political and military aspects of the nerve agent debate and on the role of public scrutiny.[17] The leading actors, including Prime Ministers, Cabinet Defence Committee members and Service Chiefs, allow for an accurate examination of the extent, level, and role of nerve agents in British defence policy. It is at this higher level of policy formulation where the major decisions were taken, and where the policy-changing debates occurred.[18] Naturally, this will include how various high-level defence committees and officials were guided and informed, which has necessitated exploring the influence and roles of mid-level advisory committees, scientists, and experts. In order to provide a holistic account of nerve agents in postwar Britain, the impacts of publicity, grass-roots movements and political opposition will be explored, as well as international influences in the form of tripartite cooperation, NATO, and the UN. The role of publicity, though, mirrors the postwar trend in other controversial issues regarding defence science in Britain, where protest movements and public criticisms, initially starved of

[17] An excellent book investigating this dimension of British CW history is Schmidt, Ulf. 2015. *Secret Science: A Century of Poison Warfare and Human Experiments*. Oxford: Oxford University Press.

[18] Naturally a core part of this is also the policy recommendations and guidance provided to more senior politicians by defence officials and Government officials.

information due to strict censorship and secrecy measures, took on new life from the late 1950s.

Since their discovery by German scientists in the 1930s and 1940s, lethal nerve agents have added a new and controversial dimension to the potential of CW. This book will analyse the trials and tribulations of British politicians and defence officials when it came to lethal nerve agent weapons. It will further develop our understanding and comprehension of the motivations, constraints, and development of policy in a highly sensitive area. At the crux of this tumultuous British experience lay pervasive, long-running and discordant debates over acquisition, deterrence and disarmament, during which conflicted politicians and defence officials attempted to adapt to the nerve agent age. These same officials struggled to balance arguments over the military justifications for acquiring nerve agent weapons against all the negative political, economic, normative, and sometimes even military drawbacks of this contentious avenue of weapons development. These divisive discussions and detailed deliberations raged for over three decades. Nerve agents in postwar Britain caused strong disagreements between the Services and government departments, attracted severe public and political criticism, and drew the ire, support, and interventions of Prime Ministers.

At the core of this book rests five overlapping themes. Together, they will help reveal the extent of British political engagement with nerve agents, the military rationale, motives and constraints involved in nerve agent acquisition, and ultimately, how far and why Britain ventured down this controversial path. Through breaking down the secret inner workings and debates over the nerve agent question, I will seek to shed light on the complex interlinking facets of secrecy, publicity, deterrence and disarmament.

The first theme is the extent to which the controversial CW field, and more specifically the lethal nerve agents, featured in British politics and defence policy. This includes exploring the motives and reasoning behind crucial policy decisions, such as acquisition, the production of nerve agent weapons, and the eventual shift towards disarmament. By investigating this aspect of British defence policy, we can reveal the degree of military support and political engagement with lethal and controversial nerve agent weapons. A central facet of this theme will thus be to establish the extent and nature of support for nerve agent acquisition and the development of policy, ranging from more junior committees and specialised researchers to Ministers and Prime Ministers. Through tracing the progression of policy, motives for change and proliferation can be identified. Issues of first-use, internal divides between government departments and Ministers, and adherence to international treaties can thus be further appreciated. Throughout the early Cold War, the level of support for nerve agent weapons is an overlooked facet of British defence policy and, at times, surprising.

Building on the extent of British engagement in the nerve agent field, the second theme of the book is secrecy and publicity. This includes the nature and ramifications of secrecy and publicity, and of politicians and defence officials attempting to exert some form of control over the flow of information. This holds key insights for historians of British politics, military historians, and even contemporary practitioners. It is also highly complementary, as it builds upon excellent existing literature which examines secrecy and controversial science in the CBW (Chemical and Biological Warfare) field, such as by Brian Balmer and Ulf Schmidt.[19] Through establishing the nature and influence of publicity, at times intense, appreciation can be given to key figures in the field, such as the influential campaigner Elizabeth Sigmund. The public protest and anti-CBW groups that emerged in the 1960s formed a crucial part in exerting pressure on government policy. This was also a multi-faceted movement and trend, including but not limited to students, members of the press, scientists, academics, campaigners and MPs. The nature of publicity and secrecy constantly evolved throughout this Cold War period. It also mirrored broader shifts in public coverage of defence science, from respect and reverence to direct criticism and questioning. Campaigners and others attempting to hold government policy to account, though, needed information on highly secret government activities in the CW field. It is an ebb and flow of information that changes over time.

Chemical warfare deterrence, itself an underexplored area and one which is heavily impacted by secrecy and publicity, is the third central theme. In both contemporary and historical assessments, the focus has tended to be placed on the much larger field of nuclear deterrence, yet this fixation has led to valuable alternative avenues of research being overlooked. One such marginalised area is British CW deterrence. Chemical weapons deterrence, and the requirement for a nerve agent compatibility, feature prominently in British decision-making. At times, the possession

[19] Balmer, Brian. 2001. *Britain and Biological Warfare: Expert Advice and Science Policy 1935–65*. Hampshire: Palgrave; Schmidt. *Secret Science*, 1–672.

and publicity of a nerve agent capability for deterrence purposes was thought more controversial than nuclear weapons, which also taps into their perceived taboo nature. Chemical weapons deterrence was open to different interpretations. Thorough defensive preparation, such as with protective gear and training, was touted as a viable means of deterrence by some government officials, whereas others, including Prime Ministers, remained keen on retaliation-in-kind. Throughout these discussions, the ambiguous nature of CW deterrence provides a fascinating and at times bizarrely contradictory account; but it is also one which features prominently in British defence policy during the Cold War.

The fourth significant theme is one that no study of British CW policy would be complete without: analysis of international cooperation, particularly that of Anglo-American and tripartite collaboration (between Britain, the United States, and Canada). The nature and depth of cooperation in the CBW field, between the three powers, is remarkable. They shared vital research findings and willingly divided the research burden at a time when nuclear weapons cooperation was struggling. Yet, while the opportunities and benefits can be readily cited, in reality, Anglo-American cooperation was not always a smooth ride. Cooperation was often strained with tensions, frustrations, and miscommunications. These tensions take on various forms and topics, such as over publicity, CW policy in NATO, British dependency, potential deals for nerve agent weapons, and chemical weapons disarmament. A key aim will thus be to reveal not just the opportunities and military benefits, but also the limits and hindrances brought about by international cooperation.

Overlapping and intertwining with considerations of international cooperation, deterrence and secrecy is the fifth major theme, that of the divides and confrontations resulting from arguments over military utility and normative values. These clashes were often stark and unyielding. Defence officials, politicians, and Foreign Office officials all represented different sides of the debate at different times. Normative factors are taken here to refer to the legal, moral, international, and political influences and thinking, which often served to discourage or counter arguments for nerve agent acquisition and use. These arguments, or variations of them, were often put forward by Foreign Office officials keen to adhere to international norms and treaties, as well as anti-CBW campaigners, politicians, and Ministers concerned by the taboo nature of this controversial form of warfare. Military utility, referring to the military and often operational benefits of nerve agent weapons, includes arguments such as their acquisition being essential for deterrence, either in Europe or the Third World, or emphasising the military effectiveness of them against Soviet forces. The latter point would give rise to discussions of first-use, and of British adherence to the 1925 Geneva Protocol. The Cold War climate and perceptions of the Soviet Union's CW capabilities, as well as fears of more widespread global proliferation, formed an integral part of arguments regarding military utility. As such, exploring the role and influence of intelligence assessments on CW policy is key to understanding why nerve agent weapons featured strongly in British defence policy and how their acquisition was justified. The British nerve agent debate is a prime exemplar of the divide between normative values and military utility. At its heart is a debate and disagreement over the legal, moral, international and political consequences vis-à-vis military requirements, deterrence and perceptions of the Soviet threat.[20]

These themes will help show that in Britain, the nerve agent question formed an important component in the development of Cold War defence policy, and that nerve agent policy was often confusing, divisive and contradictory. Nerve agents were, and in many ways still are, a controversial topic in grass-roots movements, in government, and at the international level.

In order to support the findings and exploration of the main themes, a wide net has been cast in acquiring sources. From the UK National Archives, an abundance of archival sources has allowed for a new assessment of the involvement of Prime Ministers and defence officials in CW policy. This has been greatly aided by the release of files from the Prime Minister's Office (in particular the PREM 11 series).[21] Extensive use has also been made of sources from Cabinet (abbreviated CAB, which includes files from the Cabinet Defence Committee and the Joint Intelligence

[20] Studying the divide will also partially build upon Susan Martin's excellent work through exploring the British experience, and it will provide further insights and historical case studies for those in International Relations who analyse and draw lessons from the debates between normative values and military utility. See: Martin, Susan. 2016. Norms, Military Utility, and the Use/Non-use of Weapons: The Case of Anti-plant and Irritant Agents in the Vietnam War. *Journal of Strategic Studies*, 39:3, 321–364. For the case of Syria, see: Chapman, Geoffrey, Elbahtimy, Hassan & Martin, Susan B. 2018. The Future of Chemical Weapons: Implications from the Syrian Civil War. *Security Studies*, 27:4, 704–733.

[21] Of particular use to this book has been the files from TNA, PREM 11/49, PREM 11/3099 (of great use in assessing CW policy in the 1950s), PREM 11/3464 and PREM 11/3465.

Committee), defence (DEFE, including the Chiefs of Staff), Foreign Office (FCO, later known as the Foreign and Commonwealth Office), and War Office (WO) files. This range of government sources will fuel a recasting of our understanding of the senior levels and dynamics of political and military engagement with the nerve agent question.[22] Notably, the book has significantly benefitted from recently released archival material from the 1970s, facilitating a full re-appreciation of the decisive turning point of 1976.[23] Other archives that have been explored include the United States National Archives, the Harvard-Sussex Program Archive, the George Washington University National Security Archive, the Liddell Hart Centre for Military Archives at King's College London, and the Zuckerman Archive at the University of East Anglia.

An accusation levelled against a focus on government sources is that factors outside of the traditional scope of political and defence debates, or those simply that went unrecorded, can be overlooked. To mitigate against this, when assessing normative factors and influences which were not always categorically stated or acknowledged, I have sought to read beyond the sources and to not take them purely at face value. Documents from a wide range of sources have also been collected and incorporated to enhance and deepen the analysis. To take stock of the influence of public and political opinion, significant attention has also been paid to press

[22] CAB files of particular use include CAB 131/21, CAB 131/80, CAB 131/81 and CAB 131/158 (for the assessments of the Joint Intelligence Committee). For the DEFE files, numerous sources have been utilised, including sources from the important DEFE 4 and 5 series in the 1970s, which include the reports and minutes of meetings of the Chiefs of Staff Committee (These sources are also a core thread present throughout the entire book). Other key sources include DEFE 7/700, DEFE 7/2140, DEFE 10/265, DEFE 10/445, DEFE 10/446, DEFE 10/447 (as well as many other files in the DEFE 10 series, for the Defence Policy Research Committee, among others) which tend to include important reports and minutes of meetings for mid-level defence committees). For Service interests, particularly for the Air Ministry and the War Office, select files of use include AIR 20/9440, AIR 8/1936, AIR 2/17792, WO 188/785, WO 32/20166, and the extremely useful files for the 1960s and 1970s: WO 32/21760 and WO 32/21761. Blended throughout the book is a wide-range of archival sources from TNA, of which the above if a sample of some of the most important.

[23] Key sources for the 1970s can be found in DEFE 13/1000, DEFE 13/1056, FCO 66/220, FCO 66/312, FCO 66/391, and FCO 66/497. These FCO files also provide a key insight into the rise of chemical weapons disarmament and the back and forth with defence officials over the formulation of the 1976 draft Chemical Weapons Convention.

coverage, political memoirs, parliamentary debates and grass-roots activism.[24]

In breaking down the complicated and convoluted history of nerve agents in postwar Britain, the chapters will follow a chronological structure from 1945 through to 1976. The five central themes will be included to varying degrees throughout the chapters, as will broader events in British policy and politics in the Cold War, for these often had a direct impact and influence on nerve agent policy formulation. Each chapter will also start with a brief and broad overview of the international climate, helping to frame the perspectives of senior politicians and defence officials as well as the currents of public opinion.

Chapter 2 will explore nerve agent discovery in 1945, when Allied forces uncovered a remarkable development by Nazi Germany in the CW field. Primarily, it will trace how defence officials attempted to grapple with the problematic nerve agent development. The chapter is, therefore, more of a foundational chapter, dealing with both the legacy of war and emerging Cold War, which will account for military views and internal competition for primacy in the nerve agent field. Following this, Chap. 3 will assess one of the landmarks of British Cold War defence policy, the all-important 1952 Global Strategy Paper. This chapter will bring to the fore debates and clashes over normative values, military utility, and prospects for the first-use of chemical weapons. Chapter 4, spanning 1954–1957, will then build on this through investigating the effects of the Global Strategy Paper in the ever-changing Cold War climate, and the impact of the Suez Crisis. A core part of this chapter will be an examination of the influence of Prime Minister Anthony Eden on the direction and course of British CW policy.

[24]This book has utilised a wide-range of sources and it has attempted to broaden out from purely government sources, so as to provide a more complete history of the British nerve agent experience. In this endeavour, Hansard has been used extensively to ascertain political opinion, as often parliamentary questions either reflect public concerns, or those of the politicians themselves, both of which are important avenues for consideration. In addition, newspapers have been used widely throughout the book, including sources from *The Times*, *The Guardian* and *The Telegraph*, to name a select few. Biographies and autobiographies have also been explored to add further depth to the political debates and defence policy considerations (some of which have been referred to in the literature review). This book has also made use of personal accounts by CBW protestors, one key text for grass-roots activism is by Elizabeth Sigmund, see: Sigmund, Elizabeth. 1980. *Rage Against the Dying: Campaign Against Chemical and Biological Warfare*. London: Pluto Press.

Chapter 5 covers 1958–1963, during which time British CW policy was characterised by growing support for greater interdependence with the United States, a spate of negative publicity, and, arguably most significantly, the 'Zuckerman effect'. Sir Solly Zuckerman, as Chief Scientific Advisor at the MoD after 1960, exerted considerable sway over the direction of British nerve agent policy. Chapter 6 traces and scrutinises the consequences of renewed and reinvigorated support for nerve agent weapons, and it addresses how government officials handled a tide of negative publicity. In addition, it will assess fears of chemical weapons proliferation in the Third World, which had roots in the Cuban Missile Crisis. Spanning the period from 1964–1969, the chapter brings to the fore publicity, controversial debates over CW deterrence, grass-roots activism and secrecy. Chapter 7 begins with the CS debacle and the growing influence of détente on the CBW field. It culminates with the rise of disarmament talks and the decisive year of 1976. When, military perspectives decisively shifted, and the British nerve agent debate was all but settled, with the oscillating and tempestuous course taken since 1945 finally coming to an end.

CHAPTER 2

British Defence Policy and the Nerve Agents in Postwar Britain, 1945–1950

[A] statement he made that frightened me was that we cannot afford to lag behind in bacteriological and chemical warfare preparations. It is a tragic state of affairs that a British Socialist statesman should say that at this time of day.
—Rhys Davies MP (Hansard, House of Commons, 'Defence', Vol.478, cc467–645467, 26 July 1950)

The Grand Alliance, of Britain, the Soviet Union, and the United States, stumbled and subsequently collapsed. Moving from wartime cooperation to outright hostility, relations between the United States and the Soviet Union witnessed a rapid and steady deterioration from the end of the Second World War in 1945 to the outbreak of the Korean War in 1950.[1] The tense postwar climate would see the clash of two great behemoths, bringing to the fore an ideological confrontation which would span decades and at times bring the world to the very brink of catastrophe.[2] While the origins of the Cold War have long been debated, for British defence officials it was clear from an early stage that the Soviet Union would emerge the principal threat to national security. In March 1946, as if to mark a crucial watershed,

[1] Chamberlin, Paul Thomas. 2018. *The Cold War's Killing Fields: Rethinking the Long Peace*. New York: HarperCollins, 49.
[2] Westad, Odd Arne. 2017. *The Cold War: A World History*. UK: Allen Lane, 1–3.

W. King, *Nerve Agents in Postwar Britain*, Britain and the World,
https://doi.org/10.1007/978-3-030-70474-2_2

former Prime Minister Winston Churchill stated during his famous speech in Fulton, Missouri, that an iron curtain had fallen across Europe. Later that year, members of the Foreign Office also sagely warned that the Soviets were practising 'the most vicious power politics' in pursuit of their goals.[3] In this early postwar period, British foreign and defence policy was being formulated under the perceived threat of a large, ominous and seemingly growing menace from the East. And, with the United States demobilising in the immediate postwar period, it initially seemed like Soviet power would have to be countered without Washington's active support.[4]

From early 1947, and to the relief of British defence officials, there was an evident shift as the United States increasingly committed itself to the defence and reconstruction of Western Europe. Signs of this greater engagement in the emerging Cold War were seen in January 1947 when Britain and the United States combined their zones of occupation in Germany, and in March the Truman administration pledged substantial aid to Greece and Turkey.[5] The bipolar stand-off solidified by mid-1948, as the United States began to supply vital Marshall Plan economic aid to Western Europe, and when Soviet leader Joseph Stalin tested President Truman's resolve in the Berlin Blockade crisis.[6] In isolation, Britain had little chance of holding the Soviet Union at bay. But, with this more committed United States, the British Chiefs of Staff Committee (CoS) could begin basing war plans on keeping 'the Russians in Europe as far to the east as possible'.[7]

[3] Baylis, John. 1995. *Ambiguity and Deterrence: British Nuclear Strategy 1945–1964.* Oxford: Clarendon Press, 61; Young, John W. 1991. *Cold War Europe 1945–89: A political history.* London: Edward Arnold, 3; Fry, Geoffrey K. 2005. *The Politics of Decline: An Interpretation of British Politics from the 1940s to the 1970s.* Hampshire: Palgrave Macmillan, 4–6; Varsori, Antonio. 2000. Reflections on the Origins of the Cold War. In *Reviewing the Cold War: Approaches, Interpretations, Theory*, ed. Odd Arne Westad. London: Frank Cass, 288.

[4] Bartlett, C. J. 1992. *'The Special Relationship': A Political History of Anglo-American Relations since 1945.* London: Longman, 24; Goodman, Michael S. 2007. *Spying on the Nuclear Bear: Anglo-American Intelligence and the Soviet Bomb.* Stanford, CA: Stanford University Press, 2.

[5] Hathaway, Robert M. 1991. *Ambiguous Partnership: Britain and America, 1944–1947.* New York, NY: Columbia University Press, 4–5; Young. *Cold War Europe*, 142–145; Barrass, Gordon S. 2009. *The Great Cold War: a journey through the hall of mirrors.* Stanford, CA: Stanford Security Studies, 51–53.

[6] Young, John W. 1997. *Britain and the world in the twentieth century.* London: Arnold, 154–156.

[7] TNA, CAB 131/6, 'Defence Review', Chiefs of Staff Committee, 14 September 1948.

Coinciding with this escalating and tense Cold War stand-off in Europe, the situation was also growing increasingly unpredictable and uncertain in East Asia.[8] Where, after with the defeat of Nationalist forces in China, Mao Zedong proclaimed the formation of the People's Republic of China in October 1949, bolstering the international Communist movement.[9] Extending the iron curtain with the alleged 'Bamboo curtain', the threat of Communism appeared significantly enhanced.[10] Animated by the fall-out over the so-called 'loss' of China, US fears also grew over the world-wide expansion of Communism. The sense of crisis was increased by the Soviet explosion of an atomic bomb in August 1949 and was conveyed in the national security document NSC-68, which, produced in April 1950, called for a wholesale programme of US rearmament.[11] Mistrust and suspicions, fuelled by perceived acts of aggression, continued to fester and grow. A culminating point was reached with the outbreak of the Korean War in June 1950—at the time in the West an action widely perceived as instigated by the Kremlin.[12] Britain, as a close ally of the United States and as a key regional actor in Asia, became heavily embroiled in the war sending thousands of soldiers to fight, many of whom had just five years prior survived the Second World War.[13] This British commitment to war in East Asia was to prove costly.

For British defence officials, this tumultuous postwar period was bracketed by the ending of one war and the beginning of another.[14] It is in this emerging Cold War, and in this increasingly hostile environment, that the

[8] French, David. 2012. *Army, Empire, and Cold War: The British Army and Military Policy, 1945–1971*. Oxford: Oxford University Press, 28.

[9] Jun, Niu. 2010. The birth of the People's Republic of China and the Korean War. In *The Cold War, Volume I: Origins*, ed. Melvyn Leffler and Odd Arne Westad. Cambridge: Cambridge University Press, 231.

[10] Shaw, Tony. 2001. *British Cinema and the Cold War: The State, Propaganda and Consensus*. London: I. B. Tauris, 65.

[11] Leffler, Melvyn. 2010. The Emergence of an American grand strategy, 1945–1952. In *The Cold War, Volume I: Origins*, ed. Melvyn Leffler and Odd Arne Westad. Cambridge: Cambridge University Press, 82–85.

[12] Bartlett, C. J. 1984. *The Global Conflict 1880–1970: The International Rivalry of the Great Powers*. London: Longman, 290–299.

[13] Callaghan, John. 2007. *The Labour Party and Foreign Policy: A history*. London: Routledge, 186–187.

[14] Reynolds, David. 2000. *Britannia Overruled: British policy and world power in the twentieth century*. Harlow: Pearson Education, 172–173; Hoyt, Timothy. 2016. The United States and the Cold War Arms Race. In *Arms Races in International Politics*, ed. Thomas Mahnken, Joseph Maiolo, and David Stevenson. Oxford: Oxford University Press, 152–154.

origins and foundations of British nerve agent policy were established. This chapter will first analyse the shocking discovery of nerve agents, and it will then go on to explore the secret struggle to determine their place and role in British defence policy, which culminated in crucial Cabinet Defence Committee decisions in 1950.

Discovery

When considering weapons of mass destruction (WMDs) in the immediate postwar period, it is often easy to focus almost exclusively on the impact of the atomic bomb. Yet similarly to advances in nuclear weapons, CBW was also shaken by seismic events in the field.[15] In April 1945, just weeks before the German surrender, Allied troops made a startling discovery.[16] When advancing through North-West Germany, they stumbled upon German shells strangely marked with green rings near Lübbecke.[17] These shells, alongside the uncovering of advanced German CW research at Raubkammer, revealed a remarkable development in the CW field: nerve agents.[18] While CW agents deployed in the First World War would typically choke or blister and their use could be readily detected, nerve agents were ten to fifty times more lethal than these previous chemical weapons.[19] British experts possessed no truly effective way to warn soldiers

[15] Although not uncovered by investigators until much later, one disturbing discovery of wartime activities was Imperial Japan's use of biological weapons. For details on how Britain received this information, see: King, William and Guillemin, Jeanne. 2019. The price of alliance: Anglo-American intelligence cooperation and Imperial Japan's criminal biological warfare programme, 1944–1947. *Intelligence and National Security*, 34:2, 263–277.

[16] Spiers. *Chemical Weaponry*, 5–6; Schmidt. *Secret Science*, 158. For further details on the German nerve agent development and the Allied and Soviet discovery, see: Tucker, Jonathan B. 2006. *War of Nerves: Chemical Warfare from World War I to Al-Qaeda*. New York, NY: Pantheon Books, 64–102.

[17] Sloan, Roy. 1998. *The Tale of Tabun: Nazi Chemical Weapons in North Wales*. Llanrwst, Wales: Gwasg Carreg Gwalch, 31; Hall, Charlie. 2019. *British Exploitation of German Science and Technology, 1943–1949*. Oxon: Routledge, 92–95.; Kaszeta, Dan. 2020. *Toxic: A History of Nerve Agents. From Nazi Germany to Putin's Russia*. London: C. Hurst & Co., 68.

[18] For a detailed account of the development of German CW policy in build up to and during the Second World War, see: Gellermann, Günther W. 1986. *Der Krieg, der nicht stattfand: Möglichkeiten, Überlegungen und Entscheidungen der deutschen Obersten Führung zur Verwendung chemischer Kampfstoffe im Zweiten Weltkrieg*. Koblenz: Bernard & Graefe Verlag, 38–212.

[19] Bryden, John. 1989. *Deadly Allies: Canada's Secret War 1937–1947*. Toronto: McClelland & Stewart, 196.

of their use on the battlefield. Soldiers, who had traditionally relied on gas masks for their protection, were now potentially vulnerable to this colossal upgrade in the CW threat. Anglo-American experts soon appreciated that this nerve agent development heralded a new age in the CW field, which accelerated the rush for German CW spoils.[20] In the closing stages of the European war, this dramatic unveiling of the next generation of chemical weapons highlighted not an end, but a new beginning for a revitalised CW field.

German researchers had uncovered three main nerve agents: tabun, sarin and soman.[21] While German scientists had discovered sarin and soman too late for large-scale stockpiling in the war, this was not the case for tabun.[22] As the Allies soon discovered, by the end of the war, Germany had produced over 10,500 tons of tabun.[23] Tabun is 100–1000 times more toxic than chlorine and 10–100 times more potent than 'mustard gas', phosgene or lewisite.[24] Nearly all of this substantial stockpile ended up in the British and United States zones of occupation, but the decision loomed large as to what actually to do with this new acquisition.[25] Acquiring or moving such a stockpile would require negotiation with the United States, Ministerial approval, and time to reach a decision, not to mention an extreme burden in terms of the actual logistics.

While this deliberation continued, what is clear is that the nerve agent discovery secured the future of British CW research at Porton Down, the country's principal CW research establishment, and became the focal point of postwar CW research.[26] Information from German scientists, captured materials, and documents formed the bedrock of a dangerous new venture. Reflecting this newfound drive, as noted by Ulf Schmidt, Allied soldiers soon arrived for tests involving exposure to tabun in mid-July

[20] Evans, Rob. 2000. *Gassed: British Chemical Warfare Experimentation Humans at Porton Down*. London: House of Stratus, 111; Schmidt. *Secret Science*, 176.

[21] There are variations of these and other discoveries, such as cyclosarin, but for simplicity and coherence these 'main' three have been focused on.

[22] Schmidt. *Secret Science*, 87.

[23] TNA, CAB 131/3, 'Gas – Offensive Policy and Disposal of German War Stocks', Chiefs of Staff Committee, 3 June 1946; Spiers, Edward M. 1986. *Chemical Warfare*. London: Macmillan, 84; Hall, Charlie. 2017. British Exploitation of German Science and Technology from War to Post-War, 1943–1948. PhD diss., University of Kent. 132–134.

[24] Murphy, Seán, Hay, Alastair, and Rose, Steven. 1984. *No Fire, No Thunder: The Threat of Chemical and Biological Weapons*. London: Pluto Press, 11.

[25] Spiers. *Chemical Warfare*, 84.

[26] Carter. *Chemical and Biological Defence at Porton Down*, 68–70.

1945.[27] These tests, revealing a concerning approach to both research ethics and CW research, were conducted in the utmost of secrecy, shielding researchers from public scrutiny and accountability.[28] Even with the non-use of chemical weapons in a military role in Europe, tabun and the discovered nerve agents still revolutionised the CW field. After the First World War, the future of Porton Down was in doubt; this was far from the case in 1945.[29]

Exactly why Nazi Germany did not use its advantage in CW capabilities is a difficult and debated question. Whether it was down to inflated perceptions of Allied CW capabilities, Allied aerial superiority, CW not fitting into German military doctrine or memories of the First World War, Germany ultimately did not make operational use of its nerve agent weapons.[30] The development of lethal nerve agents did not overcome substantial obstacles to their use, even during the final, desperate stages of the Second World War. Importantly, some scholars, such as Richard Price, also point towards the existence of a 'chemical weapons taboo'.[31] Price convincingly contends that the non-use of chemical weapons and treaties regarding non-use are rooted in a genuine moral rejection of this form of warfare. This is a taboo which, Price argues, remains strong to this very day despite recent challenges.[32] The 'special dread' associated with chemical weapons can be linked to unfamiliarity breading fear and enhancing the symbolic and awful nature of them.[33] This taboo did not in itself stop the fist-use of chemical weapons during the Second World War in Europe. But, it certainly was an influencing factor and one which would continue into the postwar period.[34] While this taboo is often hard to explicitly locate in government files, it does manifest in other forms, such as normative concerns over use, moral revulsion at this form of warfare and perhaps

[27] Schmidt. *Secret Science*, 179.

[28] For further details, see: Schmidt. *Secret Science*, 1–670.

[29] Carter. *Chemical and Biological Defence*, 68–69.

[30] For further details, see: Brown, Frederic. 1968. *Chemical Warfare: A Study in Restraints*. Princeton: Princeton University Press, 291–297; Lego, Jeffrey W. 1995. *Cooperation under Fire: Anglo-German Restraint During World War II*. London: Cornell University Press, 144–216.

[31] Price. *The Chemical Weapons Taboo*, 1–256.

[32] Price, Richard. 2019. Syria and the Chemical Weapons Taboo, *Journal of Global Security Studies*, 4:1, 37–52.

[33] Price. *The Chemical Weapons Taboo*, 124–125.

[34] Ibid.

even military resistance to its employment, as well as adherence to international treaties.

Resisting the initiation of CW in part took form and was embodied through adherence to the 1925 Geneva Protocol. The Geneva Protocol itself built upon prior treaties. Notable milestones were the Brussels Convention of 1874 and the Hague Declaration of 1899, which prohibited the use of projectiles the object of which was the diffusion of asphyxiating or deleterious gases.[35] Attempts were made to ban this form of warfare even before they were fully developed or stockpiled.[36] After the First World War and the horrors of gas warfare, support had mounted for a new international treaty. As noted by Catherine Jefferson, fear of the unknown had been amplified by the fear of the ways in which gas worked and its effects on the human body.[37] These fears and concerns culminated in the 1925 Geneva Protocol, prohibiting signatories from the use of chemical and biological weapons in war.[38] The treaty did not prohibit the production, acquisition, or stockpiling of CBW weapons, just their use.[39] Moreover, Britain was a reserving state. Meaning that the country reserved the right to retaliate in kind if chemical or biological weapons were used against itself or allies, and it reserved the right to use these weapons against states who had not ratified or acceded to the Protocol 'permanently'.[40] This was not unique at the time, France, for example, also imposed similar reservations allowing retaliation-in-kind. The treaty did, though, significantly raise the threshold for CW use.[41] International opinion and norms were largely against CBW use, and any justification for the first-use of such controversial weapons needed strong support. It created such a burden of proof that the only permissible situation for British officials to allow the use of chemical weapons was one of calamity or survival.[42] At such a point

[35] Murphy et al. *No Fire, No Thunder*, 7.

[36] Price. *The Chemical Weapons Taboo*, 12.

[37] Jefferson, Catherine. 2014. Origins of the norm against chemical weapons. *International Affairs*, 90: 3, 655.

[38] 'Protocol for the Prohibition of the Use in War of Asphyxiating, Poisonous or Other Gases, and of Bacteriological Methods of Warfare', Geneva, 17 June 1925.

[39] Vagts, Detlev F and Meron, Theodor. 2013. Legal Aspects of the Geneva Protocol of 1925. In *Humanizing the Laws of War: Selected Writings of Richard Baxter*, ed. Richard Baxter, Detlev F. Vagts, Theodor Meron, Stephen Schwebel, and Charles Keever, Oxford: Oxford University Press, 167–168.

[40] Ibid., 184.

[41] Price. *The Chemical Weapons Taboo*, 122.

[42] Schmidt. *Secret Science*, 153.

of desperation, Winston Churchill had then considered it 'absurd' to worry about the morality of chemical weapons use.[43]

From the perspectives of British defence officials, taboo or treaties were not, however, the main lessons learnt from non-use during the Second World War. The perceived best way to prevent the use of chemical weapons was thought to be through deterrence and the threat of retaliation-in-kind. During the war, discouraging German use of CW was an active policy for Britain.[44] Dissuading and deterring was a difficult endeavour for the Prime Minister, Winston Churchill. On the one hand, if Britain did not publicly deter Germany through its posture and with public threats of retaliation with chemical weapons, then it could have invited CW attack through silence and the absence of a deterrent.[45] On the other hand, if too many public announcements and threats of retaliation were made, German officials might have taken the view that Britain was either frightened of CW or releasing public statements as a precursor for its imminent use.[46] As Churchill was advised, deterring CW during the war entailed navigating this fine line, which by the end of the war appeared to have worked.[47] Crucially, and as a result of chemical weapons not being used in a military role in Europe, many defence officials and politicians believed wartime CW deterrence to have been a success story. Preparedness to wage CW and the occasional public warning seemingly successfully played a part in deterring the German first-use of chemical weapons.[48] Such an

[43] Ibid.

[44] Spiers, Edward M. 2010. *A History of Chemical and Biological Weapons*. London: Reaktion Books, 59–60.

[45] Britain and Churchill were not alone in making these threats, Roosevelt and Hitler also made similar remarks. See: Robinson, Julian Perry. 1981. Chemical Arms Control and the Assimilation of Chemical Weapons. *International Journal (Toronto)*, 38:3, 518.

[46] TNA, PREM 3/89, H. L. Ismay to Churchill, 24 May 1944.

[47] Ibid. Churchill's limited public statements on Britain's intent to use 'gas warfare far and wide' in retaliation to any German CW use seemed to defence officials to have played a part in deterring German use. For wartime examples see: *The Times*, 'Now is the Time to Strike', 11 May 1942, 5; *Daily Mail*, 'Gas: Churchill Warns Hitler', 11 May 1942, 1; TNA, PREM 3/89, H. L. Ismay to Churchill, 24 May 1944; Gilbert, Martin. 1986. *Winston S. Churchill, Vol. 7: Road to Victory 1941–1945*. London: Heinemann, 353. Some of Churchill's threats were actually made to Germany on behalf of the Soviet Union during the war.

[48] For some later examples of this approach to deterrence, see: TNA, DEFE 10/445, '1949 Report on Chemical Warfare', Joint Secretaries of the Chemical Warfare Sub-Committee, 17 August 1949, Attached note; Hansard, House of Commons, 'Civil Defence', Vol.529, cc1832–914, 5 July 1954; TNA, DEFE 7/700, Prime Minister's Personal Minute to the Minister of Defence, 20 June 1956.

interpretation was also in part reinforced by the nerve agent discovery. Even with this substantial capability, Germany still did not initiate CW. Brandishing a British CW capability and threatening to use it, but not too much and not too little, had seemingly been the correct course.

In the early Cold War, this perceived success of CW deterrence during the Second World War had a lasting impact on how senior British officials viewed CW deterrence, when deterring German CW use shifted to deterring Soviet CW use. An early indicator of this legacy of the Second World War in defence planning was seen in the Tizard Report of June 1945. When, Sir Henry Tizard, as scientific advisor to Churchill's wartime Coalition Government, was tasked with chairing a committee to report on the future developments in weapons and methods of war.[49] In his report, which was prepared and formed before the nerve agents were found, Tizard argued that CW research 'must continue as an insurance'.[50] In supporting this claim, the Tizard Report emphasised that it was German fears of reprisals which had mitigated against their desire to use chemical weapons during the war.[51] Chemical weapons deterrence, in defence circles, was gaining credence.

Elected in July 1945, after the surprising defeat of Winston Churchill, the new Labour Prime Minister Clement Attlee reinforced and supported this line of thinking.[52] With wartime experiences shaping his approach to CW deterrence, in November 1945, Attlee informed Parliament that: 'gas was banned before the war of 1914–1918, but it was used; and I have no doubt that if the Nazis had thought it worth while they would have used gas again'.[53] This was a rare intervention, as on the whole secrecy shrouded the CW field and political and public scrutiny and discussions were extremely limited. The public was largely unaware of the substantial nerve agent development. Indeed, Attlee only referred to poison gas, rather than

[49] Goodman, Michael. 2003. British intelligence and the Soviet atomic bomb, 1945–1950. *Journal of Strategic Studies*, 26:2, 122.

[50] TNA, CAB 80/94, 'Future Development in Weapons and Methods of War', Sir Henry Tizard's "Ad Hoc" Committee, 16 June 1945.

[51] Ibid.

[52] Bullock, Alan. 1983. *Ernest Bevin: Foreign Secretary*. Oxford: Oxford University Press, 3–4.

[53] Hansard, House of Commons, 'Foreign Affairs', 22 November 1945, Vol.416, cc601–714. This was also just a month after he had granted initial approval for British atomic energy research and development, see: Jones, Matthew. 2017. *The Official History of the UK Strategic Nuclear Deterrent: Volume I: From the V-Bomber Era to the Arrival of Polaris, 1945–1964*. Oxon: Routledge, 13.

nerve agents. Attlee also believed that the perceived military utility of chemical weapons needed to be countered, which he thought best accomplished through deterrence and the ability to retaliate. When meeting President Truman in Washington that same month, leaked reports claimed that the lesson Britain and Attlee had taken from CW and the Second World War was that:

> The decision not to use gas in World War II was not a moral decision but a military one, based on the premise that neither belligerent would undertake its use as long as the other refrained.[54]

While normative factors and the chemical weapons taboo certainly played a part, they were not acknowledged and even outrightly dismissed. The impression was given that Britain was prepared to resort to this form of warfare if needed. And, even if the moral aspect were refuted, there appeared some mutual understanding that neither side would resort to this form of warfare. By categorising it in this way, it also branded it a field for special consideration. From Attlee's perspective, or at least in what he disclosed, the prevention of CW rested upon the threat of retaliation and the removal or reduction of the military benefit of any enemy's potential use, rather than moral considerations or legal constraints.[55] His views were undoubtedly shaped by the need to strike a firm stance, itself forming a part of deterrence, as well as the harshness of the Second World War. Attlee, when part of Churchill's War Cabinet, had also defended Britain's CW posture in the House of Commons.[56]

Churchill's and Attlee's interpretations of deterring CW use through military means was also reflected in the minimal parliamentary discussions over CW deterrence in the early postwar period, with MPs observing that 'the atom bomb, like poison gas, may not be used, because of the fear of

[54] Felix Belair Jr., 'Attlee Says Bomb Cannot Be Banned; Asks A Strong UNO', *The New York Times*, 13 November 1945, 1; Fredericks, William Curtis. 1988. The Evolution of Post-World War II United States Chemical Warfare Policy. PhD diss., Oxford University, II-4.

[55] More evidence for this will be included later on, with specific examples seen in the Cabinet Defence Committee meetings of 1946 and 1950.

[56] Attlee was also often responsible during the war for defending the Government's CW posture and preparedness in the House of Commons. For example, see: Hansard, House of Commons, 'Gas Warfare', 24 February 1943, Vol.387, cc160–1.

reprisals'.[57] The perceived success of CW deterrence, through the threat of reprisals and preparedness to retaliate, was thus also seen by some as a source of hope for potential nuclear deterrence after the shock development of the atomic bombs. In the CW field, this interpretation of CW deterrence and the resulting requirement for insurance and a retaliatory capability only increased in importance after the discovery of German nerve agents. With the game-changing nerve agent development, the stakes had risen. An effective and successful CW retaliatory capability and deterrent thus now necessitated the acquisition of nerve agent weapons, without which Britain would again be at a disadvantage in the CW field and the viability of its deterrent questionable. An up-to-date retaliatory capability was imperative, especially given that British officials knew they were not the only ones to have benefitted from the nerve agent development.[58]

Sweeping in from the east, towards the end of the Second World War, Soviet forces also uncovered the startling German nerve agent development. The finding of nerve agents in Germany led to something of a race between the soon to be Cold War rivals, as the United States, Britain, and the Soviet Union all attempted to extricate from Germany as much information and materials as possible.[59] The invading armies located and, where possible, seized German CW stocks, details of production methods, equipment and even personnel.[60] In this hunt for CW gains, one of the most substantial prizes for the Soviet CW programme was the discovery of a full-scale operational German nerve agent plant at Dyhernfurth. The site had produced thousands of tons of tabun during the war.[61] This facility, with German technical assistance, was deconstructed, moved and re-built in Soviet territory.[62] In addition to the facility, captured personnel, research

[57] Hansard, House of Commons, 'Defence Policy', 4 March 1946, Vol420, cc39–146.

[58] Evans. *Gassed*, 111; Schmidt. *Secret Science*, 176.

[59] Hall. British exploitation of German Science, 132–134.

[60] Maddrell, Paul. 1998. Britain's Exploitation of Occupied Germany for Scientific and Technical Intelligence on the Soviet Union. PhD diss., University of Cambridge, 21; Schmidt *Secret Science*, 87.

[61] TNA, CAB 80/94, 'German Chemical Warfare Preparations', Inter-Service Committee on Chemical Warfare, 16 June 1945.

[62] Robinson. *The Rise of CB Weapons*, 72; Maddrell. Britain's Exploitation of Occupied Germany, 21.

findings, and CW stocks gave the Soviet CW programme a substantial boost.[63] As while Soviet officials had noted the importance of CW research since the Russian Civil War, when Western forces used chemical weapons against the Red Army, the nerve agents reinvigorated interest and triggered a substantial Soviet nerve agent programme.[64] This intense Soviet interest in the nerve agents, originating from their knowledge of German discoveries, would go on to dominate British considerations of the Soviet CW threat throughout the Cold War.

Even with this new nerve agent development and the abundance of information and sources from Germany, accurate intelligence on the nature and form of the Soviet CW threat was still hard to come by. German wartime intelligence provided British planners with a limited window into Soviet wartime CW training, delivery methods, types of stocks and defensive measures, but this was to an extent outdated after the nerve agent discovery, which drastically moved the goalposts for intelligence officials attempting to analyse the future Soviet threat.[65] Due to German nerve agents only being located towards the end of the war, much of the wartime intelligence gathered by Germany and by Britain on Soviet CW capabilities was now of limited use. Compounding this lack of accurate intelligence was also impressive levels of Soviet security and counterintelligence. These measures limited British intelligence to a 'negligible' amount of information.[66] By early 1946 British officials were almost wholly reliant on the minimal information on Soviet CW gains from Germany, rough predictions and self-mirroring (the practice of basing Soviet capabilities and progress on British or US experiences, and then including further appreciation of possible hindrances or accelerators

[63] Ibid., 21–23.

[64] Tucker. *War of Nerves*, 106–108.

[65] TNA, CAB 81/143, 'Technical Information Regarding Weapon Development in the USSR', Joint Scientific and Joint Technical Intelligence Committees, 5 December 1946; Maddrell. Britain's Exploitation of Occupied Germany, 29–31.

[66] TNA, CAB 81/132, 'Survey of Chemical Warfare Intelligence', Secretary of the Joint Intelligence Committee, 21 March 1946, Annex; Haslam, Jonathan. 2015. *Near and Distant Neighbours: A New History of Soviet Intelligence*. Oxford: Oxford University Press, 147. A significant intelligence gain came in the form of the 'Hirsch Report', which although useful was soon outdated by the rapid development of nerve agent weapons. For a summary of the report, which was published much later, see: The Harvard Sussex Program Archive, 'Soviet BW and CW Preparations and Capabilities', Col. Dr. Walter Hirsch, 19 May 1951, i–iv.

unique to the Soviet effort). These hesitant assessments led to the forecast that by 1951 the Soviet Union would be mass-producing tabun and that by 1956 the Soviet Union would be mass-producing the more lethal and complicated nerve agents such as sarin and soman.[67] Regardless of these difficulties in attaining information on the Soviet CW programme, one thing was clear for the Joint Intelligence Sub-Committee (JIC): in the near future, Britain would face a significant Soviet CW threat which was a direct result of the finding of nerve agents in Germany.[68]

The nerve agent discovery, and perceived threat of Soviet CW capabilities, would also have a more immediate impact on British CW policy. In June 1945, the Inter-Service Committee on Chemical Warfare strongly advised the Chiefs of Staff to acquire the 10,500 tons of tabun filled bombs from occupied Germany.[69] The specialist committee was responsible for CW matters, and for providing advice on the direction of British CW policy. Remarkably, even though Soviet forces had captured the Dyhernfurth facility, the majority of the German tabun filled bombs had ended up in Allied possession. Keen British military support for the acquisition of this substantial nerve agent stockpile was driven by three core factors. Firstly, as supported by British intelligence, even after German forces had bombed their own tabun plant at Dyhernfurth, the Soviet Union had still acquired detailed information on nerve agent production.[70] With concerned perceptions of Soviet intent, British planners thought it a sound tactical decision to acquire as much nerve agent material and munitions from occupied Germany as possible. Secondly, in addition to preparing for any clash with a future adversary, British officials were wary of the ongoing war in the Pacific, with concerns over possible

[67] TNA, CAB 81/143, 'Technical Information Regarding Weapon Development in the USSR'. Joint Scientific and Joint Technical Intelligence Committees, 5 December 1946; Lewis, Julian. 2003. *Changing Direction: British Military Planning for Post-war Strategic Defence, 1942–1947.* London: Frank Cass, 282.

[68] TNA, CAB 81/132, 'Survey of Chemical Warfare Intelligence', Secretary of the Joint Intelligence Committee, 21 March 1946, Annex.

[69] TNA, CAB 80/94, 'Disposal of German Chemical Warfare Stocks', Inter-Service Committee on Chemical Warfare, 16 June 1945; TNA, CAB 81/67, 'Offensive Gas Policy', Inter-Service Committee on Chemical Warfare, 20 March 1946.

[70] TNA, CAB 80/94, 'German Chemical Warfare Preparations', Inter-Service Committee on Chemical Warfare, 16 June 1945; Maddrell, Paul. 2006. *Spying on Science: Western Intelligence in Divided Germany 1945–1961.* Oxford: Oxford University Press, 273.

Japanese use of chemical weapons in the closing stages of the fighting.[71] If CW was initiated in the Pacific Theatre, then German nerve agent stocks would provide a 'useful and flexible' capability with which to retaliate.[72] United States officials also supported this stance, as they recommended, just weeks before the dropping of atomic bombs on Hiroshima and Nagasaki, that German tabun stocks be shipped to the Far East for possible retaliatory use against Imperial Japan.[73] In the closing stages of the war, continued adherence to retaliation-only did further reveal some tacit support amongst senior defence officials for the Geneva Protocol. As despite Churchill and officials contemplating the possibilities of use during the most desperate stages of the war, Britain adhered to the Protocol throughout the conflict.

A third driving factor for acquisition was a technical one, which tied into how and when the nerve agent weapons could and would be used. If British officials wanted a temporary, economic and short-term capability, then regardless as to whether this was for use against Japan or to deter the Soviet Union, it would have to come from captured German stocks.[74] This was due to the complexity and difficulty involved in producing nerve agents in bulk. Britain, it was estimated, would be unable to produce nerve agent weapons in less than three to four years.[75] Officials also recognised that tabun, while extremely effective, was the lesser of the nerve agents. The question that thus arose was whether Britain should settle for this lesser nerve agent when it had captured information on how to produce more advanced nerve agents such as sarin and soman. This observation further boosted the argument for acquiring German tabun stocks, as while sarin and soman were desirable, they would take longer to produce. If one of these more advanced nerve agents were to become a chosen CW agent for mass-production, then Britain would still need some form of

[71] It should also be noted here that British officials were doubtful as to whether Japan actually knew of the German nerve agent discovery, a suspicion which was to prove correct.

[72] TNA, CAB 80/94, 'Disposal of German Chemical Warfare Stocks', Inter-Service Committee on Chemical Warfare, 16 June 1945.

[73] Ibid.; Carter, Gradon and Balmer, Brian. 1999. Chemical and Biological Warfare and Defence, 1945–90. In *Cold War Hot Science: Applied Research in Britain's Defence Laboratories 1945–1990*, ed. Robert Bud and Philip Gummett, Amsterdam: Harwood, 136.

[74] TNA, CAB 80/94, 'Disposal of German Chemical Warfare Stocks', Inter-Service Committee on Chemical Warfare, 16 June 1945.

[75] Ibid.

temporary nerve agent capability to fill the interim period.[76] Deterrence demanded an effective retaliatory capability. German tabun stocks therefore not only represented an economical and expedient nerve agent capability for Japan and the nascent Cold War, but they also acted as a stop-gap in allowing British scientists to focus on producing and researching preferable nerve agents such as sarin.[77]

Reflecting growing fears over the Soviet CW and conventional threat, and in addition to the tabun stocks, the Inter-Service Committee on Chemical Warfare further recommended in March 1946 that Britain retain a substantial quantity of wartime chemical weapons.[78] This precautionary measure reveals the emphasis and importance attached to maintaining some form of stop-gap CW capability and a credible CW deterrent in the emerging Cold War. As despite fears that existing chemical weapons, such as those containing sulphur mustard or phosgene, were 'outmoded' by the nerve agent development, their retention was still recommended.[79] This was, though, no small endeavour. By the end of the war Britain had stockpiled around 35,171 tons of sulphur mustard, 6744 tons of phosgene, and 1383 tons of 'other gases'.[80] If these substantial figures are added to the acquisition of the 10,500 tons of tabun from Germany, then at the end of the Second World War Britain possessed over 50,000 tons of CW agents.[81] Of this substantial amount, the CoS was advised to retain enough sulphur mustard and phosgene to maintain a significant 'war reserve', and to maintain around 30% of Britain's total wartime production capacity for CW agents.[82] This British war reserve came to 11,700 tons of sulphur mustard,

[76] TNA, CAB 80/94, 'German Chemical Warfare Preparations', Inter-Service Committee on Chemical Warfare, 16 June 1945. For while Germany had mass-produced tabun, British officials observed that Germany had only established a small pilot plant for sarin, and that it had not moved on to the mass-production of soman.

[77] TNA, CAB 80/94, 'Disposal of German Chemical Warfare Stocks', Inter-Service Committee on Chemical Warfare, 16 June 1945; TNA, CAB 81/67, 'Offensive Gas Policy', Inter-Service Committee on Chemical Warfare, 20 March 1946.

[78] Ibid.

[79] Ibid.

[80] TNA, PREM 3/89, Norman Brook to Churchill, 6 June 1945, Table C; Carter and Balmer. Chemical and Biological Warfare and Defence, 295.

[81] This number was still in fact less than German stockpiles of chemical weapons at the end of the Second World War, which was estimated to be around 70,000 tons. Spiers. *Chemical Warfare*, 79.

[82] TNA, CAB 81/67, 'Offensive Gas Policy', Inter-Service Committee on Chemical Warfare, 20 March 1946, Annex IV.

2900 tons of phosgene, and all 10,500 tons of tabun from occupied Germany.[83] A substantial stop-gap capability was in place.

In June 1946, and in light of the emerging Soviet threat, the considerable technological advancement which the nerve agents represented and the need for a stop-gap capability, both the CoS and Attlee's Cabinet Defence Committee approved these major proposals.[84] Britain would acquire German tabun stocks and maintain a significant war reserve for the Cold War. Attlee, as Prime Minister and Chair of the Cabinet Defence Committee, had in effect again reiterated his commitment to deterrence through a viable retaliatory capability, by approving the retention of wartime stocks and the acquisition of lethal nerve agent weapons from Germany.

This approval for the acquisition of a CW capability was to prove all the more significant for the CoS, given that, as they put it in June 1946, 'it is unlikely that atomic or biological weapons will be available for our use on a large scale for at least five years and we must therefore rely on our existing weapons in this period'.[85] Intensive research was still underway in the field of biological weapons, and a British atomic weapon would take time and substantial resources to develop. The nerve agent capability would, therefore, act as an essential stop-gap WMD capability. The importance of nerve agent weapons was also not purely confined to this postwar stop-gap WMD role though, as even after nuclear and biological weapons were developed, the CoS and the Defence Committee both agreed that Britain should remain 'in a position to wage chemical warfare from the start of hostilities'.[86]

Once Ministerial confirmation was given for the acquisition of German tabun stocks, actually acquiring and moving 10,500 tons of tabun, which if including casings weighed around 18,000 tons, was found to be no easy

[83] TNA, CAB 131/3, 'Gas – Offensive Policy and Disposal of German War Stocks', Chiefs of Staff Committee, 3 June 1946.

[84] Ibid. Although the explicit confirmation and minutes from the Cabinet Defence Committee are not in the CAB 131 files, later confirmation and reference to this approval can be found in: TNA, DEFE 5/37, 'Chemical and Biological Warfare', Chairman of the Defence Research Policy Committee, 6 March 1952, Annex.

[85] TNA, CAB 131/3, 'Gas – Offensive Policy and Disposal of German War Stocks', Chiefs of Staff Committee, 3 June 1946.

[86] Ibid.

feat.[87] For in addition to the immense problem of the sheer weight involved, there was also the possibility of nerve agent leakage in transit, fears of the press or members of the public finding out, and the fact that over half the tabun bombs were stored in the zone of Germany occupied by the United States. Just monitoring and looking after the already filled bombs would require the attention of 60 full-time personnel and storage at a special site at an airfield in Llandwrog, Wales.[88] Due to the tabun being stored in German bombs, which were unsuitable for carriage on RAF aircraft and due to them having fuses already fitted (making them more dangerous for transportation and storage), British forces also lacked an effective means of delivery.[89] Notwithstanding all these drawbacks, the nerve agents were still deemed simply too valuable. The CoS believed that disposing of them would be 'imprudent' and that their retention would give Britain a 'commanding lead in the field'.[90]

After negotiations with the United States, whose zone contained around 5700 tons of tabun, Britain secured the vast majority of the world's existing nerve agent stocks from occupied Germany.[91] Surprisingly the United States willingly accepted this agreement, on the condition that they be allowed a select few tabun bombs for trials.[92] A key reason for this was that the United States Chemical Warfare Service (becoming the Chemical Corps in 1946) had its eye on something far more substantial, the mass-production of sarin. On both sides of the Atlantic, sarin was emerging as the ideal CW agent for mass-production. Its perceived military utility and slightly easier means of production compared to soman rendered it the preferred option.[93] For the United States, the ability to retaliate with their growing stockpile of atomic weapons, Britain being an ideal storage location for nerve agent deployment in a European war, and the fact that they were able to secure sarin samples from Germany influenced

[87] TNA, CAB 81/67, 'Disposal of Stocks of German Nerve Gases', Inter-Service Committee on Chemical Warfare, 19 February 1946.

[88] TNA, CAB 131/3, 'Gas – Offensive Policy and Disposal of German War Stocks', Chiefs of Staff Committee, 3 June 1946; McCamley, N. J. 2006. *Secret History of Chemical Warfare*. Barnsley: Pen & Sword Military, 136–137.

[89] Kaszeta. *A History of Nerve Agents*, 86–87.

[90] TNA, CAB 131/3, 'Gas – Offensive Policy and Disposal of German War Stocks', Chiefs of Staff Committee, 3 June 1946.

[91] Ibid.

[92] Tucker. *War of Nerves*, 92.

[93] Ibid., 100–123. For later British considerations of sarin, see: TNA, DEFE 10/445, 'Operational Value of Nerve Gas', Chemical Warfare Sub-Committee, 15 November 1949.

the decision. As did concerns over the logistics of shipping and repurposing the German tabun bombs. All these factors played a part in this willingness to allow British acquisition of thousands of tons of lethal nerve agent weapons. As a result, from October 1946, 10,500 tons of the German tabun filled bombs were secretly shipped to Wales, in what was astutely called Operation Dismal.[94]

How Britain was able to ship this substantial quantity of nerve agents, without public attention, was thanks in large part to the D-Notice system.[95] The system was a 'sharp tool' to encourage the press not to cover issues that may damage national security.[96] Alongside the occasional threat of prosecution, the D-Notice system was, in essence, a collaborative effort between the press and the government which sought to limit press coverage of sensitive information, and thus the potential acquisition of this information by hostile actors. This tacit support of the D-Notice system contrasted starkly with the outpouring of scientific information in the postwar period.[97] New technologies in medicine, energy generation, transport and communications were produced during the war.[98] But, coverage of much of the activities in military science was deemed to not be in the public interest, such as with certain aspects of the atomic bomb.[99] The transportation of lethal nerve agents, heralding a remarkable new threat, was unsurprisingly also kept under wraps and placed under the D-Notice system.

When it came to public disclosures, strict secrecy was applied. Despite the substantial nerve agent find and its ramifications, in Britain, public awareness of the nerve agent development was minimal in the immediate postwar period. Perhaps to an even greater degree than the limited disclosures on postwar nuclear weapons—with the clear and horrific display of the weapons potential forcing some admission as to its existence and

[94] TNA, CAB 131/3, 'Gas – Offensive Policy and Disposal of German War Stocks', Chiefs of Staff Committee, 3 June 1946; Sloan. *The Tale of Tabun*, 38–43; Tucker. *War of Nerves*, 93.

[95] Wilkinson, Nicholas. 2009. *Secrecy and the Media: The Official History of the United Kingdom's D-Notice System*. Oxon: Routledge, 513.

[96] Jenks, John. 2006. *British Propaganda and News Media in the Cold War*. Edinburgh: Edinburgh University Press, 2.

[97] Gregory, Jane and Miller, Steve. 1998. *Science in Public: Communication, Culture, and Credibility*. New York: Plenum Press, 3, 37–38.

[98] Ibid., 37–38.

[99] Hill, Christopher. 2018. *Peace and Power in Cold War Britain: Media, Movements and Democracy, c.1945–1968*. London: Bloomsbury, 60–61.

nature—CW was kept under tight censorship, with a few rare exceptions seeping out. This was not unique to CW, as BW research also progressed in a similar vein—which, as outlined by Brian Balmer, actually hindered efforts to expand postwar BW research and draw more scientists into the field.[100] One early revelation had come soon after the nerve agents were located in June 1945, with *The Times* reporting that Germany had produced a 'new gas in great quantity'.[101] Information was, however, extremely limited, and little mention was made of the nerve agents. There were only vague references to a 'new gas'.[102] This shroud of secrecy surrounding the nerve agent discovery was again fleetingly pierced when senior Nazi scientists were prosecuted at Nuremberg, with horrific stories emerging of mass-killings with gas and of human experimentation in concentration camps.[103] Emerging from this ongoing process was the testimony and trial of Albert Speer, who, as well as disclosing a slightly more bizarre story about his plan to kill Hitler by introducing poison gas into the Chancellery ventilation system, revealed in his trial the German development of 'two new terrible poison gases', that of tabun and sarin.[104] Either this remarkable revelation was seemingly not fully picked up by the press, or adherence to secrecy and the D-Notice system prevailed, as one

[100] Balmer. *Secrecy and Science*, 25–31.

[101] *The Times*, 'Germany's Secret Weapons' 29 June 1945, 5; Sloan. *The Tale of Tabun*, 32. The *Daily Telegraph* also released the story of the disposal of substantial quantities of German gases, which also referred to the 'newest poison gas'. See: *Daily Telegraph*, 'Sea Grave of Poison Gas', 7 August 1945, 3. Also see the later reference: *The Times*, 'German Poison Gas Plans', 26 September 1947, 3.

[102] Ibid.

[103] For examples and further information, see: *The Times*, 'All Jews Destined For Gas Chambers', 6 February 1946, 3; *The Times*, 'Guilt Of German Leaders', 30 August 1946, 4; Baumslag, Naomi. 2005. *Murderous Medicine: Nazi Doctors, Human Experimentation, and Typhus.* London: Praeger, xxiii–xxix, 64; Earl, Hilary. 2013. Prosecuting genocide before the Genocide Convention: Raphael Lemkin and the Nuremberg Trials, 1945–1949. *Journal of Genocide Research*, 15:3, 317–319; Guillemin, Jeanne. 2017. *Hidden Atrocities: Japanese Germ Warfare and American Obstruction of Justice at the Tokyo Trial.* New York, NY: Columbia University Press, 53–54. For a thorough and highly informative account of the Doctor's Trial, medical ethics and the Nuremberg trials of scientists, see: Schmidt, Ulf. 2004. *Justice at Nuremberg: Leo Alexander and the Nazi Doctors' Trial.* Hampshire: Palgrave Macmillan, 1–297.

[104] *The Times*, 'Speer's Strange Story', 21 June 1946, 4; *The Times*, 'Frank Evidence by Speer', 22 June 1946, 4; *Daily Telegraph*, 'Speer's Plan to Gas Hitler in his Chancellery', 4 December 1946, 4; Borkin, Joseph. 1978. *The Crime and Punishment of I.G. Farben.* London: Free Press, 133.

of the first public confirmations of the nerve agents garnered little traction or public attention. In Britain, these early postwar years were thus characterised by intense levels of secrecy. Only extremely limited information on the nerve agents made it into the public domain.[105]

While the Defence Committee and the CoS had secretly approved the acquisition of German tabun bombs and the retention of wartime stocks, they recognised that both of these measures were only temporary solutions. A core part of Attlee's and the Defence Committee's 1946 policy centred not only on existing stocks, but on planning for the future creation of a 'new offensive policy' based on nerve agents.[106] As such, the CoS concluded that British research 'must continue with the object of discovering…further major advances in chemical warfare'.[107] Alongside the potentialities of mass-producing and weaponising sarin and soman, British officials were therefore also hopeful of discovering further developments. The nerve agent discovery held open the option of even further potential finds. Only once this had been explored, and greater consideration given to mass-production, would Britain begin the domestic production of nerve agent weapons and completely shift its CW policy to embrace the new nerve agent age. After these new nerve agent weapons had either been discovered or developed, they would provide the foundation and bedrock of Britain's new offensive CW policy for the Cold War.

An Emerging Role

From early 1947, demand and military support for nerve agent weapons continued to grow, with a clear role for them slowly emerging and consolidating. In May 1947, the Defence Research Policy Sub-Committee (DRPC) produced a core set of guidelines for CW policy, in which the committee further emphasised the role of, and strengthened the

[105] Further revelations from Nuremberg were also later triggered by the trial of Karl Brandt in mid-1947; Brandt had engaged in human CW experiments. Much of the information was again kept under tight secrecy, but there was some public awareness of human experiments. For more details, see: Schmidt, Ulf. 2007. *Karl Brandt: The Nazi Doctor, Medicine and Power in the Third Reich*. Cornwall: Hambledon Continuum, 284–296, 354, 382; *The Times*, 'Sentences on Nazi Doctors', 21 August 1947, 3.

[106] TNA, CAB 131/3, 'Gas – Offensive Policy and Disposal of German War Stocks', Chiefs of Staff Committee, 3 June 1946.

[107] Ibid.

requirement for, British nerve agent weapons.[108] The DRPC, responsible for providing advice on and overseeing aspects of British defence policy, was highly influential and operated under the CoS.[109] Believing that Britain needed chemical weapons in the Cold War, especially as they could prove of 'great tactical use', the DRPC urged acquisition and development.[110] Here, even from an early stage, the nerve agents were assigned a 'tactical' role, for targeted or limited use, rather than a strategic role. While revealing the early stage of policy formulation and still the potential of nerve agent weapons, it is clear the DRPC was deliberately placing nerve agents in a separate category from other WMDs. Nuclear weapons were 'strategic', as were some biological weapons, but this was deemed not the case for these lethal new nerve agents.[111] Any overlaps with biological or nuclear weapons were avoided. By labelling them as tactical, the DRPC thus in a way removed a source of direct competition for the acquisition of nerve agent weapons as well as showing early appreciation of the practical difficulties of dispersing thousands of tons of nerve agents over a wide area. A tactical nerve agent capability, especially if used on the ground by the Army in artillery shells, could support a strategic aerial nuclear weapons capability, and be used in tandem, whereas a strategic aerial nerve agent capability would encroach on the increasingly dominant field of nuclear weapons.

The DRPC recommended, in May 1947, that Britain begin the domestic production of nerve agents for a tactical role in Europe. Although production was initially envisaged on a small scale, this was thought a necessary pre-cursor to mass-production.[112] The escalation in the scale of production would facilitate invaluable learning and training experiences for scientists and engineers, but also not commit Britain fully to mass-production. While this recommendation would not be immediately acted

[108] TNA, DEFE 10/18, 'Future Defence Policy', Defence Research Policy Committee, 1 May 1947, Annex. For further details on the DRPC, see: Agar, Jon and Balmer, Brian. 1998. British Scientists and the Cold War: The Defence Research Policy Committee and Information Networks, 1947–1963. *Historical Studies in the Physical and Biological Sciences*, 28:2, 209–252.

[109] Ibid., 209–213.

[110] TNA, DEFE 10/18, 'Future Defence Policy', Defence Research Policy Committee, 1 May 1947, Annex; Agar and Balmer. British Scientists and the Cold War, 217, 219.

[111] Balmer. *Britain and Biological Warfare*, 61–65.

[112] TNA, DEFE 10/18, 'Future Defence Policy', Defence Research Policy Committee, 1 May 1947, Annex.

upon, as the role and place of nerve agents in British defence policy was still consolidating and thus to a degree uncertain, it would reappear with renewed vigour in 1949–1950. The DRPC also reiterated earlier fears, noting that British forces still had no effective means of detecting the nerve agents, and as such CW defensive measures, including methods of detection, were also assigned a high priority.[113]

In this formative period, and still drawing from their wartime experience, British officials increasingly appreciated that a core part of preventing or mitigating Soviet CW use was through CW deterrence. This legacy of wartime CW deterrence also influenced the development of ideas about nuclear deterrence, when it was assumed that the Soviet Union would be working to develop its own atomic bomb to break the US monopoly. In April 1947, when the CoS had turned to debate and discuss the possible first-use of weapons of mass destruction in the Cold War, it was to the CW field which they looked to for past examples and guidance.[114] This reflected the fact that despite the nerve agent development, chemical weapons were actually the most familiar of all the weapons of mass destruction.[115] The CoS took the case of CW deterrence in the Second World War as an example which showed that if they were strong and prepared to retaliate on a like-for-like basis, then that form of warfare would be less likely to be used. They thus believed that the only way to prevent Soviet use of weapons of mass destruction was through facing 'her with the threat of large scale damage from similar weapons' and that this threat of retaliation was to 'be a most effective deterrent to war'.[116] Deterring different types of warfare, whether chemical biological or nuclear, thus demanded acquiring that same capability to retaliate in kind. This logic, reinforced by the wartime CW experience, would come to dominate deterrence considerations in the Cold War. In terms of CW policy, this same deterrence frame of mind was reflected in the retention of the substantial war reserve and the tabun stocks from Germany, which were both thought to have some deterrence value. For officials based at Porton Down, the nerve agent weapons were

[113] Ibid.

[114] TNA, DEFE 5/4, 'Future Defence Policy', Joint Secretary of the Chiefs of Staff Committee, 23 April 1947, Annex II, Appendix II.

[115] Ibid.

[116] TNA, DEFE 6/2, 'Future Defence Policy', Joint Planning Staff, 7 May 1947, Annex.

an important 'bargaining chip'.[117] They represented an important stop-gap deterrent to discourage Soviet first-use.[118]

The perceived Soviet CW threat and the need for a reliable, effective and long-term nerve agent capability and deterrent also required informed policy decisions based on accurate intelligence assessments and further integration in trilateral cooperation. In the intelligence sphere, in May 1948, the JIC outlined the priorities for SIGINT (signals intelligence).[119] Alongside atomic weapons intelligence as a priority I was CW intelligence.[120] The JIC correctly feared that German advances in the CW field had stimulated Soviet research. It observed that due to the 'relative backwardness' of the Soviet Union it might be forced to settle for the nerve agents in the short-term, before developing a nuclear weapons capability.[121] Despite this high importance attached to CW intelligence, results were poor. The JIC accepted that it had very limited information on the Soviet CW threat.[122] Intelligence assessments were reliant on wartime information and scraps of new findings. The lack of verifiable intelligence meant that in this formative period, British nerve agent weapons policy was being formulated in the dark. Defence officials had little idea about what the actual Soviet CW threat was, the level of research and development it had achieved, or if Soviet planners were considering first-use.[123] These limitations meant that the JIC could provide little direct assistance or guidance in the early creation of postwar British nerve agent policy, in establishing how useful a nerve agent capability would be against Soviet forces, or in revealing how urgent or substantial the Soviet CW threat was.

[117] TNA, DEFE 5/4, 'Future Defence Policy', Joint Secretary of the Chiefs of Staff Committee, 23 April 1947, Annex II, Appendix II.; Tucker. *War of Nerves*, 101.

[118] TNA, DEFE 5/4, 'Future Defence Policy', Joint Secretary of the Chiefs of Staff Committee, 23 April 1947, Annex II, Appendix II.

[119] Aldrich, Richard and Coleman, Michael. 1989. The Cold War, the JIC and British Signals Intelligence, 1948. *Intelligence and National Security*, 4:3, 538–539.

[120] TNA, CAB 158/3, 'Russian Interests, Intentions and Capabilities', Joint Intelligence Committee, 23 July 1948, Annex II; Hennessy, Peter. 2007. *Cabinets and the Bomb*. Oxford: Oxford University Press, 74–77.

[121] For a brief summary and review of CW intelligence in this postwar period, see: TNA, CAB 158/3, 'Russian Interests, Intentions and Capabilities', Joint Intelligence Committee, 23 July 1948, Annex II.

[122] TNA, CAB 158/3, 'Increase of the Likelihood of a Major War', Secretary of the Joint Intelligence Committee, 29 January 1948, Annex; TNA, CAB 158/3, 'Russian Interests, Intentions and Capabilities', Joint Intelligence Committee, 23 July 1948, Annex II.

[123] Ibid.

A far more fruitful avenue for British CW policy was trilateral cooperation, where the remarkable level of collaboration between Britain, the United States and Canada had continued to grow. During the war, the United States, Canada and Britain had all cooperated extensively in CW research, with each country having representatives in the relevant research and intelligence committees of the other two.[124] In contrast to other areas of Anglo-American defence cooperation, all parties continued this unusual level of collaboration into the immediate postwar period. Their programmes operated so 'closely in step as to be virtually integrated'.[125] In the early Cold War, yearly trilateral meetings exemplified this level of cooperation. During these senior meetings, defence officials and scientists from all three national research programmes discussed CBW research, the sharing of technical information, the division of research tasks and the pooling of resources and findings.[126] From August 1948, this level of trilateral cooperation grew yet further, when the British Air Ministry consulted with colleagues in the United States and Canada on the potential use of chemical weapons against the Soviet Union.[127] The participants aimed to establish a standardisation procedure which would further bridge the research efforts of all three countries and further integrate their respective CW policies, leading to a more cohesive and unified common practice.

These consultations did achieve an impressive degree of standardisation within the CW field. For as all three countries agreed, they would seek to 'obtain the greatest possible economy in the use of our combined effort and resources', and there would be minimal obstacles to 'full cooperation'.[128] This was especially the case in assessments of the effectiveness of certain types of chemical weapons and with delivery methods. Branching out from this initial tripartite agreement, the trilateral Standardisation Working

[124] For examples, see: TNA, CAB 121/100, 'United States representative on Inter-Service Chemical Warfare Committee', Inter-Service Chemical Warfare Committee, 27 May 1943; Carter, Gradon and Pearson, Graham S. 1996. North Atlantic Chemical and Biological Research Collaboration: 1916–1995. *Journal of Strategic Studies*, 19:1, 78.

[125] Gowing, Margaret. 1974. *Independence and Deterrence: Britain and Atomic Energy 1945–52, Volume 1: Policy Making*. London: Palgrave Macmillan, 94.

[126] For example, see: TNA, DEFE 10/264, 'Collaboration between Great Britain, Canada and United States', Secretary of the Chemical Warfare Sub-Committee, 5 August 1947, Annex I.

[127] A summary of the meeting and the results attained can be found in: TNA, DEFE 10/445, 'Liaison with the United States and Canada', Joint Secretary of the Chemical Warfare Sub-Committee, 16 August 1949, Attached report.

[128] Ibid.

Party for CW built upon and developed the cohesive practices amongst the three countries. This working party was not just for CW research, but also for the standardisation of the types of weapons stockpiled, their delivery methods and even target selection.[129] With tightening resources, this harmonisation with the United States and Canada was a substantial asset to British defence policy. Close trilateral relations influenced the direction of British nerve agent research, but it ensured that British research maintained its advanced level.[130] In isolation, British CW research, which was feeling the strain of broader postwar economic cutbacks in defence, would have struggled to keep pace with significant developments in the field.[131]

Much of this trilateral cooperation was, as with British policy and the nerve agent discovery itself, still kept hidden in a highly secretive environment, concealed from outside observers and the domestic population.[132] As seen in the scarcity of direct parliamentary questions and newspaper reports, there continued to be no real public comprehension or awareness of the nerve agents in 1947–1948. After the previous fleeting mentions of new gases, in March 1948, in a rare instance when CW was mentioned in any capacity in Parliament, cryptic references were made to 'advances in the field of chemical warfare' and to 'new forms of chemical warfare'.[133] These comments gave some slight indication that there had indeed been a development in the CW field, but the nature and form of it remained undisclosed. This minimal information meant that the majority of the public remained largely unaware of the substantial nerve agent development. Strict adherence to secrecy was also reflective of wider Government defence policy in scientific research and development, with it accepted that little could be said publicly about defence research and development, for

[129] Ibid.

[130] For the immediate postwar increase in resources and funding for Porton Down, see: Carter. *Chemical and Biological Defence*, 67–68.

[131] British liaison was not only exceptionally close with the United States and Canada, but relations with other Commonwealth countries also proved beneficial to British CW research. One area of particular importance was that of scientific staff. Dr Rivett, a chemist from South Africa, and two physiologists from Australia proved useful contributions to the British programme.

[132] Balmer. *Secrecy and Science*, 5, 7–8, 59–61.

[133] Hansard, House of Commons, 'Defence', Vol.448, cc40–167, 1 March 1948; Hansard, House of Commons, 'Mr. Shinwell's Statement', Vol.448, cc1012–1931012, 9 March 1948.

much of it 'must remain secret' in the increasingly hostile and confrontational Cold War climate.[134]

More senior oversight and influence did, though, manifest itself in other ways, particularly when it came to shaping broader defence policy. Substantial questions over continental defence, and whether Britain should be willing to fight a protracted ground war in the defence of Western Europe, had a lasting impact on British CW policy. The debate over continental defence had raged since the end of the war, with tensions mounting between the Service Chiefs.[135] A flashpoint in this broader debate and a crucial turning point for CW policy can be seen in 1948–1949, with even greater United States involvement in the defence of Western Europe. This growing commitment by the United States to Western European defence, as seen with the Marshall Plan, NATO, and the Berlin Blockade, made continental defence far more feasible.[136] In turn, British defence policy shifted much more favourably towards continental defence, and the priority for land-based weapons for use against Soviet forces increased.[137]

This growing British commitment to continental defence had direct ramifications for CW policy. Chemical weapons were in part envisaged to deter Soviet first-use of chemical weapons in a continental war. A nerve agent capability could also deter Soviet aerial attacks with chemical weapons. But, the sheer magnitude of Soviet ground forces, as well as the logistical burden of aerial nerve agent weapons, led to emphasis increasingly being placed on a land-based capability.[138] Without firm commitments to continental defence, British CW policy would have lost a major potential role and it may have struggled to fit into the much broader framework and reorientation of defence policy at the time. However, as defence officials were increasingly shifting towards continental defence, support for chemi-

[134] TNA, PREM 8/960, 'Statement on Defence 1949', Presented by the Minister of Defence to Parliament, February 1949.

[135] TNA, CAB 131/5, Minutes of Meeting, Cabinet Defence Committee, 8 January 1948; Kent, John and Young, John. 1992. The "Western Union" concept and British defence policy, 1947–8. In *British Intelligence, Strategy and the Cold War, 1945–51*, ed. Richard Aldrich. Oxon: Routledge, 172.

[136] Young. *Britain and the World*, 154–156.

[137] Dockrill, Saki. 1992. Britain's strategy for Europe: must West Germany be rearmed? 1949–51. In *British Intelligence, Strategy and the Cold War, 1945–51*, ed. Richard Aldrich. Oxon: Routledge, 193; Young. *Cold War Europe 1945–1989*, 5, 144; Reynolds. *Britannia Overruled*, 150.

[138] For example, see: TNA, CAB 81/67, 'Offensive Gas Policy', Inter Service Sub-Committee on Chemical Warfare, 20 March 1946.

cal weapons gained momentum and strengthened alongside this broader transition in British defence policy. The role and place of nerve agent weapons was further established alongside Britain's defence policy and Cold War outlook.

This approach to CW deterrence, focusing on a ground-based continental deterrent, solidified during the crucial and formative years of 1948–1949. During which the War Office, the primary beneficiary of the shift in favour of continental defence, would begin to take the driving seat in the formulation of British nerve agent policy. The War Office began to exert considerable influence over the direction and role of a British nerve agent capability and deterrent. Coinciding with the War Office's increasing interest in the possibilities of chemical weapons for continental defence were consistent arguments by advocates of CW, also mainly in the War Office, who built upon the insights of the DRPC and continued to successfully separate the role of nerve agents from other weapons of mass destruction.[139] As defence officials argued, 'CW is specifically suitable for use against the Russians' in a tactical localised role.[140] Although tactical, rather than strategic, may at first appear to limit nerve agent policy, in reality, this continued to free it from direct and damning comparisons to atomic weapons.

In late 1949, after further trials and testing, it was fully recognised that the nerve agents were most effective against troops and against tanks, rather than in the targeting of Soviet cities on a strategic scale.[141] Nerve agents were thought particularly potent against the T-series of Soviet tanks, which officials had branded 'one of our greatest menaces'.[142] The Soviet Union had shown during the Second World War the advanced nature of its tank design.[143] It was feared that with massed and modern tanks it could launch an unstoppable advance across Western Europe.[144]

[139] TNA, DEFE 10/445, 'Chemical Warfare Policy', Joint Secretaries of the Chemical Warfare Sub-Committee, 2 December 1949, Annex.

[140] Ibid.

[141] TNA, DEFE 10/445, 'Operational Value of Nerve Agents', Chemical Warfare Sub-Committee, 15 November 1949.

[142] TNA, DEFE 10/446, 'Chemical Warfare Policy', Chemical Warfare Sub-Committee, 19 May 1950, Annex.

[143] Maiolo, Joe. 2010. *Cry Havoc: The Arms Race and the Second World War 1931–1941*. London: John Murray, 7.

[144] Castaldi, Carolina and Nuvolari, Alessandro. 2009. Chariots of Fire: The Evolution of Tank Technology, 1915–1945. *Journal of Evolutionary Economics*, 19:4, 551.

After further research to better understand the nerve agents, War Office officials began to fully recognise the benefits that they could bring in countering this form of Soviet warfare. A single nerve agent shell could remove the threat of a Soviet tank. The nerve agent could reach the crew through any gaps in the tank's armour. In addition, this form of nerve agent use was thought to have a substantial psychological effect on ground troops and tank crews.[145] Regardless of the moral or ethical qualms involved in their use in a continental war, from a purely military utility perspective, the nerve agents were increasingly seen as a remarkably effective weapon for countering numerically superior Soviet troops and tanks.

This consolidating tactical role for nerve agents all but removed any potential overlap with atomic and biological weapons, which continued to be thought of as strategic weapons for aerial delivery against targets within the Soviet Union.[146] This separation was all the more critical given that, while atomic weapons were the undisputed prime of weapons of mass destruction, in the postwar period biological weapons were seen as second. As revealed by Brian Balmer, British BW researchers had been exploring the potentialities of anthrax bombs, which would have been useable against Soviet cities.[147] The overlap between atomic weapons and biological weapons meant that they occupied a similar position in the strategic sphere: the targeting of cities and the striking at the Soviet core.[148] This clash would later harm BW policy, for without a truly unique role and with them not reciprocating on their substantial promise and investment, the DRPC began to increasingly turn to chemical weapons as the preferred option for a 'second-scale' weapon of mass destruction.[149] Importantly, rather than compete with nuclear weapons, chemical weapons were

[145] TNA, DEFE 10/445, 'Operational Value of Nerve Agents', Chemical Warfare Sub-Committee, 15 November 1949; TNA, DEFE 10/445, 'Chemical Warfare Policy', Joint Secretaries of the Chemical Warfare Sub-Committee, 2 December 1949, Annex.

[146] TNA, CAB 81/67, 'Offensive Gas Policy', Inter Service Sub-Committee on Chemical Warfare, 20 March 1946; TNA, DEFE 10/445, 'Chemical Warfare Policy', Joint Secretaries of the Chemical Warfare Sub-Committee, 2 December 1949, Annex. Lewis. *Changing Direction*, 214; Balmer. *Britain and Biological Warfare*, 65–67.

[147] Ibid., 65–68.

[148] Lewis. *Changing Direction*, 219; Balmer, Brian. 2006. The UK Biological Weapons Programme. In *Deadly Cultures: Biological Weapons Since 1945*, ed. Mark Wheelis, Lajos Rózsa, and Malcolm Dando. London: Harvard University Press, 52–53.

[149] Agar and Balmer. British Scientists and the Cold War, 219–220.

increasingly seen as complementary.[150] The two forms of warfare could be used in tandem in a continental war, nerve agent weapons against the brunt of the Soviet ground offensive in Europe and atomic weapons against the core of the Soviet Union. Nuclear weapons and nerve agent weapons would also be employed by different sections of the Armed Forces—strategic and tactical—and they would have two very different roles. These differences were crucial for defining and securing the place of nerve agent weapons in Cold War British defence policy.

Where CW did overlap with nuclear weapons, its role was often curtailed and limited. An example of this was seen with Air Ministry requests for an aerial strategic CW capability. Here CW policy did not just overlap with atomic weapons, but also even biological weapons. In early 1950, the DRPC unsurprisingly quashed the requirement for strategic chemical weapons, as it would have overlapped too heavily with atomic weapons.[151] Further limiting CW to a tactical role, Britain lacked the correct CW agent for a strategic capability. For although Britain possessed tabun, and could produce nerve agents such as sarin, these CW agents were non-persistent.[152] Once released, a non-persistent nerve agent did not remain lethal for an extended duration of time, meaning that they were far more effective at achieving quick results, for example against troops or tanks, rather than for rendering cities or large-scale infrastructure inhospitable for an extended period of time.[153] There was also the issue of scale. Covering a substantial area with nerve agents, one comparable to an atomic bomb or even an anthrax bomb, would have required numerous aircraft in a significant operation, likely entailing frequent renewal operations to maintain the nerve agent coverage.

The rejection of a strategic nerve agent capability further bolstered the dominance of the War Office over CW policy. In early January 1950, this

[150] The focus here is on Soviet troops, for it was also recognised that a strategic chemical weapon was as effective when compared to other methods. For strategic atomic weapons see: Baylis. *Ambiguity and Deterrence*, 36–37. For strategic biological weapons see: Balmer. *Britain and Biological Warfare*, 65–67. For the tactical and strategic role of nerve agents see: TNA, CAB 81/67, 'Offensive Gas Policy', Inter-Service Sub-Committee on Chemical Warfare, 20 March 1946.

[151] TNA, DEFE 5/19, 'Chemical Warfare Research and Development Policy', Defence Research Policy Committee, 11 January 1950; TNA, DEFE 5/19, 'Chemical Warfare Research and Development Policy', Vice Chief of the Air Staff, 11 February 1950.

[152] Spiers. *Chemical Warfare*, 2, 5–7.

[153] Dunn, Michael and Sidell, Frederick. 1989. Progress in Medical Defense Against Nerve Agents. *Journal of the American Medical Association*, 262:5, 649–652.

was particularly apparent when the DRPC significantly downgraded the importance of possessing an aerial delivery capability, regardless as to whether it was persistent or non-persistent.[154] When ranking research priorities, the DRPC accorded the highest priorities to defensive equipment, detection methods and nerve agent weapons for the Army. This was despite the Air Ministry attaching 'outstanding' importance to its acquisition of a new aerial nerve agent capability.[155] The direction of CW policy was thus becoming shaped by the growing dominance of the Army's ideas on use, by the perceived military utility of tactical non-persistent nerve agents against Soviet ground forces, and by the fact that nerve agent weapons were increasingly viewed as complementary to atomic weapons, rather than as a competitor.

The impact of CW policy turning more towards the Army, and away from the Air Force, actively shaped how British defence officials viewed CW deterrence in 1949 and early 1950.[156] This problem of deterrence was taking on even greater urgency as Second World War era CW stocks were gradually deteriorating, existing delivery methods were 'unsatisfactory', and it would take a number of years until a replacement nerve agent capability could be domestically mass-produced.[157] Both the War Office and Air Ministry agreed that, due to the poor state of the war reserve, Britain was now unable to wage CW effectively from the outset of war and it would soon be without an efficient CW deterrent.[158] Even after detailed planning and research, Britain was struggling to fulfil the Defence Committee's directive of 1946 that the country should be prepared to wage CW at the outset of hostilities. However, addressing this issue, and deciding how best to deter the Soviet Union, evoked considerable disagreement between the

[154] TNA, DEFE 5/19, 'Chemical Warfare Research and Development Policy', Defence Research Policy Committee, 11 January 1950; TNA, DEFE 5/19, 'Chemical Warfare Research and Development Policy', Vice Chief of the Air Staff, 11 February 1950. Whereas Army projectiles could be targeted against localised forces, such as tanks or troops, it was also difficult to find an effective and accurate means of dispersal via the Air Force.

[155] TNA, DEFE 5/19, 'Chemical Warfare Research and Development Policy', Defence Research Policy Committee, 11 January 1950; TNA, DEFE 5/19, 'Chemical Warfare Research and Development Policy', Vice Chief of the Air Staff, 11 February 1950.

[156] TNA, DEFE 10/445, Minutes of Meeting, Chemical Warfare Sub-Committee, 1 November 1949.

[157] Ibid.

[158] Ibid.

Air Ministry and the War Office. Both Services had very different interpretations over how best to deter the Soviet Union.[159]

For the Air Ministry, the poor state of the war reserve was ‘unlikely to deter an aggressor if he considers that the use of CW will be to his advantage’.[160] Despite this apparent weakness, the Air Ministry was keen to point out that deterrence should not be based on the immediate capacity to wage CW, as the ‘ultimate threat’ was actually the total potential CW development and production capacity held by Britain and her allies.[161] This led to Air Ministry officials concluding in November 1949 that ‘the only real deterrent will undoubtedly be the fear of ultimate reprisals on a very large scale’.[162] They believed that emphasis should thus be placed on the total potential production capacity which Britain could bring to bear in the CW field throughout a prolonged war, instead of stressing an immediate retaliatory capability. This total potential capability implied a protracted war, yet it was deemed the most effective form of deterring Soviet CW use as it threatened substantial reprisals. In addition, the Air Ministry proposed abandoning certain ineffective CW munitions held in the war reserve, such as phosgene bombs, in order to focus on the nerve agents and the total potential production capacity.[163] Freeing up resources, the Air Force perhaps had an eye on revitalising its aerial nerve agent capability at a later stage.

These arguments also conveniently coincided with the state of play in CW policy. The Air Ministry was unlikely to possess a retaliatory nerve agent capability in the near future with the DRPC questioning its importance and role, and with priority assigned to nerve agent weapons for the Army. As a result, it had little vested interest in supporting the acquisition of an immediate retaliatory capability for deterrence. The Air Ministry had little incentive to support interpretations of deterrence based on an

[159] TNA, DEFE 10/445, ‘1949 Report on Chemical Warfare’, Joint Secretaries of the Chemical Warfare Sub-Committee, 17 August 1949, Attached note; TNA, DEFE 10/265, ‘Chemical Warfare Policy’, Joint Secretaries of the Chemical Warfare Sub-Committee, 11 October 1949, Annex; TNA, DEFE 10/445, Minutes of Meeting, Chemical Warfare Sub-Committee, 1 November 1949.

[160] TNA, DEFE 10/265, ‘Chemical Warfare Policy’, Joint Secretaries of the Chemical Warfare Sub-Committee, 11 October 1949, Annex.

[161] TNA, DEFE 10/265, ‘1949 Report on Chemical Warfare’, Joint Secretary of the Chemical Warfare Sub-Committee, 11 October 1949, Annex.

[162] Ibid.

[163] TNA, DEFE 10/445, Minutes of Meeting, Chemical Warfare Sub-Committee, 1 November 1949.

immediate retaliatory capability. Its support of a longer-term alternative interpretation of CW deterrence, and questioning the need for an immediate retaliatory nerve agent capability, reveals a growing divergence in approach to CW policy by the Air Ministry and the War Office. It also shows the Air Ministry somewhat deviating from the more widely shared interpretation of CW deterrence, which was that the ability and threat to retaliate immediately with chemical weapons prevented its outbreak.

For the War Office, an immediate retaliatory capability and maintaining existing stocks was the clear and unquestionable way of best deterring Soviet first-use, as they believed that the:

> Knowledge that we possess them [chemical weapons] would impose on any potential enemy the inconvenience of ensuring the protection of his own troops, and act as a deterrent to any enemy from initiating CW.[164]

Even after the protestations of the Air Ministry, but in a sign of its expanding influence over CW policy, it was this deterrent argument championed by the War Office which prevailed. Stocks of CW agents were maintained. Emphasis was placed on acquiring an immediate nerve agent retaliatory capability for the Army.[165] Existing stocks were now deemed to possess 'some deterrent value', notwithstanding their inefficiency and being outmoded.[166] Any disposal of these existing stocks was dubbed a 'most imprudent' step.[167] Nerve agent stocks acquired from the continent would therefore continue to be used to deter the outbreak of CW on it, as even though the existing war reserve and the tabun stocks could not be deployed effectively, the knowledge that Britain possessed them was still thought to play some part in deterring Soviet first-use. These discussions over the future role of nerve agent weapons in British nerve agent policy had taken almost four years.

[164] TNA, DEFE 10/445, '1949 Report on Chemical Warfare', Joint Secretaries of the Chemical Warfare Sub-Committee, 17 August 1949, Attached note.

[165] TNA, DEFE 10/445, Minutes of Meeting, Chemical Warfare Sub-Committee, 1 November 1949.

[166] Ibid.

[167] Ibid.

'A Tragic State of Affairs'[168]

In late 1949, after they had initially disagreed over what to do with existing stocks and how best to deter Soviet CW use, the War Office, Air Ministry, the Chemical Warfare Sub-Committee (CWSC) of the CoS, and the DRPC all eventually agreed on the path forward. The consensus in defence circles, taking around four years since the finding of German nerve agents, coincided with growing Cold War tensions and came shortly after Communist victory in the Chinese Civil War. British defence officials now agreed that the domestic production of nerve agent weapons was a necessary step for defence policy in the Cold War.[169] If deterrence were to work, then British capabilities needed to be credible and effective, rather than deteriorating and borderline unusable.[170] Possession of a substantial nerve agent capability was therefore dubbed 'one of the most powerful deterrents to the initiation of CW by an enemy'.[171] A nerve agent capability was envisaged to begin with the domestic production of one ton of sarin per week, but after the appropriate techniques and methods had been mastered, the site would be massively expanded to produce 50 tons of sarin per week.[172] This considerable expansion was believed a necessary step to meet Britain's requirement for a nerve agent capability and to deter Soviet first-use of chemical weapons. By selecting sarin as the prime CW agent, and similarly to the findings of the United States CW programme, officials also took the safer middle path. Of the three main German nerve agent discoveries, tabun, as already recognised in 1945–1946, was less lethal than both sarin and soman. Soman, however, was extremely difficult to mass-produce, leaving sarin, with its high lethality and comparatively

[168] Hansard, House of Commons, 'Defence', Vol.478, cc467–645467, 26 July 1950.

[169] TNA, DEFE 10/445, '1949 Report on Chemical Warfare', Joint Secretary of the Chemical Warfare Sub-Committee, 17 August 1949, Annex; TNA, DEFE 10/265, '1949 Report on Chemical Warfare', Joint Secretary of the Chemical Warfare Sub-Committee, 11 October 1949, Annex; TNA, DEFE 10/445, 'Operational Value of Nerve Gas', Chemical Warfare Sub-Committee, 15 November 1949. The Chemical Warfare Sub-Committee advised the CoS on CW policy.

[170] TNA, DEFE 10/446, 'Chemical Warfare Policy', Chemical Warfare Sub-Committee, 9 May 1950.

[171] Ibid.

[172] TNA, DEFE 10/445, 'Operational Value of Nerve Gas', Chemical Warfare Sub-Committee, 15 November 1949.

easier means of mass-production, as the preferred nerve agent for mass-production.[173]

The focus and emphasis on the nerve agent weapons, however, had not yet trickled down to funding for CW research, which only stood at around £300,000 per annum in 1949.[174] This funding did not include additional costs such as the construction of a sarin facility, which would cost an additional several million pounds.[175] While funding was limited, British CW research at Porton Down had survived the postwar period. Cuts to staff numbers, in line with other cutbacks in defence, were substantial but not fatal, dropping from 1000 to 800.[176] The nerve agents secured funding for Porton Down's continued CW research, yet they were not enough on their own to equate with or override postwar reductions in defence expenditure as a whole.

The shift to supporting the development of a domestic nerve agent capability was given greater impetus with the CoS Global Strategy Paper of 1950, which firmly placed and confirmed the role of continental defence at the top of British defence requirements.[177] The document endorsed the role of the Army and committed Britain to the defence of mainland Europe. As noted by the DRPC, the emphasis on defending Western Europe was significant, with it having a 'considerable effect on our research and development policy'.[178] In confronting the Soviet numerical advantage, allied forces needed to be supplied with high-quality equipment for dealing with, amongst other threats, large numbers of troops and heavily armoured tanks.[179] The number one priority was for a nerve agent weapon

[173] For similar findings in the United States, see: Tucker. *War of Nerves*, 122–123. For British considerations of sarin, see: TNA, DEFE 10/445, 'Operational Value of Nerve Gas', Chemical Warfare Sub-Committee, 15 November 1949.

[174] TNA, CAB 131/7, 'Defence Research and Production Programmes, 1949–50', Report by Defence Production Committee, 10 March 1949, Annex I, Appendix B.

[175] For later costs on the expansion of the Nancekuke facility, see: TNA, DEFE 5/37, 'Chemical and Biological Warfare', Chairman of the Defence Research Policy Committee, 6 March 1952, Annex; TNA, DEFE 5/43, 'Chemical Warfare Policy', Deputy Chairman of the Defence Research Policy Committee, 15 December 1952.

[176] Carter. *Chemical and Biological Defence*, 70

[177] TNA, CAB 131/9, 'Defence Policy and Global Strategy', Chiefs of Staff Committee, 7 June 1950.

[178] Ibid.

[179] TNA, DEFE 4/35, Minutes of Meeting, Chiefs of Staff Committee, 23 August 1950, Confidential Annex; TNA, DEFE 4/35, Minutes of Meeting, Chiefs of Staff Committee, 13 September 1950, Confidential Annex.

with an anti-tank role. The second priority was already for a nerve agent weapon to use against troops.[180] British nerve agent policy thus directly mirrored and benefitted from ongoing defence debates, with the rise and confirmation of continental defence securing the role and funding of British nerve agent weapons research and development.

Alongside these important developments, the impressive levels of secrecy surrounding British CW policy during this period meant that there still remained extremely limited coverage, criticism and accountability. Again, only minor revelations continued to drip into the public sphere. One such example was seen in July 1950, when the Labour Minister for Defence Manny Shinwell revealed Government interest in CW research when addressing defence policy.[181] Shinwell, as part of a much larger statement on defence, included the observation that 'we cannot afford to lag behind' in researching CBW defensive measures.[182] An astute Labour MP, Rhys Davies, pounced upon on this rare reference to CBW policy, which revealed Government activity in the CBW field. Davies, proclaiming that the reference to CBW had 'frightened' him, lambasted his fellow Labour MP by declaring that 'it is a tragic state of affairs that a British Socialist statesman should say that' CBW research was necessary.[183] This incident was an early indicator of backbench Labour MPs views, opposition and outright hostility towards British involvement in CBW research.

By September 1950, behind closed doors and still adhering to stringent levels of secrecy, British CW policy now had a clear purpose: the defence of continental Europe, the deterring of Soviet first-use, and the countering of Soviet troops and tanks. As such, the DRPC passed its conclusions and the recommendations for research, development and production of nerve agent weapons up to the CoS, which now approved the requests.[184] The Minister of Defence, Manny Shinwell, also supported the immediate construction of a nerve agent facility at Nancekuke, near Portreath, and the production of sarin.[185]

[180] TNA, DEFE 10/446, 'Chemical Warfare Policy', Chemical Warfare Sub-Committee, 9 May 1950.

[181] Hansard, House of Commons, 'Defence', Vol.478, cc467–645467, 26 July 1950.

[182] Ibid.

[183] Ibid.

[184] TNA, DEFE 4/35, Minutes of Meeting, Chiefs of Staff Committee, 23 August 1950, Confidential Annex.

[185] Ibid.; TNA, CAB 131/9, 'Chemical Warfare Policy', Chiefs of Staff Committee, 4 September 1950.

On 4 September 1950, with the Korean War raging, the CoS updated the Cabinet Defence Committee. And, it clarified and reinforced the emerging interpretation of nerve agent weapons and CW deterrence, by stating that:

> As proved in the last war, the ability to retaliate effectively, and immediately, is one of the most powerful deterrents to the initiation of Chemical Warfare by an enemy.[186]

While again showing the legacy and impact of the Second World War on British conceptions of CW deterrence, outright and clear CoS backing was a pivotal moment for advocates of a nerve agent capability. Though the Korean War undoubtedly spurred on the decision to acquire a nerve agent capability, it is worth noting that there had already been very strong and mounting military support for such a policy change. In order to meet the deterrent requirement and in line with the suggestions of defence officials, the CoS informed the Defence Committee of the decision to begin the immediate construction of the one ton per week sarin production facility at Nancekuke.[187] The CoS also re-affirmed the dominance of the Army in the nerve agent field, by confirming that the number one priority for nerve agent weapons production was for the development of anti-tank nerve agent weapons for the Army.[188] The second priority was for a nerve agent artillery shell for use against Soviet troops, and an aerial nerve agent bomb was a distant third.[189] These nerve agent requirements, of domestic mass-production and delivery methods, were thought essential to British defence policy. As noted by the CoS, the nerve agents had 'tremendous potentialities'.[190]

[186] Ibid. This continuation of the legacy of wartime CW deterrence would also seem to run in tandem with the strengthening of moral aversion to the CW. For the development of the moral aspect and non-use, see: Price. *The Chemical Weapons Taboo*, 174–176.

[187] The Minister of Defence, Emmanuel Shinwell, had already given the green light for nerve agent production in August 1950. The plant was to be located at Portreath, Nancekuke. In September this policy was clarified, and the entire Cabinet Defence Committee was informed of the decision. See: TNA, DEFE 10/37, Minutes of Meeting, Defence Research Policy Committee, 12 April 1949; TNA, CAB 131/9, 'Chemical Warfare Policy', Chiefs of Staff Committee, 4 September 1950.

[188] Ibid.

[189] Ibid.

[190] Ibid.

On 27 September 1950, alongside the escalating Korean War and shortly after the CoS and Shinwell's approval, the Defence Committee, again chaired by Attlee, met to confirm the future of British CW research. They agreed with and supported the decision to domestically produce sarin.[191] Little consideration was explicitly given to moral or legal norms, military necessity and the fearful Cold War climate dominated considerations. For the third time in the postwar period, Attlee again revealed his tacit support of, and commitment to, CW deterrence through the threat of reprisals. The Defence Committee, concerned by the Soviet threat and the poor state of Britain's CW preparedness, also supported the future expansion of Nancekuke to a substantial 50 tons of sarin per week and the continued research and development of new nerve agent weapons.[192] In a time of war, without any public statement or parliamentary acknowledgement, Britain thus again approved the domestic production of chemical weapons.[193] With the endorsement of the Attlee Government, the production of sarin and the development of nerve agent weapons was accepted and supported at the highest level.[194]

Notably, all these changes and expenditures were recommended despite the worsening state of the economy. Nerve agent weapons were deemed a necessary endeavour. Unfortunately for defence officials though, and as recognised by the CoS and Labour ministers, Britain's mass-produced sarin capability would not be ready until 1957, whereas the Soviet Union was thought capable of mass-producing nerve agents from mid-1951.[195] This perceived six year time lag was brought about by Britain relying on German tabun stocks and war reserves of sulphur mustard and phosgene as an interim capability, whilst the Soviet Union was believed to have moved straight on to the domestic production of nerve agents, like the United States.[196] In the immediate postwar period, Britain possessed the

[191] TNA, CAB 131/8, Minutes of Meeting, Cabinet Defence Committee, 27 September 1950.

[192] Ibid.

[193] Had the Korean War not been raging, political approval for the production of lethal and controversial nerve agent weapons would almost certainly have been far less likely.

[194] TNA, CAB 131/8, Minutes of Meeting, Cabinet Defence Committee, 27 September 1950.

[195] TNA, CAB 131/9, 'Chemical Warfare Policy', Chiefs of Staff Committee, 4 September 1950; TNA, CAB 131/8, Minutes of Meeting, Cabinet Defence Committee, 27 September 1950.

[196] Tucker. *War of Nerves*, 122–130.

bulk of German nerve agents, whereas in 1950 it had fallen seriously behind the superpowers in the CW field, with the CoS fearful of the 'formidable' Soviet nerve agent threat and unable to effectively retaliate in kind.[197] This perceived Soviet advantage added a real sense of urgency to British CW policy. For as the CoS revealed, another key reason for Army requirements being the top priority in nerve agent weapons was also the speed with which they could be produced.[198] An effective aerial nerve agent weapon and its suitable delivery method would have required greater technical sophistication, and taken more time to develop; time which Britain no longer had. The early lead could soon be lost.[199]

Another avenue to address this perceived growing imbalance in capabilities vis-à-vis the Soviet Union continued to be through Anglo-American cooperation. Unlike atomic cooperation, which was undermined by the McMahon Act in 1946, CW cooperation between the two countries remained unhindered. This level of cooperation rendered possible the potential British acquisition of nerve agents produced in the United States, which would have provided Britain with one of the most advanced chemical weapons known at the time and have acted as a stop-gap deterrent while domestic production caught up.[200] Ruefully, however, defence officials recognised that this option was not yet viable. The United States would only be mass-producing sarin from 1952, a target date which British experts thought overly optimistic. Because of this, British officials believed that the focus should be on producing 'our own weapons' with greater alacrity, rather than relying on the United States for a stop-gap nerve agent capability in the Cold War.[201]

The feared deficit in immediate CW capabilities and with a lack of viable short-term options, the Defence Committee pushed the Air Ministry to explore re-purposing the tabun bombs captured from Germany.[202] These bombs still contained some useable tabun, it was just the delivery mechanism which was inefficient. Extending the life-span of captured tabun

[197] TNA, CAB 131/9, 'Chemical Warfare Policy', Chiefs of Staff Committee, 4 September 1950.

[198] Ibid.

[199] Ibid.

[200] TNA, DEFE 4/35, Minutes of Meeting, Chiefs of Staff Committee, 23 August 1950, Confidential Annex.

[201] Ibid.

[202] TNA, CAB 131/8, Minutes of Meeting, Cabinet Defence Committee, 27 September 1950.

stocks, and their role as a stop-gap deterrent was, therefore, another option which British politicians and defence officials explored.[203] The desperate situation led to a search for a 'quick fix'.

Fears of the Soviet threat manifested itself in other ways. Attlee, the CoS and the rest of the Defence Committee were particularly concerned over the nerve agent threat to British civilians.[204] With the Korean War raging there was great uncertainty over Soviet intent, and tensions were high both is East Asia and in Europe. Perceptions of the danger of Communist aggression in Europe directly impacted civil defence planning, with concerns raised over British vulnerability to a Soviet aerial offensive with nerve agent weapons. Similarly to atomic weapons, the population density of London and its proximity to the Soviet Union made it a prime target for an attack with nerve agents.[205] Despite having reservations over the viability of strategic aerial delivery, British CW experts alarmingly reported that if the Soviet Union used nerve agent weapons in an attack against London, there would be thirty times more deaths than with conventional bombing.[206] Given this uncertainty over Soviet intent, method of use and the looming nerve agent deficit, the need for effective civilian defensive equipment was thought all the more important.[207]

Civil defence preparations for CW were significantly accelerated. A core part of this increased focus on defensive measures and civilian preparations was the stepping up of the mass-production of new respirators.[208] These new respirators would be far more effective in mitigating against the nerve

[203] TNA, CAB 131/9, 'Chemical Warfare Policy', Chiefs of Staff Committee, 4 September 1950; TNA, CAB 131/8, Minutes of Meeting, Cabinet Defence Committee, 27 September 1950.

[204] TNA, CAB 131/9, 'Chemical Warfare Policy', Chiefs of Staff Committee, 4 September 1950; TNA, CAB 131/8, Minutes of Meeting, Cabinet Defence Committee, 27 September 1950.

[205] Lewis. *Changing Direction*, 223.

[206] TNA, DEFE 10/445, 'Effects of an Attack with Nerve Gas on Central London', Joint Secretaries of the Chemical Warfare Sub-Committee, 20 June 1949.

[207] TNA, CAB 131/8, Minutes of Meeting, Cabinet Defence Committee, 27 September 1950. Some scientific advisors in 1950, such as J. W. Martin a member of the Scientific Advisers Branch of the Home Office, also argued that the main way to stop Soviet CW use was through deterrence. See: TNA, HO 228/11, 'The Potentialities of Nerve Gas as a CW Agent', J. W. Martin, 12 April 1950.

[208] TNA, CAB 131/8, Minutes of Meeting, Cabinet Defence Committee, 27 September 1950.

agent threat, as they would be more advanced and better fitting.[209] Taking shelter with a new gas mask was believed to give a good degree of protection, as remaining inside would mitigate against nerve agent exposure through skin contact and the new gas masks would protect against inhalation. Although the substantial increase in defensive equipment was approved, neither the CoS nor the Defence Committee was impressed by the relatively poor state of affairs and of Britain's distinct vulnerability. The Defence Committee disapprovingly accused the Ministry of Supply and defence officials of seeking 'too high a standard of perfection' at the cost of valuable time. Britain was vulnerable.[210] Without an effective nerve agent capability and with limited defensive equipment, the country was indeed in a poor state to deter Soviet use of chemical weapons and unable to properly defend against an attack. The 1950 Cabinet Defence Committee decision and expansion of CW policy was, therefore, an amalgamation of attempts to mitigate this concerning imbalance regarding the Soviet threat and a recognition of the perceived military value and utility of nerve agent weapons for both use and deterrence.

By the end of 1950, in secret and behind closed doors, British CW policy had thus significantly accelerated. Even though in 1945 British CW research was spurred into action after the hugely significant nerve agent development and even with key figures such as Clement Attlee supporting acquisition for deterrence, there was no clear role or place for these new nerve agent weapons in British defence policy. Years were spent secretly investigating and analysing the possible military uses of the nerve agents, with scientific research continuing at pace. Understanding the ramifications of the nerve agent development took time. A clear and agreed-upon role required some form of consensus in a period initially characterised by competing interests. The consideration of acquisition centred on whether nerve agent weapons were needed, how or if they could deter Soviet first-use, counter Soviet numerical advantages in conventional forces, or be used for strategic bombing. Towards the end of the 1940s, amid uncertainty and despite the divergent attempts of the Air Ministry, it was the Army who emerged as the chief advocate of nerve agent weapons. This

[209] TNA, CAB 131/9, 'Chemical Warfare Policy', Chiefs of Staff Committee, 4 September 1950; TNA, CAB 131/8, Minutes of Meeting, Cabinet Defence Committee, 27 September 1950.

[210] TNA, CAB 131/8, Minutes of Meeting, Cabinet Defence Committee, 27 September 1950.

growing dominance of the Army, which was supported by military assessments, the continental shift in defence policy, the escalating Cold War, and by the findings of the DRPC and the CWSC, meant that deterring CW came to be seen as predominantly achievable through land-based weapons. This focus on a tactical chemical weapon had the substantial benefit of distinguishing nerve agent weapons from other weapons of mass destruction.

After a clear role had emerged, it was however soon realised that even with Britain possessing the bulk of the world's nerve agents after the occupation of Germany, after 1950 it would soon be lagging behind the Soviet Union and the United States. British dependence on wartime stocks of sulphur mustard, phosgene and captured tabun from Germany had, in fact, placed it at a significant disadvantage in the long-run. The Soviet Union, which had not discovered substantial stocks in Germany, but had acquired the methods, techniques, and equipment for mass-production, was forced to explore mass-production immediately. The shock realisation that Britain could be around five years behind the Soviet Union in nerve agent production, and without a truly effective deterrent, pushed Attlee's Labour Government to secretly accelerate and approve the production of defensive equipment, the development of nerve agent weapons and the domestic production of sarin. The urgency attached to nerve agent production, alongside the acceptance of its military utility and the consolidation of its deterrence based and retaliatory roles, would have significant ramifications for British CW policy in the Cold War, especially when it came to the landmark 1952 Global Strategy Paper.

CHAPTER 3

A Step Too Far: The Nerve Agents and the Global Strategy Paper, 1951–1953

The United Kingdom is at present committed by the Geneva Convention not to use [chemical and biological weapons] except in retaliation... The new nerve gases can, however, be used tactically to great advantage and would provide the Allies with weapons of real value against an enemy who relies on massed formations.
Chiefs of Staff Committee (TNA, PREM 11/49, 'Defence Policy and Global Strategy', Chiefs of Staff Committee, 9 July 1952)

The attention of the United States, Britain, and the Soviet Union had rapidly shifted from Europe to East Asia, in 1950, with the Korean War. Although the exact causal factors and responsibility for the outbreak of the war are still debated to this day, the ramifications of the conflict are clear.[1] Labelled a 'great calamity', the Korean War represented the extension of the Cold War to East Asia.[2] It was a conflict which would have a legacy far beyond the immediate postwar period, and even the Cold War. The ramifications of the war were not confined to East Asia; they were global. Leading to a greater commitment to combat perceived Soviet aggression abroad, the United States pledged to provide

[1] Park, Myunglim. 1998. The '"American Boundary", Provocation, and the Outbreak of the Korean War. *Social Science Japan Journal*, 1:1, 31.
[2] Westad. *The Cold War*, 159.

W. King, *Nerve Agents in Postwar Britain*, Britain and the World,
https://doi.org/10.1007/978-3-030-70474-2_3

more land forces to NATO in Western Europe in early 1951 and adopted the significant rearmament measures advocated in NSC-68.[3] Britain, conversely, was struggling to mirror this rapid escalation in capabilities and expenditure.

The outbreak and outcome of the Korean War also bore witness to an increasingly awkward and challenging role for Britain. Even though Prime Minister Winston Churchill was reunited with his former wartime leaders, President Harry Truman and Soviet leader Joseph Stalin, after the Conservatives won the October 1951 general election, the relationships and dynamics between the three countries had changed entirely.[4] Churchill returned to the top of British politics during yet another war. But, this time, he had remarkably different responsibilities.[5] His predecessor, Clement Attlee, had launched a substantial re-armament programme, including approving nerve agent production, weakening the already parlous state of Britain's economy.[6] The economic strains of the Korean War combined with the mounting costs of the growing welfare state and of continued postwar recovery. By mid-1951 Britain was in significant financial difficulty.[7] Churchill, rather than being in a position to tackle issues and the war head-on, was instead tasked with addressing economic pressures, reducing inflated defence expenditure and assuaging political uncertainty and alarmism.

The outbreak of the Korean War triggered a period of intense anxiety about the future security of Europe and the reliability of the United States.[8] Confrontational United States foreign policy in this tumultuous period led to some disagreement between the Anglo-American partners over the very nature of the Soviet threat, and on the dangers of the atomic

[3] Leffler. The emergence of American grand strategy, 1945–1952, 82–86.

[4] Bartlett. *The Special Relationship*, 51; Jones. *UK Strategic Deterrent, Volume I*, 21.

[5] Fry. *The Politics of Decline*, 114–116.

[6] Lowe, Peter. 1989. The Significance of the Korean War in Anglo-American Relations, 1950–1953. In *British Foreign Policy 1945–1956*, ed. Michael L. Dockrill and John W. Young. London: Macmillan, 126; Baylis, John. 1989. *British Defence Policy: Striking the Right Balance*. New York, NY: Palgrave Macmillan, 74.

[7] Rosecrance, R. N. 1968. *Defence of the Realm: British Strategy in the Nuclear Epoch*. New York, NY: Columbia University Press, 154–155.

[8] Deighton, Anne. 1992. "Arming the key battleground": German rearmament, 1950–1955. *Journal of Diplomacy & Statecraft*, 3:2, 344–345.

age.[9] In the United States, the 'red menace' became an obsession for many.[10] While British experts took the Soviet threat seriously, United States assessments of a global, aggressive and expansionist Soviet Union were thought exaggerated. Believing that the Soviet Union would not risk global war due to the overwhelming US superiority in nuclear weapons, the British CoS disagreed with their counterparts in the United States.[11] Given this uncertainty, pressures on the defence budget generated by rearmament, and the development by the Soviet Union of its own nuclear capabilities (when the UK had yet to test its first atomic bomb), the Churchill Government sought to clarify British defence policy in a groundbreaking review in 1952.

The importance of the 1952 Defence Policy and Global Strategy Paper (GSP), formulated by the British CoS under the instruction of the Churchill Government, is widely debated. For while the GSP confirmed the centrality of nuclear weapons and nuclear deterrence to British defence policy, it also stated that the immediate threat of war was unlikely.[12] The GSP outlined how British defence policy should shift from purely addressing and preparing for the threat of immediate war, to more of a long-term focus.[13] Some historians, for example Baylis, Clarke and Wheeler, question the importance and impact of the paper. They cite its economic focus and its re-iteration of what was in effect already existing policy.[14] Certain United States officials were also underwhelmed by the GSP. They damningly concluding that Britain was merely 'rearranging their strategic estimate to fit their economic situation'.[15] In contrast, other commentators, for example, Rosecrance, argue that the GSP was the most influential British defence paper in the postwar period, and Pierre labels it a 'classic among military documents'.[16] The importance of the paper can be debated

[9] Clark, Ian and Wheeler, Nicholas. 1989. *The British Origins of Nuclear Strategy 1945–1955*. Oxford: Clarendon Press, 136.

[10] Reynolds. *Britannia Overruled*, 172.

[11] Clark and Wheeler. *The British Origins of Nuclear Strategy*, 146.

[12] Jones. *UK Strategic Deterrent, Volume I*, 21.

[13] Twigge, Stephen and Scott, Len. 2000. *Planning Armageddon: Britain, the United States and the Command of Western Forces 1945–1964*. Amsterdam: OPA, 148.

[14] Clark and Wheeler. *The British Origins of Nuclear Strategy*, 170; Baylis. *Ambiguity and Deterrence*, 149.

[15] Johnston, Andrew. 1998. Mr Slessor Goes to Washington: The Influence of the British Global Strategy Paper on the Eisenhower New Look. *Diplomatic History*, 22:3, 379.

[16] Rosecrance. *Defence of the Realm*, 159; Pierre, Andrew. 1972. *Nuclear Politics: The British Experience with an independent Strategic Force, 1939–1970*. Oxford: Oxford University

in terms of what it meant for nuclear strategy and the defence budget, but its various nuances are often underappreciated and overlooked.

For British CW policy, the 1952 GSP represents a crucial turning point. This importance is not just illustrated by the final text of the 1952 GSP. What was left out at the drafting stage, and what the CoS actually wanted to include, hold illuminating and alarming insights. This chapter will analyse the build-up to the GSP, following on from the growing and consolidating military interest seen in the previous chapter. It will then explore the debate over the inclusion of CW, and end by assessing the significant impact the GSP had on British nerve agent policy in the Cold War.

A Weapon for War

After September 1950, when the Cabinet Defence Committee approved the production of nerve agents in Britain and the creation of a nerve agent facility, the CoS moved quickly and secretly. Fears were growing over the Soviet CW threat, and a consensus was emerging over the perceived military role and value of nerve agent weapons in British defence policy.[17] Within a year of granting political approval, the location of Britain's nerve agent facility was finalised. Construction had begun. Nancekuke, on the north coast of Cornwall, was confirmed as the ideal location, despite concerns that it was vulnerable to air, coast and seaward attack—especially in the form of small raids or acts of sabotage.[18] Even though this location was not militarily the soundest, the site had good railway and communications links, and it was close to vital chemical industries in South Wales, which could supply chemicals necessary for nerve agent production. The coastal location would also allow for the discharge of 'dangerous effluent' into the sea.[19] This issue of disposal and the environment in CW policy, although extremely controversial, attracted comparatively little public attention at

Press, 86–88; Baylis, John and Stoddart, Kristan. 2015. *The British Nuclear Experience: The Role of Beliefs, Culture, and Identity*. Oxford: Oxford University Press, 44; Anthony Eden, then Foreign Secretary at the time, even informed Churchill he thought it a 'remarkable state paper'. See: TNA, PREM 11/49, Anthony Eden to Winston Churchill, 18 July 1952.

[17] TNA, CAB 131/8, Minutes of Meeting, Cabinet Defence Committee, 27 September 1950.

[18] TNA, DEFE 4/47, Minutes of Meeting, Chiefs of Staff Committee, 5 October 1951.

[19] Ibid.

the time and appeared to raise few moral qualms. Although the plans to discharge chemical waste into the sea were kept secret, in the House of Commons, just months before, it was revealed that since 1945 Britain had dumped over 100,000 tons of chemical weapons, including weapons casings and some captured and outmoded German stocks, into the Atlantic.[20] British CW policy and this massive dumping of surplus chemical weapons triggered only minor public interest, again revealing the stringent levels of secrecy and the comparative dearth of publicly available information. While snippets of information on British CW activities and the disposal of CW munitions slipped into the public sphere, the nerve agent facility, and the substantial scale envisaged for it, were kept secret. The D-Notice system was still limiting press coverage of sensitive topics. Even five years after their finding, many in Britain remained oblivious of the nerve agents.

The services and much of the groundwork for the full 50 ton per week nerve agent facility were installed by late 1950.[21] This groundwork and core infrastructure would facilitate the rapid expansion of the site to a level which would meet British nerve agent requirements for the Cold War.[22] In this effort, assistance and advice from the United States were thought to be of significant value in easing technical and production difficulties. Producing the nerve agents in bulk was no easy feat.[23] In 1951 alone the United States spent around $3.6 million on nerve agent research. It had almost completed a substantial production facility at the substantial cost of around $79 million, which could produce 9000 tons of sarin per year.[24] These United States actions, and the sheer scale and ambition of its nerve agent programme, coincided with the growing momentum behind, and support for, a nerve agent capability in British military circles.

British officials were increasingly keen on not just owning an effective CW deterrent, but the military utility of nerve agent weapons was growing in appeal. From the early 1950s, key British defence committees

[20] Hansard, House of Commons, 'Equipment (Dumping)', Vol.486, cc1656–7, 17 April 1951.

[21] TNA, DEFE 4/47, Minutes of Meeting, Chiefs of Staff Committee, 5 October 1951.

[22] Ibid.

[23] Ibid.

[24] NARA II, RG 330, Entry 241, Box 486, 'Funding of Chemical Corps G Agent Research and Development Program', Office of the Chief Chemical Officer, 25 January 1952; NARA II, RG 218, Central Decimal File 1951–1953, Box 152, 'Priority for Chemical and Biological Warfare Facilities', Chief of Staff, United States Army, 25 February 1952.

increasingly advocated a more significant, active, and prominent role for nerve agent weapons in defence planning. This rising tide of support for nerve agent weapons was seen in the reports of the newly formed sub-committee on Anti-Tank Defence Measures, which operated under the influential DRPC, as well as the CWSC, which operated under the CoS. All of these committees, either directly or indirectly, were responsible for guiding and informing British defence policy at the highest levels. Their collective findings represented a mounting consensus in defence policy. Their assessments would hold significant ramifications for both British CW policy and defence policy in the Cold War.

Reflective of the grave concerns British defence officials had over Soviet tanks, and of the need for new weapons in the Cold War, the Anti-Tank Defence Measures Sub-Committee was established in June 1950.[25] Even from the 1930s, there had been clear indicators of the frightening potential of Soviet military capabilities, with the rise of 'red militarism'.[26] The Second World War had compounded these fears, with rapid Soviet mobilisation and the mass-production of tanks. In 1950, the Joint Intelligence Committee estimated that the Soviet Union possessed 20 tank divisions, each one counting several hundred tanks, and that it was able to produce 335 tanks every single month.[27]

The committee, tasked with addressing this unenviable problem, had a broad remit: to discover and recommend for development new anti-tank weapons. Its findings played a crucial part in adding to the emerging consensus that nerve agent weapons had a significant role to play in any war with the Soviet Union.[28] Just two months after its creation, the committee assigned a nerve agent anti-tank weapon a ranking of 10+, the highest possible.[29] In October 1951, after continued research and field trials, it was further appreciated that nerve agent shells would be extremely effective

[25] TNA, DEFE 10/30, 'Development of Anti-Tank Weapons in the Long Term', Sub-Committee on Anti-Tank Defence Measures, 22 October 1951.

[26] Maiolo. *Cry Havoc*, 7–22.

[27] TNA, CAB 131/9, 'Military and Economic Strength of the Soviet Union', Chiefs of Staff Committee, 29 November 1950, attached report.

[28] TNA, DEFE 10/30, 'Development of Anti-Tank Weapons in the Long Term', Sub-Committee on Anti-Tank Defence Measures, 22 October 1951.

[29] TNA, DEFE 10/420, 'Range of Army Anti-Tank Weapons for 1953/54', Sub-Committee on Anti-Tank Defence Measures, 29 August 1950. Note: The ATDMC also referred to the role of LVT-1, which had an incendiary affect that was very much greater than napalm.

against Soviet tanks.[30] Experts now strongly believed, as had been tentatively predicted in the late 1940s, that it would take only a single shell, filled with around a pint of a nerve agent, to incapacitate a tank crew before they were even aware of the CW attack.[31] Such a nerve agent shell, once mass-produced, could be readily deployed by British forces. Trials had shown that it could be delivered effectively from a pdr-25, which was the dominant British field gun during the Second World War, in Malaya, and in the Korean War.[32] This ease of implementation meant that the vast majority of British artillery units could use and deploy nerve agent weapons against Soviet tanks with relative ease.

To a degree reinforcing earlier military assessments, nerve agent use was also beneficially thought to have 'very great' psychological effects on Soviet tank crews and personnel.[33] Paranoid of nerve agent exposure and hindered by cumbersome defensive equipment, Soviet tank crews would be significantly less effective in battle. Although little was mentioned on the morality of resorting to nerve agents, the recognition that trained and experienced Soviet tank crews would be disturbed, frightened and have their combat ability seriously undermined by its use, does reveal some appreciation of the horrors of such a method of war.

Considerations of alternative dimensions of nerve agent use were to a degree reflected in assessments of British commitments under the Geneva Protocol of 1925, which defence officials thought a hindrance to the development of an anti-tank nerve agent weapon.[34] It was viewed as an obstacle. The Geneva Protocol acted as an effective constraint and another hurdle to overcome for military planners. The Geneva Protocol prohibited the first-use in war of asphyxiating, poisonous and other gases, as well as their liquid form, but it did not contain any punishment for

[30] TNA, DEFE 10/30, 'Development of Anti-Tank Weapons in the Long Term', Sub-Committee on Anti-Tank Defence Measures, 22 October 1951. The interim report can be found at: TNA, DEFE 10/420, 'Report by No. 4 Working Party on Potential Lethality of Future Anti-Tank Weapons', Sub-Committee on Anti-Tank Defence Measures, 6 June 1951.

[31] TNA, DEFE 10/30, 'Development of Anti-Tank Weapons in the Long Term', Sub-Committee on Anti-Tank Defence Measures, 22 October 1951.

[32] Ibid.; Henry, Chris and Fuller, Mike. 2002. *The 25-pounder Field Gun 1939–72*. Oxford: Osprey Publishing, 38.

[33] TNA, DEFE 10/30, 'Development of Anti-Tank Weapons in the Long Term', Sub-Committee on Anti-Tank Defence Measures, 22 October 1951.

[34] For detailed coverage of the origins of the Geneva Protocol of 1925, see: Spiers. *Chemical Warfare*, 34–61; Spiers, Edward M. 2006. Gas disarmament in the 1920s: Hopes confounded. *Journal of Strategic Studies*, 29:2, 281–300.

non-compliance, nor did it ban the stockpiling of chemical weapons or their use in retaliation by reserving countries.[35] This limitation of retaliatory use, though, did partly push British experts towards considering other weapons which would not be limited to a purely retaliatory role. One such option was 'LVT-1', a substance that ignited on contact with oil or water.[36] Due to defence officials believing that this potential weapon was not covered by the Geneva Protocol, despite its toxic after-effects, they placed it as the top priority for research and development for an anti-tank weapon. Yet, even after recognising the limitations of the Geneva Protocol, for the Anti-Tank Defence Measures Sub-Committee a nerve agent shell had such great military utility that it remained one of the most important areas for research and development. The findings, including the strong support for a nerve agent anti-tank weapon, were passed upwards to the DRPC and the CoS who both endorsed and supported the conclusions.[37]

The military desire to fully embrace nerve agent weapons gained further credence with the strong backing of the CWSC. In terms of the possibility of a nerve agent anti-tank weapon, the CWSC agreed with the findings of the Anti-Tank Defence Measures Sub-Committee, and fully supported the acquisition of a nerve agent anti-tank weapon. In many other areas, the CWSC went much further. It advocated and supported the need for an effective CW deterrent, the mass-production of nerve agents and even tentatively addressed their potential first-use on the battlefield.[38] For defence officials, this latter point would have solidified and confirmed the role and place of nerve agents in British defence policy, whilst removing doubts over whether the nerve agents would actually be used in a future war with the Soviet Union.

The main objectives of policy were therefore figuring out how best to discourage Soviet first-use, retaliate against any actual Soviet use, and gain

[35] Spiers. The Geneva Protocol, 327; Spiers. *Chemical Warfare*, 45–47.

[36] TNA, DEFE 10/30, 'Development of Anti-Tank Weapons in the Long Term', Sub-Committee on Anti-Tank Defence Measures, 22 October 1951.

[37] TNA, DEFE 10/30, 'Development of Anti-Tank Weapons in the Long Term', Chairman of the Defence Research Policy Committee to the Secretary of the Chiefs of Staff Committee, 10 November 1951; TNA, DEFE 4/49, Minutes of Meeting, Chiefs of Staff Committee, 28 November 1951; TNA, DEFE 5/37, 'Chemical and Biological Warfare', Chairman of the Defence Research Policy Committee, 6 March 1952, Annex.

[38] For examples, see this line of military thinking in: TNA, DEFE 10/447, 'Annual Review 1950', Joint Secretary of the Chemical Warfare Sub-Committee, 10 August 1951, Appendix A, Section III; TNA, DEFE 10/447, 'Review of Chemical Warfare 1950–1951', Joint Secretaries of the Chemical Warfare Sub-Committee, 11 December 1951, Attached report.

an advantage over Soviet CW capabilities. In terms of deterring Soviet first-use, the CWSC reiterated the view that the ability to retaliate, as seen during the Second World War, was 'one of the most powerful deterrents to the initiation of Chemical Warfare by an enemy'.[39] The Second World War continued as the prime case study for successful CW deterrence, even in 1951. This interpretation supported and coincided with the arguments over developing a nerve agent capability. The deterrence line of argument also played down the horrors of actually resorting to CW. Their role was to deter and be held in reserve, rather than be employed at the outbreak of war. This retaliation-only role was beginning to be challenged, though, for the CWSC not only urged CW preparedness for CW deterrence but it also supported nerve agent acquisition because the weapons were thought to be so militarily effective against Soviet forces.[40] It can be argued that a weapon needs to be effective in order to deter, but the interest of the CWSC in nerve agent weapons went far beyond just finding a viable deterrent, it delved into considerations of British first-use.

In August 1951, in addition to recognising the value of an anti-tank weapon, the CWSC surmised that nerve agent weapons had 'tremendous potentialities' against concentrations of troops.[41] Importantly, nerve agent weapons could give small forces the power to inflict heavy casualties. They could 'offset the great numerical superiority of the Russians'.[42] This numerical imbalance of conventional forces was a huge area of concern for British defence officials, as while nerve agent weapons were not the only solution to this problem, for the CWSC they certainly represented one of the most viable means of countering it. Even outside of the nerve agents, chemical weapons were judged to provide an advantage if used. Defence officials thought that sulphur mustard could be used to cover a retreat on the continent, slowing down the Soviet advance while nuclear weapons devastated the Soviet Union.[43] In light of all these perceived advantages in

[39] TNA, DEFE 10/447, 'Annual Review 1950', Joint Secretary of the Chemical Warfare Sub-Committee, 10 August 1951, Appendix A, Section III.

[40] Ibid.

[41] Ibid.

[42] TNA, DEFE 10/447, 'Review of Chemical Warfare 1950–1951', Joint Secretaries of the Chemical Warfare Sub-Committee, 11 December 1951, Attached report.

[43] TNA, DEFE 10/447, Minutes of Meeting, Chemical Warfare Sub-Committee, 25 October 1951. Note: There was some debate over the value of sulphur mustard. In 1953 the RAF was also in the process of potentially producing 1000lb bombs of sulphur mustard gas, as a stop-gap measure for a CW capability.

British forces using a combination of chemical weapons, the CWSC concluded that it was actually beneficial for Britain to resort to CW; as CW use was judged, on balance, to favour the Western Powers rather than the Soviet Union.[44] Even with dire warnings from Anglo-American intelligence agencies and the poor state of British civil defence, this belief in the perceived advantages of using chemical weapons gained traction.

In terms of the intelligence picture, the JIC confirmed that the only real CW threat of note came from the Soviet Union. Throughout the early 1950s, the British JIC warned that large-scale Soviet nerve agent production could have started in 1951, that the Soviet Union could possess enough nerve agent weapons for operational use in 1953 and have a considerable capacity available from 1956.[45] While this Soviet threat was branded 'formidable', it was still believed that Soviet defensive measures were poor, leaving their forces exposed to Allied CW use.[46] British experts also believed that 'good anti-gas training requires a high degree of intelligence in the individual', which was dubbed more applicable to NATO soldiers than their Soviet counterparts.[47] In defending against a CW attack, the sheer scale of Soviet ground forces was thought to be a disadvantage and an added burden. Equipping millions of troops with the required defensive equipment was an enormous task, and it was an endeavour presumed to be beyond the capabilities of the Soviet Union.[48]

Despite these claims, intelligence on Soviet CW capabilities was itself of very poor quality. It was not the most influential factor when it came to guiding policy. During this period, British CW policy was often conducted in partial to complete ignorance of the true nature of the Soviet CW

[44] Ibid.

[45] TNA, DEFE 10/447, 'Annual Review 1950', Chemical Warfare Sub-Committee, 10 August 1951, Appendix a, Section III. For intelligence forecasts in 1951 see: TNA, DEFE 10/498, 'Russian Research and Development', Directorate of Scientific Intelligence and Joint Technical Intelligence Committee, 20 December 1951. For 1952 see: TNA CAB 158/14, Part I, 'Russian Research and Development, Joint Intelligence Committee, 27 February 1952. For 1953 see: TNA, CAB 158/15, Part I, 'Soviet and Satellite War Potential, 1953–1956', Joint Intelligence Committee, 10 April 1953.

[46] For the 'formidable' threat see: TNA, DEFE 7/700, 'Review of Chemical Warfare Development up to the End of 1951', Chemical Warfare Sub-Committee, January 1952. For defensive measures see: TNA, DEFE 10/447, 'Review of Chemical Warfare 1950–1951', Note by the Joint Secretaries of the Chemical Warfare Sub-Committee, 11 December 1951, Attached report.

[47] Ibid.

[48] Ibid.

threat, as intelligence could provide little verifiable evidence on Soviet nerve agent developments. This led to the CWSC lamenting, in August 1951, that 'practically no post-war information on [Soviet] CW has been secured'.[49] Soviet security measures were cited as the primary reason for this absence of accurate intelligence.[50] The level of security was deemed so thorough that a large-scale CW installation could exist without any real possibility of it becoming known to British intelligence.[51] Of particular concern, when considering the potential use of chemical weapons in war, British defence officials were highly uncertain as to how the Soviet Union would respond. Cautious officials noted that if the Soviets adopted the same approach they had to minefields, 'namely, to march their troops straight through and accept casualties', then absolutely no reliance could be placed on certain chemical weapons for slowing down a Soviet advance across Western Europe.[52] This was a core aspect and rationale to the military justification for the deployment of chemical weapons, as although the nerve agents would not be used solely in this role, Britain's stocks of sulphur mustard would be. Unanswered questions also lurked in the background—what if the Soviet Union responded in kind with vast stockpiles of chemical weapons, or perhaps worse still, with other WMDs?

Ultimately, as a result of poor intelligence, British officials lacked the means to accurately assess the CW threat they faced. Instead, judgements were made which relied on information now several years old. The CWSC believed that a significant nerve agent threat existed which needed to be deterred and countered. But, they had little verifiable intelligence as to how Soviet officials approached CW, what form Soviet use of CW would take, in what quantity it would be used, or how the Soviets would respond.

Even with the paucity of accurate intelligence the Soviet nerve agent threat was still thought significant. For the CWSC, this perceived Soviet threat had significant ramifications for civil defence, which was still feared

[49] TNA, DEFE 10/447, 'Annual Review 1950', Chemical Warfare Sub-Committee, 10 August 1951, Appendix a, Section III.

[50] TNA, DEFE 10/171, 'Report on Russian Research and Development', Directorate of Scientific Intelligence and Joint Technical Intelligence Committee, 20 December 1951.

[51] Ibid.

[52] TNA, DEFE 10/447, Minutes of Meeting, Chemical Warfare Sub-Committee, 25 October 1951.

unsatisfactory.[53] The quandary was that effective civil defence necessitated circulating information on the nerve agents. If civil defence were to mitigate against Soviet CW use in the nerve agent age, doctors, regional medical staff and civil defence officials needed to be made aware of the existence, threat, and nature of the nerve agents.[54] Public information on the nerve agents had remained threadbare up to 1952. Few knew that this significant development in CW had actually occurred. Attempts to address domestic vulnerability, therefore, entailed informing a far greater number of people and of letting the previously tight control over information on the nerve agents slip ever so slightly. A partial revelation was now needed.

Information on the nerve agents was rarely reported on due to close adherence to tight secrecy, but on 8 August 1952, the *Daily Mail* published an article titled 'New Gas Destroys Nerves'.[55] Remarkably, seven years after British discovery, and around 16 years after German scientists originally uncovered tabun, the article warned of dangerous 'new' nerve agents. It reported that they were almost odourless, colourless and that unlike typical CW agents they produced no easily recognisable symptoms of exposure, such as itching skin or sore eyes. Most distressingly, the *Daily Mail* reported on the substantial dangers and loss of life that would result from Soviet air raids using nerve agents.[56] This warning mirrored the findings of the secret CWSC report in 1949, which had explicitly warned of the dangers of an aerial nerve agent attack on cities such as London. The release of this information was not, however, some accidental slip-up, or a result of pioneering investigatory journalism. It was a deliberate and calculated effort by government officials to slightly lift the shroud of secrecy and raise public awareness. In light of the nerve agent threat and the uncertain global environment, stringent levels of secrecy had to be deliberately weakened for the benefit of defensive preparedness. This revelation was also made despite serious concerns, which spanned the Atlantic, over the adverse publicity which would result from any public mention of the nerve agents.[57] The need for a strong and informed civil defence effort had

[53] TNA, DEFE 7/700, 'Review of Chemical Warfare Development up to the End of 1951', Chemical Warfare Sub-Committee, 12 March 1952.

[54] Ibid.

[55] *Daily Mail*, 'New Gas Destroys Nerves', 8 August 1952, 3.

[56] Ibid.

[57] For the United States, see: National Security Archive, George Washington University, Chemical and Biological Warfare, Box 2, Statements of Policy and Directives on Biological Warfare, Secretaries of the Joint Strategic Plans Committee, 27 May 1952. These United States fears also coincided with Chinese accusations of United States BW use in the Korean

thus necessitated a public intervention, which came in the form of Sir Harry Garner, Chief Scientist at the Ministry of Supply.[58] As reported in the *Daily Mail*, Garner confirmed the existence of the lethal nerve agents, revealed 'detailed facts' about them, and raised public awareness as to how best mitigate against their use.[59] This increased civil defence effort represented a significant point in postwar British nerve agent policy. The existence of nerve agents was acknowledged, and the threat was deemed substantial enough to warrant the deliberate loosening of extremely tight levels of secrecy.

Even with this apparent civilian vulnerability and a lack of accurate and verifiable intelligence on the Soviet CW threat, defence officials still remained committed to the view that Britain held the upper hand in any war with chemical weapons.[60] Perceived Soviet defensive weakness combined with the substantial military advantages offered by nerve agent use fuelled this positive outlook. Using nerve agents was deemed a militarily sound option. The CWSC remained committed to the production of nerve agent weapons for use in war, deterrence through an immediate retaliatory capability, and the expansion of the sarin facility at Nancekuke.[61] This drive by the CWSC for a larger nerve agent role played a significant part in altering the direction of British nerve agent research and development. When the CoS came to review CW policy in secret, it would heavily rely on the findings and advice of the sub-committee on Anti-Tank Defence Measures and the CWSC, as well as those of the DRPC.

In March 1952, the DRPC's support for nerve agent weapons, whilst reflective of growing military interest in the utility and benefits of using nerve agent weapons against Soviet forces, had become slightly more cautious than the sub-committee on Anti-Tank Defence Measures and the CWSC. In a strategic role, against populated cities or production centres, as before the DRPC recognised that chemical weapons would pale in comparison to atomic weapons. In a tactical role, it noted that nerve agent

War. For Canadian concerns see: TNA, CAB 129/55, 'Chemical Warfare Policy', Secretary of the Cabinet, 7 October 1952, Attached minute.

[58] *Daily Mail*, 'New Gas Destroys Nerves', 8 August 1952, 3.

[59] Ibid.

[60] TNA, DEFE 10/447, 'Review of Chemical Warfare 1950–1951', Note by the Joint Secretaries of the Chemical Warfare Sub-Committee, 11 December 1951, Attached report; TNA, DEFE 7/700, 'Review of Chemical Warfare Development up to the End of 1951', Chemical Warfare Sub-Committee, 12 March 1952.

[61] Ibid.

weapons were 'extremely important', and it even reconsidered its prior dismissal of a nerve agent aerial bomb, which was now deemed 'impressive'.[62] The language here had again shifted. An aerial bomb, which beforehand was deemed strategic, was now also thought of as tactical, sparing it from harmful comparisons to biological or atomic weapons.[63] The use of key terms to separate chemical weapons from atomic weapons had thus continued, and its use expanded to in part justify and rationalise the production and development of additional weapons. This also reflected the Air Ministry changing the required and defined role of nerve agents. After their interpretation of CW deterrence was rebuffed in the late 1940s, the Air Ministry had thus continued to press for some form of nerve agent capability. A strategic capability was not viable, so they had shifted to a tactical one, which was now endorsed. From attacking Soviet cities with nerve agents to Soviet soldiers on the front line, the Air Ministry had adapted to the tactical definition and the growing consensus in defence circles as to how nerve agent weapons should best be used. Avenues and possibilities for nerve agent acquisition and military use were opening and strengthening.

But, one seemingly unassailable obstacle remained. The growing consensus and momentum in defence circles for a nerve agent capability had still to tackle a particular issue regarding use. As the DRPC warned, and defence officials feared, the limitations placed on CW use by the 1925 Geneva Protocol continued to cast a significant shadow over whether researching, developing and deploying this type of weapon was truly a worthwhile endeavour due to the legal constraints and pressures of international norms. Coupled with this was the impact of economic cutbacks on scientific research and development funding under the Churchill Government.[64] This combination, of international treaties barring first-use and economic restraints, led to the DRPC erring on the side of caution. At a time when the front of Korean War was now relatively stable, the

[62] TNA, DEFE 5/37, 'Chemical and Biological Warfare', Chairman of the Defence Research Policy Committee, 6 March 1952, Annex.

[63] TNA, DEFE 10/30, 'Development of Anti-Tank Weapons in the Long Term', Sub-Committee on Anti-Tank Defence Measures, 22 October 1951. For a later appreciation see: TNA, DEFE 7/700, 'Development of Manufacturing Capacity for Nerve Gas', N. Brownjohn to First Sea Lord, CIGS and CAS, 13 January 1953, Attached note by General Sir Kenneth Crawford.

[64] TNA, DEFE 5/37, 'Chemical and Biological Warfare', Chairman of the Defence Research Policy Committee, 6 March 1952, Annex.

committee questioned the decision to invest heavy capital, of around £2 million, into the expansion of the Nancekuke facility and the mass-production of sarin when there were no assurances that nerve agent weapons would be used in a future war.[65] The urgency of war, which had partially characterised Clement Attlee's support for nerve agent mass-production, had started to slightly wane with the relative stalemate. As a result, the DRPC advised the CoS that, despite the strong support of other committees, there still needed to be assurances over the use of nerve agent weapons for them to warrant the investment and commitment that mass-production required.[66]

In April 1952, just before public revelations for civil defence and the completion of the 1952 Global Strategy Paper, yet after receiving the analysis and recommendations of the Anti-Tank Defence Measures Sub-Committee, the CWSC and the DRPC, the CoS favourably assessed British CW policy. Three key committees had provided it with detailed assessments. Nerve agent weapons could counter Soviet tanks and troops, act as an effective deterrent, and add to Britain's military capabilities, keeping it at the top table alongside the Soviet Union and the United States. Based on the advice it received, for the CoS, the military utility of nerve agent weapons was to trump all other concerns. The CoS believed that development should proceed. The nerve agent plant at Nancekuke should be expanded with haste to mass-produce sarin. And, Britain must be in a position to wage CW at the start of hostilities.[67] Political approval and confirmation now constituted the final step. Without nerve agent production, the CoS warned that Britain would be 'entirely dependent on the United States', which was thought an unacceptable last resort.[68] It also alleged that if the Soviets were to discover any Western weakness in the CW field, they would engage in CW to offset Western atomic dominance.[69] As such, maintaining a strong CW posture with a viable nerve agent capability was increasingly important. Without an updated nerve agent capability, Britain would be unable to retaliate with 'any worthwhile

[65] Ibid.

[66] Ibid.

[67] TNA, DEFE 41/157, 'Chemical Warfare', Joint Secretaries of the Chemical Warfare Sub-Committee, 1 May 1952. Decision over the actual mass-production of chemical weapons would be taken at a later date, when the weapons were fully developed.

[68] TNA, DEFE 4/53, Minutes of Meetings, Chiefs of Staff Committee, 1 April 1952, Confidential Annex.

[69] Ibid.

gas'.[70] In the nerve agent age, the CoS thus decided that Britain needed to expand its production 'as soon as possible'.[71] These new nerve agent weapons would provide a 'considerable' deterrent and military capability.[72]

However, this mounting support for nerve agent weapons coincided with the 1952 GSP. And, even with keen military support and the Attlee Government's prior political approval in 1950, the role and place of nerve agent weapons was at the mercy of the sweeping 1952 GSP.[73] Caught up in this politically driven review of all of defence policy, nerve agent development, in an economically harsh environment and as forewarned by the DRPC, was in reality in a precarious position. This was despite strong support from defence officials. Other actors would become involved with the GSP, and have their say on nerve agent policy, such as the Foreign Office and Government Ministers. If the outcome of the GSP towards CW policy were positive, then the Nancekuke facility could indeed be rapidly expanded to meet Britain's Cold War nerve agent requirements. For a positive outcome, in a period characterised by the Churchill Government's cutbacks in defence expenditure, defence officials needed to show that nerve agent weapons were essential to defence. Crucially, proving that they were essential morphed into proving that they would be used in the event of war, which for chemical weapons is very dangerous territory indeed. Under the Geneva Protocol, it would have been illegal for Britain to initiate CW, and thus there was no guarantee that nerve agent weapons could or would be used in a future conflict, casting doubt over their viability. This did not lead to military officials or the CoS backing down in their desire for nerve agent weapons; rather, they sought to change Britain's interpretation of the Geneva Protocol of 1925.

The 1952 GSP

As John Baylis and Alan Macmillan have correctly noted, the section on CBW within the GSP was altered after feedback from government departments; unfortunately, they were unable to find the original wording.[74] The

[70] Ibid.; TNA, DEFE 4/53, Minutes of Meetings, Chiefs of Staff Committee, 23 April 1952, Confidential Annex.

[71] Ibid.

[72] Ibid.

[73] Ibid.

[74] Baylis, John and Macmillan, Alan. 1993. The British global strategy paper of 1952. *The Journal of Strategic Studies*, 16:2, 211.

original draft copy of the GSP, now available, holds substantial implications for British CW policy and defence policy. It displays just how strongly the CoS and defence officials felt about nerve agent weapons. What is significant regarding CW policy, in this original version of the GSP, is that the CoS attempted to change British policy for the first-use of lethal chemical weapons, and distance the country from its commitments under the Geneva Protocol.[75] In their version of the 1952 GSP, the CoS elaborated on their earlier support for nerve agent weapons and wanted the Government's approval for the following:

> We consider that the Allies should be prepared to use these [CBW] weapons in war *when they think it to their advantage to do so*, and that this should be reflected in their public attitude to the employment of these forms of warfare.[76]

This substantial shift would have taken the onus away from a policy of retaliation-only. It would have allowed military planners to use lethal chemical weapons when they thought it beneficial to do so. Strikingly, this policy referred to not just chemical weapons, but also biological weapons. If the Geneva Protocol were to be side-lined, then defence planners wanted to go all the way. As reflected in existing military thinking, this potential British use of nerve agent weapons would likely have been immediate rather than hypothetical. If the option were on the table in 1952, the perceived advantages of use, military utility of nerve agent weapons and the numerical imbalance vis-à-vis Soviet forces all but assured British first-use on the European battlefield.

The extra condition regarding the 'public attitude' is a particularly potent point. It deviated greatly from the prior practice of keeping information on actual nerve agent policy highly secret. This condition implied that if Britain were to renounce its no-first-use policy, then it should publicly acknowledge doing so. The requirement also had possible roots in the perceived need for a bold CBW deterrent, which defence officials thought required some degree of public recognition or rhetoric. In justifying this

[75] TNA, PREM 11/49, Ian Jacob to Winston Churchill, 10 July 1952, Attached expurgated version.

[76] Ibid. [Italics added by author] The CoS seemingly wished to have available for possible use a full range of CBW weapons at the start of hostilities. This also mirrored earlier debates in the United States over first-use, see: Fredericks. United States Chemical Warfare Policy. II, 22–24.

remarkable request, with it spurning both moral and legal qualms, the CoS claimed that military utility, the scale of the Soviet threat, developments in nuclear weapons and peculiarly the CW policy of the United States all justified changing Britain's adherence to the legally binding and internationally accepted Geneva Protocol.

A crucial part of the rationale for this change in first-use policy was the assessment that if war did break out Soviet forces would easily overrun Western Europe, and that Soviet aircraft would attempt to destroy United States bases in the United Kingdom.[77] Such a Soviet assault on Western Europe would likely have triggered a United States nuclear response and, as Churchill warned, this atomic retaliation would then further spur on Russian forces to over-run Western Europe as rapidly as possible.[78] The Red Army would roll relentlessly forward. Hindering this rapid Soviet advance was therefore where chemical weapons came to the fore in the original GSP. A dire situation seemed to permit, and possibly require, CW use. This stance was remarkably similar to the one taken during the Second World War. The nerve agents could be used to hamper Soviet progress across the continent and exert a considerable toll on Soviet ground forces, whilst atomic weapons struck at the heart of the Soviet Union. This argument over the scale of the Soviet threat chimed with the military utility of chemical weapons, for in justifying a shift in first-use policy the two arguments were intertwined for the CoS: the exceptional scale and level of the Soviet threat required an exceptional military solution. Britain's apparently grim situation remained a pressing concern.

Although recognising the nerve agents as weapons suitable for an exceptional circumstance, compared to the atomic bomb, the CoS deemed them morally no worse. The Service Chiefs argued that 'the moral objections to chemical warfare can surely be no greater than to atomic warfare'.[79] While revealing an awareness of the moral and ethical dilemmas involved in CW, this comparison was actually further used to justify the use of chemical weapons in war. In the logic of the CoS, if atomic weapons were to feature in a future war, then why not chemical weapons as well? In a total global war, all of Britain's military capabilities would have to be thrown at the

[77] Baylis. *Ambiguity and Deterrence*, 143.

[78] TNA, DEFE 32/2, Minutes of Meeting, Chiefs of Staff Committee, 18 June 1952, Confidential Annex.

[79] TNA, PREM 11/49, Ian Jacob to Winston Churchill, 10 July 1952, Attached expurgated version.

Soviet juggernaut. This again closely reflected prior wartime considerations—if the survival of the realm was at stake, arguments over morality or international law would not prevent the use of chemical weapons.

In terms of the scale of destruction, at this stage an atomic bomb undoubtedly far outweighed chemical weapons. The CoS thus queried why chemical weapons should be any different from that of atomic warfare, and why there were such moral objections to the use of chemical weapons.[80] By comparing chemical weapons to atomic weapons, the CoS hoped to rationalise their use, even if this massively distorted and overlooked the moral aspects of resorting to such a form of warfare. Alongside legal constraints, the CoS was therefore also attempting to put forward its case for nerve agent weapons being morally no worse than other methods of warfare. It also focused on the immediate casualties and destruction, akin to comparing the means of destruction to a large bomb or 'normal' means of war, rather than the harrowing nature of either weapon. The effects of nuclear fallout and radiological weapons, more comparable manifestations and forms of warfare, were unsurprisingly omitted.[81]

In justifying the first-use of chemical weapons, the CoS emphasised not just the potential United States use of nuclear weapons, but also its CW policy. As alleged by the CoS, 'the Americans have not adhered to the Geneva Convention and will certainly not hesitate to use either Bacteriological Warfare or Chemical Warfare if they consider it advantageous to do so'.[82] In requesting British chemical weapons for use at the start of hostilities, the CoS gave the impression that they were harmonising British policy with that of the United States. Due to the potential use of CBW by the United States, it was feared that Britain, regardless of its stance, could be dragged into a conflict with chemical weapons. Any first-use of chemical weapons by the United States would likely have resulted in Soviet reprisals against NATO forces, triggering the retaliation-only policy of the Geneva Protocol and lifting the legal constraints on British CW use. Conversely, this would have rendered any British change to its interpretation of the Geneva Protocol unnecessary. But, this was not a guaranteed scenario, and so defence officials still pressed for the option to use chemical weapons first.

[80] Ibid.

[81] For the British history of radiological weapons, see: King, William. 2021. A weapon too far: The British radiological warfare experience, 1940–1955. *War in History*, [online first], 1–23.

[82] TNA, PREM 11/49, Ian Jacob to Winston Churchill, 10 July 1952, Attached expurgated version.

For British officials, the likelihood of the United States resorting to CW seemed to be given even greater credence by the ongoing Korean War. As outlined by Julian Perry Robinson, senior defence officials in the United States, alongside front-line Commanders, had been pressing to use chemical weapons during the war.[83] The Chiefs of Staff of both the United States Army and Air Force were recommending changing first-use policy to make possible the use of chemical weapons when it was deemed advantageous to do so.[84] Previously, the United States had self-imposed a retaliation-only policy, in line with the Geneva Protocol. The proposed first-use shift was pushed for despite protests from a resistant Navy and even with concerns that a unilateral change might alienate the United Kingdom.[85] This latter point reveals the influence of Anglo-American relations, with United States officials taking British CW policy into account when formulating their own approach. Conversely, it also shows the relative isolation and confusion resulting from their respective debates over military utility and first-use policy; with both countries seemingly misreading, or at least highly uncertain, as to the actual first-use policy and intent of the other.

What is apparent in the experiences of Britain and the United States is the degree of uniformity in how both countries viewed the military utility of chemical weapons, and how certain British and United States defence officials wanted to be able to use nerve agent weapons at the outset of war. The first-use dilemma was not a decision that could be taken in isolation. Defence officials in the United States were hesitant to change policy unilaterally, and the British CoS was keen to bring its policy in line with that of the United States, or at least in line with what the United States was predicted to do in a global war. The CoS, in their original version of the GSP, thus used the uncertainty of United States CW policy, and its potential first-use, as another reason and rationale for changing Britain's adherence to an internationally recognised treaty banning the first-use of chemical weapons.[86]

[83] Robinson. *The Rise of CB Weapons*, 158.

[84] Coffey, Patrick. 2014. *American Arsenal: A Century of Waging War*. Oxford: Oxford University Press, 163.

[85] Tucker. *War of Nerves*, 127.

[86] TNA, DEFE 32/2, Minutes of Meeting, Chiefs of Staff Committee, 18 June 1952, Confidential Annex; TNA, PREM 11/49, Ian Jacob to Winston Churchill, 10 July 1952, Attached expurgated version.

Unfortunately for the CoS, despite all their justifications, British adherence to the Geneva Protocol was resolutely resistant to change. The substantial support for a change in first-use policy by defence officials was to clash with a champion of the Geneva Protocol, the Foreign Office. As before their version of the GSP was sent for Cabinet approval in mid-1952, it was first circulated to other government departments. In this process, the Foreign Office pushed back against Britain publicly reneging on an internationally recognised treaty, a treaty which Britain had adhered to even during the Second World War.[87] Representing more normative based arguments emphasising the rule of law, the Foreign Office crucially modified the text of the GSP, arguing that it should now read: 'the Allies should not take up a position which would deprive them of their ability to use chemical and bacteriological warfare *in retaliation*'.[88] Explicitly countering the wishes of the CoS, the Foreign Office re-draft reiterated that Britain was a signatory of the Geneva Convention and would only use chemical or biological weapons in retaliation.[89] In addition to international commitments, other likely factors further driving this resistance include prestige, avoiding a tarnished reputation and the subsequent Soviet PR coup, and even self-interest—if Britain were to renege publicly, what would stop rivals from doing the same? After the strong rebuttal by the Foreign Office, the updated version of the GSP was drastically different from the original. Any reference to the United States and its potential use of CBW, moral comparisons between CW and atomic warfare, and any mention of British first-use were all removed from the text of the GSP.[90] The input of the Foreign Office proved a costly and bruising experience for CW advocates. Their envisaged policy for the first-use of nerve agent weapons was substantially watered down and overridden.[91]

In July 1952, aware of the hugely divergent stances of the CoS and the Foreign Office, Churchill put the topic of Britain's first-use of CBW to the Cabinet's Defence Committee for review.[92] This decision by Churchill to hold a high-level meeting was also spurred on by concerned British BW experts, who had caught wind of the proposed changes in first-use policy put forward by the CoS. Here there appears some disconnect between

[87] TNA, PREM 11/49, Ian Jacob to Winston Churchill, 10 July 1952.

[88] Ibid., Attached expurgated version. [italics added by author]

[89] Ibid.

[90] Ibid.

[91] Baylis and Macmillan. The British Global Strategy Paper of 1952, 211.

[92] TNA, CAB 131/12, Minutes of Meeting, Cabinet Defence Committee, 9 July 1952.

CW and BW, for many British CW experts had strongly supported this change in first-use policy, yet for many BW experts it caused alarm and warranted the seeking of Ministerial clarification.[93] This concern and the decision by Churchill to review the proposed changes, led to the Defence Committee directly addressing first-use.

In July 1952, political oversight and input was to prove a crushing blow for CW advocates. From the outset members of the Defence Committee were highly sceptical of any proposed change in Britain's retaliation-only policy for CBW. The Foreign Secretary, Anthony Eden, was absent from the meeting. Had he been present British CW policy may have taken a very different turn.[94] Without Eden's presence, the Secretary of State for the Colonies, Oliver Lyttelton, lambasted the desired first-use policy change as 'impossible' to justify. He argued that such a policy change could only ever be enacted 'when we were fighting for our lives'.[95] Lyttelton, clearly impassioned and reflecting the taboo nature of CW, was not alone in his damning assessment. The rest of the Defence Committee agreed that CBW first-use could only be considered at a time of war and in the direst of situations. It was too controversial a form of warfare to warrant publicity or a change in first-use policy. Even Churchill, a prior advocate of CW policy, shied away from the dramatic shift put forward by the CoS. He was especially cautious over the publicity aspect of revealing anything about British CW policy. The global situation was severe, but apparently not severe enough to warrant a change in first-use policy; only war itself could bring about such a change. This was all the more remarkable given that just a decade before he had publicly threatened to 'carry out gas warfare on the largest possible scale' against Germany.[96]

Chemical and biological weapons were seen as so controversial and morally dubious that only in the most severe, desperate situation, would their first-use be considered. The exact origins of this moral aversion are hard to trace, and it is certainly not unique to British politicians. As argued by Price, the taboo nature of chemical weapons reflects a genuine and

[93] Ibid; TNA, PREM 11/49, 'Defence Policy and Global Strategy', Secretary of the Chiefs of Staff Committee, 10 July 1952; TNA, PREM 11/49, Ian Jacob to Winston Churchill, 10 July 1952.

[94] TNA, CAB 131/12, Minutes of Meeting, Cabinet Defence Committee, 9 July 1952. For more details on Eden's involvement with CW policy, see Chap. 4: The last man standing.

[95] TNA, CAB 131/12, Minutes of Meeting, Cabinet Defence Committee, 9 July 1952.

[96] *The Daily Telegraph*, 'Premier's Poison Gas Warning to Hitler', 11 May 1942, 1; Spiers. *Chemical Warfare*, 76–77; Spiers. *A History of Chemical and Biological Weapons*, 59.

widespread moral rejection of this means of modern warfare, with CW contravening what was expected of technological states of the 'civilised' world.[97] The chemical weapons taboo, a social and political construction, operates as a norm in international society.[98] In the British case, perceptions of how a great power should act played a part in political aversion towards nerve agent weapons. A cultural element was also a particularly strong factor in this aversion, with the legacy of the First World War, the shock value of the nerve agents themselves, and stories of Nazi death camps and human experiments all still within recent memory.[99] However, regardless of the exact driver of political aversion, in the absence of a global war, CW was thought such a controversial topic that now even Churchill shied away from publicly acknowledging a change in CW policy, let alone indicating British willingness to resort to nerve agent use. As a result, Churchill and the rest of the Defence Committee decided, in July 1952, that no public announcement on CW should be made unless they were obliged to, and they agreed with the Foreign Office that Britain should adhere to the Geneva Protocol.[100]

Following the Defence Committee's verdict, and after the Foreign Office review, the final 1952 GSP stated that:

> The United Kingdom is at present committed by the Geneva Convention not to use [chemical and biological weapons] except in retaliation... The new nerve gases can, however, be used tactically to great advantage and would provide the Allies with weapons of real value against an enemy who relies on massed formations.[101]

[97] Price. *The Chemical Weapons Taboo*, 164.

[98] Ibid., 168–169.

[99] This cultural aspect is more in line with Cole's work, who argues that the taboo nature of chemical weapons has roots in both biology, that of innate human aversion to toxic substances, and cultural, with the evolution of repugnance and societal contempt towards this form of warfare. See: Cole, Leonard. 1998. The Poison Weapons Taboo: Biology, Culture, and Policy. *Politics and the Life Sciences*, 17:2, 119–132. While this book has attempted to do some justice to this avenue of moral aversion through its exploration of the direction, development and causality of policy, there is certainly more work to be done on the British experience, particularly in regard to public perceptions.

[100] TNA, CAB 131/12, Minutes of Meeting, Cabinet Defence Committee, 9 July 1952.

[101] TNA, PREM 11/49, 'Defence Policy and Global Strategy', Chiefs of Staff Committee, 9 July 1952.

While this final version of the GSP acknowledged the tactical value of the nerve agents against Soviet forces, it was a far cry from what the CoS had wanted it to say. Instead of granting the ability to use chemical weapons at the outset of war, and decide when to initiate their use, British politicians had re-affirmed the country's commitment to the retaliation-only policy of the Geneva Protocol. The arguments put forward by the CoS had fallen on deaf ears, and even irritated some, with ministers on the Defence Committee finding the CoS request a step too far. This rebuff would have substantial consequences for British CW policy, with it significantly changing Britain's relationship and engagement with nerve agent weapons and the direction of CW policy in the Cold War.

An About-turn

The failed attempt to overcome the perceived limitations of a no-first-use policy in the GSP would have a considerable impact on British CW policy. It would prove a damning blow to any hopes of mass-producing nerve agents. A. E. Childs, the Director of the Chemical Defence Experimental Establishment at Porton Down, ruefully noted that the outcome of the GSP was likely to be a 'dampening effect, in that however good toxicological weapons might be, they could not be integrated with early and immediate war plans'.[102] With no change to Britain's interpretation of first-use, advocates of a British nerve agent capability had thus fallen at the Geneva Protocol hurdle. The Foreign Office-led rebuttal transpired even after the military utility of nerve agents had been established and prior political approval had been granted for production by the Attlee Government. The backlash and fallout from this failed attempt to alter first-use policy also coincided with, and was further exacerbated by, the alarming deterioration of Britain's CW stockpile and the Churchill Government's economic cutbacks to defence expenditure.

In late July 1952, shortly after the GSP decision, it was fully recognised that Britain was without a viable or truly effective CW capability. For the GSP had not only highlighted the military desirability of nerve agent weapons, but also revealed the urgency and importance military planners attached to acquiring a nerve agent capability and deterrent in the Cold War. The GSP had still supported a nerve agent capability in some form,

[102] TNA, WO 188/802, 'A History of Porton Down', Ministry of Defence, 1960, Appendix B; Carter. *Porton Down*, 56; Carter and Pearson. North Atlantic chemical and biological research collaboration, 85.

just not for first-use, which in essence limited it to a marginal and doubtful role. But, Britain still only possessed outmoded forms of chemical weapons which did not offer a meaningful deterrent capability.[103] As the Air Ministry warned, leftover stores of sulphur mustard bombs were no longer satisfactory, and phosgene bombs contained an 'obsolescent agent'.[104] Even with this damning conclusion, it was recommended that these outmoded and ineffective weapons still be retained, for they represented the country's only real CW retaliatory capability, with captured German tabun bombs still largely unusable.[105] The situation was so severe for defence officials that they clung to an obsolete, deteriorating, and inefficient CW capability, which could only realistically be used in an absolute 'last-ditch' effort against the Soviet Union. Reflective of this desperation, and with doubts over the fate of a nerve agent capability after the GSP setback, the Air Ministry even tentatively approved the production of 19,000 × 1000lb sulphur mustard bombs.[106] British CW capabilities were in a poor state. And, the nerve agent option was seemingly disappearing from sight.

In the Cold War nerve agent age, Britain was increasingly falling behind the superpowers. The situation was bleak for defence officials, as not only did the GSP quash a change in first-use policy, it also provided no concrete political guidance on how Britain's retaliatory CW capability was to be fulfilled, or what form it would take.[107]

Given the concerning situation, the Air Ministry searched for a potential remedy for the nerve agent shortage, with the new sulphur mustard bombs acting as a temporary, if questionable, stop-gap.[108] The Air Ministry still maintained that Britain needed a nerve agent capability, even if just for deterrence and retaliation, rather than for first-use. One potential solution was to again revisit the possibility of modifying captured but ageing German tabun bombs to better suit British means of delivery; this would have at least provided a slightly more credible retaliatory capability with nerve agents.[109]

[103] TNA, DEFE 41/157, 'Chemical Warfare Reserve Policy', Air Ministry, 30 July 1952.
[104] Ibid.
[105] Attlee's 1950 Cabinet Defence Committees decision over the re-purposing of stocks was also seemingly not acted upon, with policy implementation halted until the completion of the 1952 GSP.
[106] TNA, DEFE 41/157, 'Chemical Warfare Reserve Policy', Air Ministry, 30 July 1952.
[107] TNA, DEFE 5/43. 'Policy for Chemical Warfare and Biological Warfare Research and Development', Chairman of the Chemical Warfare Sub-Committee, 11 December 1952.
[108] TNA, DEFE 41/157, 'Chemical Warfare Reserve Policy', Air Ministry, 30 July 1952.
[109] Ibid.

The second option suggested by the Air Ministry was the sounding out and revival of a potential nerve agent deal with the United States.[110] This was thought a particularly promising avenue. Britain was in a poor economic state, and the expansion of the Nancekuke facility to 50 tons per week was now believed to require an investment of around £4 million, along with a further £5 million for the chemical industry.[111] Strangely though this proposal was suggested despite Britain not being alone in its poor CW preparedness. The United States was also struggling to develop nerve agent weapons in mid-1952.[112] It was, however, struggling for very different reasons. The United States had started the mass-production of sarin, but technical difficulties were hampering the level of output. While the United States was therefore grappling with the mass-production of sarin, after the GSP, Britain was left scrambling resources together to acquire some form of CW capability, even if it was only in the form of sulphur mustard bombs or through acquiring nerve agents from the United States.

This disparity in effort and scale was further compounded with the souring of military enthusiasm for nerve agent weapons in Britain, and the introduction of harsh economic cutbacks in defence expenditure. Reflective of Childs' concerns, without any guarantee over the use of nerve agent weapons, previous keen military interest in the nerve agents waned. Former advocates of nerve agent weapons, such as the DRPC, the War Office and the Air Ministry, began to question the nerve agent requirement and domestic mass-production. The path forward was again uncertain.

In January 1953, when the CoS met to discuss the future of CW policy, the tone was markedly different.[113] The optimism and enthusiasm surrounding the acquisition of nerve agent weapons, which characterised the

[110] Ibid.

[111] TNA, DEFE 5/43, 'Chemical Warfare Policy', Deputy Chairman of the Defence Research Policy Committee, 15 December 1952. For other estimates, see: TNA, DEFE 41/157, 'Chemical Warfare Reserve Policy', Air Ministry, 30 July 1952; TNA, DEFE 5/43,'Chemical Warfare Policy', Deputy Chairman of the Defence Research Policy Committee, 15 December 1952; TNA, PREM 11/3099, 'Policy for Chemical Warfare', N. Brownjohn to Winston Churchill, 4 May 1953.

[112] Tucker. *War of Nerves*, 130–134.

[113] TNA, DEFE 4/59, Minutes of Meeting, Chiefs of Staff Committee, 6 January 1953, Confidential Annex; TNA, DEFE 10/32, 'Development of Manufacturing Capacity for Nerve Gas', Secretary of the Defence Research Policy Committee, 9 January 1953, Annex.

period from the late 1940s up to the GSP, had faded. The Service Chiefs oscillated and changed course very quickly in the aftermath of the GSP. Field Marshal Sir John Harding, Chief of the Imperial General Staff, held a particularly negative view regarding British nerve agent policy.[114] Harding, representing the Army and in a substantial break from his predecessors, argued that Britain should confine itself to purely defensive research, abandon aspirations for a nerve agent capability, and scrap offensive research altogether.[115] This reversal was all the more remarkable given that it was the Army which had previously been the main advocate and central driving force behind nerve agent weapons acquisition. Harding's measures would have severely debilitated British CW policy. He believed that in order to address this potential shortcoming, Britain should willingly become entirely reliant on the nerve agent capability of the United States in order to save 'large sums of money'.[116]

The second key advocate of nerve agent weapons, the Air Ministry, also changed course after the GSP failure. Although not as drastic as Harding, Air Chief Marshal Sir William Dickson, Chief of the Air Staff, believed that instead of forsaking nerve agent weapons altogether, Britain should seek to purchase nerve agent weapons from the United States.[117] Such a step represented a possible solution to Britain's nerve agent requirements. It was a half-way measure, allowing a potential financial saving from abandoning domestic production yet avoiding the worst-case scenario of casting aside all aspects of offensive CW policy.[118]

This turn against a domestically produced British nerve agent capability also coincided with substantial advances in British nuclear weapons capabilities; Britain tested its first atomic bomb in October 1952.[119] After this key juncture, and with the advent of the thermonuclear age, even greater attention, support and funding would need to be allocated to the field.[120] This primacy of atomic weapons came at the cost of other areas of defence,

[114] TNA, DEFE 4/59, Minutes of Meeting, Chiefs of Staff Committee, 6 January 1953, Confidential Annex.

[115] Ibid.

[116] Ibid.; TNA, DEFE 10/32, 'Development of Manufacturing Capacity for Nerve Gas', Secretary of the Defence Research Policy Committee, 9 January 1953, Annex.

[117] TNA, DEFE 4/59, Minutes of Meeting, Chiefs of Staff Committee, 6 January 1953, Confidential Annex.

[118] Ibid.

[119] Jones. *UK Strategic Deterrent, Volume I*, 25–26.

[120] Ibid.

and it undoubtedly played a part in defence officials revoking their previously strong support for a nerve agent capability.

In light of the GSP and this shift in military opinion, the DRPC concluded with 'great reluctance' that plans for the expansion of Nancekuke should be abandoned.[121] The groundwork and a smaller facility would remain, but any aspirations for mass-production were scrapped. Financial constraints, the dominance of nuclear weapons, concerns over the Geneva Protocol, the adverse impact of prioritising nerve agent weapons over other 'more urgent' weapons, and the United States potentially soon providing a substantial nerve agent deterrent all combined to undermine arguments for domestic nerve agent mass-production.[122] The actions of the United States were a particularly potent factor. Production in the United States, despite setbacks, was making some headway. And, members of the DRPC observed that 'as with the atom bomb, nerve gas would prove a powerful deterrent in the hands of the United States, even if we ourselves did not possess it'.[123] With there being no certainty of use and with the costs involved, the DRPC, Air Staff, and Army all revoked their support and backing for domestic mass-production.

After reviewing policy, the CoS concluded in January 1953 that 'in light of this country's present financial position', Britain could not 'afford to provide the means of retaliation with modern CW nerve gas weapons'.[124] The CoS, in tune with the DRPC and the change in military opinion, also counselled that as with the atomic bomb, nerve agent would indeed provide a powerful deterrent in the hands of the United States. With the United States providing the future CW deterrent, and a potential avenue for future acquisition, Britain did not need to mass-produce its

[121] TNA, DEFE 10/32, 'Development of Manufacturing Capacity for Nerve Gas', Secretary of the Defence Research Policy Committee, 9 January 1953, Attached report. This assessment also formed a part of wider cuts in expenditure in the defence field, where the DRPC was tasked with investigating the impact of a 10% reduction, see the later conclusion: TNA, DEFE 10/32, 'Review of the Research and Development Programme', Defence Research Policy Committee, 22 April 1953.

[122] TNA, DEFE 10/32, 'Development of Manufacturing Capacity for Nerve Gas', Secretary of the Defence Research Policy Committee, 9 January 1953, Attached report.

[123] Ibid.

[124] For the earlier discussion and motivations involved, see: TNA, DEFE 4/59, Minutes of Meeting, Chiefs of Staff Committee, 6 January 1953, Confidential Annex. For the implementation and confirmation, see: TNA, DEFE 7/700, 'Policy for Chemical and Biological Warfare', Secretary of the Chiefs of Staff Committee to First Sea Lord, CIGS and CAS, 20 January 1953, Attached report.

own nerve agent weapons. This decision was now thought a necessary sacrifice. Other areas of defence 'of even greater importance' needed funding.[125] During a period of economic cutbacks and with the ascendency of nuclear weapons in defence policy, funding had to go towards weapons which were vital to defence and which would be used, rather than those that only might be used.[126] Military interest in biological weapons was also following a similar course. Coinciding with the backlash from the GSP failure and wider shifts, British BW policy was to focus on defensive measures, the creation of an experimental plant was to be postponed, and collaboration with the United States continued. In the postwar BW field, there had previously been hopes for an offensive capability, but these were also starting to drift and diminish.[127]

The substantial policy reversal for the CW field, put forward by the CoS, was met with widespread acceptance. One exception was the specialised CW sub-committee of the CoS. Under the Chairmanship of Sir Kenneth Crawford, the sub-committee put up bitter resistance. Crawford, in one of his final acts before retirement, condemned the direction of British CW policy. He strongly criticised the stance of Harding and Dickson, and bemoaned the poor state of British CW preparedness.[128] Crawford argued that the 'nerve gases, used as tactical weapons in the field, would be of a potency far in excess of anything previously known' and that they would prove 'a substantial deterrent'.[129] Reminiscent of the CoS arguments in the original GSP, Crawford lobbied for nerve agent weapons, believing that they would be hugely important for the defence of Western Europe in the Cold War.

[125] TNA, DEFE 4/59, Minutes of Meeting, Chiefs of Staff Committee, 6 January 1953, Confidential Annex.

[126] Ibid.

[127] Balmer. *Britain and Biological Warfare*, 139–141.

[128] TNA, DEFE 7/700, 'Development of Manufacturing Capacity for Nerve Gas', N. Brownjohn to First Sea Lord, CIGS and CAS, 13 January 1953, Attached note by General Sir Kenneth Crawford; *The Daily Telegraph*, 'New Munitions Chief', 2 September 2 1953, 6.

[129] TNA, DEFE 7/700, 'Development of Manufacturing Capacity for Nerve Gas', N. Brownjohn to First Sea Lord, CIGS and CAS, 13 January 1953, Attached note by General Sir Kenneth Crawford. Crawford put up strong resistance in the CoS meeting, which is where the 'substantial deterrent' quote can be found, see: TNA, DEFE 4/59, Minutes of Meeting, Chiefs of Staff Committee, 6 January 1953, Confidential Annex.

Crawford further warned that once atomic weapons were used, nerve agents would not remain 'excluded for long'.[130] He thus dubbed first-use a non-issue. In an all-out war no weapon would be barred, and the Americans would use everything in their arsenal. Here, he can be seen to be mirroring earlier and repeated arguments that, if the existence of the realm were at stake, moral and legal arguments would cease to be pressing factors, allowing Britain to use nerve agent weapons. While emphasising their utility and benefits, Crawford, through comparing them to nuclear weapons, was slightly downplaying the ramifications of use. The use of chemical weapons, particularly in terms of scale of casualties and destruction, paled in comparison to that of nuclear weapons. Why should chemical weapons not be acquired when Britain was proceeding with its endeavour to develop nuclear weapons? The two fields were, to a degree, connected. His comments point towards a belief that if nuclear weapons were used, all other weapons would be allowed shortly after. This was a point not widely discussed or investigated at the time in senior military circles when deciding CW policy. Certainly, the survival of the realm was cited as a potential breakthrough for use, as was the use of nuclear weapons, but a detailed exploration of how this would manifest was notably absent in high-level policy. Many questions remained unanswered. For example, would one atomic bomb allow CW use in a retaliatory attack, or was it reserved for either a total war or retaliation-in-kind? Investigating the potential scenarios for the use of chemical weapons in detail could have helped consolidate or define a role for nerve agent weapons after the setback of the GSP. Any such effort, though, was also undermined by a lack of verifiable intelligence on Soviet intent. Defence planners did not know how the Soviet Union would respond. Crawford was left clutching at straws in an attempt to salvage nerve agent policy.

A willingness to accept reliance on the United States was also slammed by Crawford. He argued that prior Anglo-American cooperation was based on a degree of reciprocity, not complete dependence.[131] If Britain chose such a position of dependence, then he cautioned that the depths and levels of Anglo-American cooperation would diminish. Britain would be unable to supply the United States with information of value in the CW

[130] TNA, DEFE 7/700, 'Development of Manufacturing Capacity for Nerve Gas', N. Brownjohn to First Sea Lord, CIGS and CAS, 13 January 1953, Attached note by General Sir Kenneth Crawford.

[131] Ibid.

field.[132] Crawford concluded that Britain needed to maintain its contribution to the 'common fund of knowledge' to warrant continued access to the United States programme.[133] The Chairman of the CWSC therefore roundly condemned the direction and perceived decline of British CW policy. He recommended that given the economic situation, Britain should not abandon nerve agent weapons, but simply seek the most economically viable means of possessing them.

Testament to his influence, which amalgamated with lingering military hopes over acquiring some form of a nerve agent weapon, Crawford's harsh rebuke did not go completely unheeded. After his intervention, while the CoS accepted that the planned 50-ton-a-week nerve agent facility should indeed be abandoned, contrary to Harding they agreed that advanced British research into offensive CW must continue, particularly as the perceived Soviet threat remained 'formidable'.[134] Crawford's desire for a more economical nerve agent capability also chimed with the Air Ministry, which had been advocating a potentially more economically viable nerve agent capability since the GSP failure. This would, though, entail further reliance on the United States. The Air Ministry believed that Britain could purchase nerve agents from the United States, whilst continuing its own offensive CW research, maintaining Britain's contributions to the partnership. This compromise gained traction, especially since after the GSP the Service Chiefs had reduced the quantity of nerve agent weapons they required for the outbreak of war; increasing the likelihood that the United States could fulfil Britain's request. Before the GSP, the War Office and the Air Ministry had desired around 12,000 tons of sarin for military use. This would have proved a tall order for United States officials. However, immediately after the GSP setback, this figure was drastically curtailed to 2500 tons.[135] Given this comparatively low amount, and

[132] Ibid.

[133] TNA, DEFE 5/43, 'Policy for Chemical and Biological Warfare Research and Development', Chairman of the Chemical Warfare Sub-Committee, 11 December 1952, Annex.

[134] TNA, DEFE 7/700, 'Policy for Chemical and Biological Warfare', Secretary of the Chiefs of Staff Committee to First Sea Lord, CIGS and CAS, 20 January 1953, Attached re-drafted report.

[135] TNA, DEFE 5/44, 'Policy for Chemical Warfare', Ministry of Supply, 17 February 1953. Of this amount, 1600 tons were intended for the army and 900 tons for the Air Ministry. See: TNA, DEFE 7/700, Minutes of Meeting, Chemical Warfare Sub-Committee, 7 November 1952.

even with the stuttering nature of the United States CW programme, British officials were hopeful for a deal.[136] As the Ministry of Supply observed, 'the overheads required for us to produce our modest requirements seem out of proportion when considering the huge effort being expended in the United States'.[137] Edging towards further dependence, nerve agents would have to be purchased from the United States, but advanced British nerve agent research would be preserved.

Despite still not possessing any substantial nerve agent capability in early 1953 as a result of technical difficulties, the United States was rapidly approaching sarin mass-production.[138] This was greatly aided by the fact that after the outbreak of the Korean War, the United States CW budget was tripled. This left Crawford to ruefully conclude that 'the American effort is such as to make ours puny by comparison'.[139] When the optimal level of United States nerve agent mass-production was reached, it would be producing 15,000 to 20,000 tons of sarin a year.[140] British officials in the Ministry of Supply still thought this rapid expansion of nerve agent production highly ambitious, and reflective of earlier concerns they cautioned that it would be 'surprising' if these plans actually met the schedule envisaged.[141] But, when mass-production was reached, British officials were keen to benefit. Owing to the degree of Anglo-American CW cooperation, British research did already directly benefit from United States nerve agent research and production programme in other ways. British experts were allowed access to production facilities, given information on production techniques and they were regularly appraised of the progress

[136] TNA, DEFE 5/44, 'Policy for Chemical Warfare', Ministry of Supply, 17 February 1953; Schmidt. *Secret Science*, 189.

[137] TNA, DEFE 5/44, 'Policy for Chemical Warfare', Ministry of Supply, 17 February 1953.

[138] NARA II, Record Group 218, Central Decimal File 1951–1953, Box 152, 'Chemical (toxic) and Biological Warfare Readiness', Joint Strategic Plans Committee, 31 August 1953.

[139] TNA, DEFE 5/43, 'Policy for Chemical and Biological Warfare Research and Development', Chairman of the Chemical Warfare Sub-Committee, 11 December 1952, Annex; Fredericks. United States Chemical Warfare Policy, II-25; Coffey. *American Arsenal*, 163.

[140] Harris, Robert and Paxman, Jeremy. 1982. *A Higher Form of Killing: The Secret of Gas and German Warfare*. London: Chatto & Windus, 182; Kirby, Reid. 2006. Nerve Gas: America's Fifteen-Year Struggle for Modern Chemical Weapons. *CML Army Chemical Review*, Jan-June, 42–43.

[141] TNA, DEFE 10/447, 'Development of Manufacturing Capacity for Nerve Gas', Ministry of Supply, 4 July 1951.

of production.[142] Due to domestic, economic, and political difficulties however, British officials now wanted far more than to just learn about methods of production from the United States, they also wanted the output.

In 1951, the idea of acquiring nerve agents supplied by the United States had been rejected by British CW planners, yet by early 1953, British officials had shifted their position.[143] Throughout 1953 advanced discussions took place between the United States Department of State, the Department of Defense and the British Ministry of Defence (MoD) over whether a nerve agent deal was viable, and over how much the nerve agents would cost.[144] The potential deal had the backing of the United States Joint Chiefs of Staff and the British CoS. But, there were serious reservations on both sides and substantial obstacles to overcome. A major problem was that United States officials estimated that 2500 tons of sarin would cost Britain around $10 million.[145] Another was that the United States did not produce suitable weapons or casings for nerve agent use by British forces—meaning that if Britain wanted nerve agent weapons, it would have to supply the delivery method of the weapons, whilst the United States supplied the actual nerve agent filling.[146] A novel solution was proposed. Highlighting the remarkable depth of trilateral cooperation in the CW field at the time, Canada would produce British-designed casings for weapons which would then be shipped to the United States, filled

[142] For an example of the depth of cooperation see: TNA, DEFE 7/700, 'Review of Chemical Warfare Development up to the End of 1951', Chemical Warfare Sub-Committee, 12 March 1952.

[143] For 1951 see: TNA, DEFE 10/447, 'Development of Manufacturing Capacity for Nerve Gas', Memorandum by the Ministry of Supply, 4 July 1951; TNA, DEFE 5/33, 'Development of Manufacturing Capacity for Nerve Gas', Chemical Warfare Sub-Committee, 20 September 1951. For 1953 see: TNA, DEFE 5/44, 'Policy for Chemical Warfare', Ministry of Supply, 17 February 1953; NARA II, RG 218, Chairman's File: General Bradley 1949–1953, Box 07, 'Chemical Warfare', Memorandum for General of the Army Omar Bradley, 18 June 1953.

[144] NARA II, RG 218, Chairman's File: General Bradley 1949–1953, Box 07, 'Chemical Warfare', Memorandum for General of the Army Omar Bradley, 18 June 1953.

[145] NARA II, RG 218, Central Decimal File 1951–1953, Box 152, Aide-Memoire, Department of State, 16 April 1953.

[146] Ibid.

with sarin, and then these sarin-filled weapons would be shipped to Britain.[147]

A second stumbling block to any deal was that Britain would have to wait for United States production to reach the required level, as in early 1953 production was not sufficient to meet the United States' military requirements, let alone modest British ones as well.[148] It was predicted that only by mid-1956 would the United States be ready for the large-scale use of nerve agents, thus opening the opportunity of potential British acquisition.[149] This delay though caused some consternation on the British side. Concerns were still simmering over whether the country should voluntarily become 'entirely dependent' on the United States for a 'potentially very powerful weapon'.[150] It again brought to the forefront British vulnerability in the CW field, a pressing concern for Clement Attlee in 1950 and one which remained unresolved.

As discussions advanced, exploration of the nerve agent deal gained momentum, and with it emerged senior political interest in the form of the British Minister of Defence, Field Marshal Harold Alexander, and the Prime Minister, Winston Churchill. In March 1953, Alexander agreed with the CoS that a deal should be pursued.[151] He believed that Britain's nerve agent requirements were 'comparatively small' when placed next to the substantial scale of the United States CW programme.[152] Of the potential 2500 tons to be acquired from the United States deal, it was also confirmed that 1600 tons would go to the Army, with the rest going to the RAF.[153] After this green light by Alexander, discussions over a potential deal gained even further authority and weight.

[147] TNA, DEFE 13/265, 'Policy for Chemical Warfare', N. Brownjohn, 25 March 1953. Canada would produce British designed weapons as the United States and Britain used different delivery methods for chemical weapons.

[148] NARA II, RG 218, Central Decimal File 1951–1953, Box 152, Aide-Memoire, Department of State, 16 April 1953.

[149] NARA II, RG 218, Central Decimal File 1951–1953, Box 152, 'BW, CW and RW Annex to Joint Strategic Objectives Plan', Joint Chiefs of Staff, 10 July 1953, Attached report. This was coincidentally the same year in which United States intelligence feared that the Soviet Union would also be prepared to wage war on a substantial scale with lethal nerve agents.

[150] TNA, DEFE 13/265, 'Policy for Chemical Warfare', N. Brownjohn, 25 March 1953.

[151] TNA, CAB 21/3912, 'Policy for Chemical Warfare', Minister of Defence, 6 March 1953.

[152] Ibid.

[153] Ibid.

However, the complexity of transporting nerve agents and the sheer costs involved began to take a toll on negotiations. With divides emerging in March-April 1953, Churchill attempted to take a more active role in the nerve agent negotiations.[154] To procure nerve agent weapons for Britain, Churchill sought to use his relationship with General Walter Bedell Smith, the United States Under-Secretary of State and a close ally of Eisenhower, to spur on negotiations.[155] The proposed letter shows Churchill attempting to intervene at the highest level to secure nerve agent supplies for Britain.[156] It also reveals that Churchill did not only want 2500 tons of sarin for British forces, but a United States commitment to meet all of Britain's nerve agent requirements should war break out. This would have been a substantial undertaking.[157] Churchill's direct intervention never came to fruition. The letter was scuppered by the very same person who had drafted it. Alexander actively discouraged the Prime Minister from intervening, warning that it might be 'straining' his contacts within the United States administration, especially since Bedell Smith was not even directly involved with the negotiations.[158] Churchill, heeding Alexander's advice, agreed to take a back seat.[159]

While negotiations stumbled on, without Churchill's intervention but with his support, plans emerged for the potential use of BBC (bromo benzyl cyanide) teargas during the 'Malayan Emergency'.[160] An influx of people, the Second World War, and the rise of communism in Southeast Asia, had thrown British control of the region into doubt.[161] From 1948, British forces had confronted the communist insurgency and sought to maintain control. As noted by David French, Britain confronted an

[154] TNA, DEFE 13/265, 'Policy for Chemical Warfare', N. Brownjohn, 4 May 1953.

[155] Snyder, W.P. 1984. Walter Bedell Smith: Eisenhower's Chief of Staff. *Journal of Military Affairs*, 48:1, 12.

[156] TNA, PREM 11/3099, 'Nerve Gas', Harold Alexander to Winston Churchill, 8 May 1953, Attached draft letter.

[157] Ibid.

[158] TNA, PREM 11/3099, 'Nerve Gas', Harold Alexander to Winston Churchill, 8 May 1953.

[159] Ibid.

[160] Mankoo, Alexander Chanan. 2019. A Historical Sociology of Teargas in Britain and the Empire, 1925–1965. PhD thesis, University College London, 54.

[161] Ucko, David H. 2019. Counterinsurgency as armed reform: The political history of the Malayan Emergency. *Journal of Strategic Studies*, 42:3–4, 50.

insurgency that pursued a 'Maoist strategy'.[162] In 1953 this conflict was still ongoing, and in an effort to restore control alternative methods were explored. Alexander Mankoo observes that, in April 1953, Porton Down officials conducted experimental trials to determine whether the BBC teargas might be used in the conflict.[163] On Tenggol, an island situated near Malaya, volunteers were exposed to the teargas. But, due to limited stocks and fears of potential publicity and criticism, the decision was made to not pursue this option further.[164] As with the nerve agents, the episode reveals an awareness that all forms of chemical weapons are controversial—the degree of the chemical weapons taboo varies based on the agent, but even with teargas it is still present and a pressing concern. The case of BBC teargas in Malaya and its non-use on a military scale also reveals an acknowledgement of international norms, and concern over a potential public backlash should information on use, or intent and stockpiling, reach the public domain. If this were a concern for teargas, it was one which could only be magnified for the nerve agents. Although this strand of CW policy operated in a slightly different sphere with different limitations and constraints than the nerve agents, it holds important insights and certain overlaps that are present in the entire field.

Shortly after the experimental trials and Churchill's attempted intervention, in May 1953, negotiations encountered their most significant obstacle. The CoS had become increasingly concerned by the burden of having to pay outright for the 2500 tons of sarin, rather than receiving it on more favourable terms through US defence aid. Summarising the findings of the CoS, Alexander informed Churchill that the proposed deal, in its current form, was simply not a viable alternative to domestic mass-production.[165] The expansion of the British nerve agent facility at Nancekuke was predicted to cost in the region of £4 million. The cost of purchasing nerve agents from the United States was to be £2.2 million, creating a saving of around £1.8 million.[166] This attractive saving, in a

[162] French, David. 2011. *The British way in counter-insurgency, 1945–1967*. Oxford: Oxford University Press, 248.

[163] Mankoo. A Historical Sociology of Teargas, 54, 212–222.

[164] Ibid.

[165] TNA, PREM 11/3099, 'Nerve Gas', Harold Alexander to Winston Churchill, 8 May 1953.

[166] This figure is derived from information used in deciding the efficiency of the deal. Domestic cost of production was predicted as £4m, total predicted savings, not including transport, were £1.8m. Calculated cost of purchasing nerve agents, around £2.2m. Figures

period of defence cutbacks, did not however include the costs of transporting the nerve agents. It was soon realised, after detailed negotiations and planning, that this saving of £1.8 million would actually have been largely offset by the costs of filling the ammunition and transporting the weapons across the Atlantic.[167] As a result of this seemingly insurmountable obstacle, Alexander informed Churchill that the nerve agent deal should be rejected in its current form. The deal did not fulfil its primary purpose of providing Britain with an economically viable nerve agent capability.[168]

While these negotiations were ongoing, Churchill was again drawn into nerve agent policy. On 7 May 1953, he was informed of British nerve agent experiments on volunteer soldiers.[169] These experiments were a part of a much more extensive research programme, with nerve agent experiments carried out thousands of times a year at Porton Down. Although these trials were often on animals such as guinea pigs and rabbits, rather than humans, from 1949, scientists at Porton had thought it essential to begin testing small amounts of nerve agents on volunteer soldiers.[170] This was deemed a necessary scientific step for the accurate measurement of the effects of nerve agent exposure in humans. In total, from 1949 to May 1953, around 1500 soldiers were subjected to tests with lethal nerve agents.[171] Churchill was remarkably informed that these trials were 'exceedingly mild'.[172] As uncovered by Ulf Schmidt, British scientists may have genuinely believed this, given that they were using smaller doses than experiments conducted in the United States.[173] But, a concerning feature of these trials was the increasing level of risk involved, as an increasing number of soldiers were exposed to increasing amounts of sarin.[174]

Churchill was notified of these activities because, on 6 May 1953, a British volunteer soldier had died from nerve agent exposure after an

found in: TNA, PREM 11/3099, 'Policy for Chemical Warfare', N. Brownjohn to the Winston Churchill, 4 May 1953.

[167] Ibid.

[168] TNA, DEFE 13/265, 'Nerve Gas', Harold Alexander to Winston Churchill, 8 May 1953.

[169] TNA, PREM 11/3099, JS [full name unreadable] to Winston Churchill, 7 May 1953.

[170] Schmidt. *Secret Science*, 219–227.

[171] TNA, PREM 11/3099, JS [full name unreadable] to Winston Churchill, 7 May 1953.

[172] Ibid.

[173] Schmidt. *Secret Science*, 215.

[174] Ibid., 263.

experiment at Porton Down.[175] In the trial, researchers had exposed Leading Aircraftman Ronald Maddison to a small quantity of sarin.[176] After a short amount of time, he suffered from violent convulsions which resulted in a traumatic death. This experiment on Maddison was carried out despite, in the months preceding his death, numerous other instances of soldiers suffering seriously negative effects from the trials.[177] In addition, many of the soldiers involved in these experiments were completely unaware that they were being tested with lethal nerve agents.[178] These trials conducted at Porton, and the death of Maddison, did not come to light at the time. The true nature of his death was concealed and kept secret.[179] The subsequent Home Office inquest was also kept secret, and Maddison's death was labelled as accidental.[180] It was not until 1960 that the *Daily Mail* reported that scientists at Porton Down had experimented on humans, but even then the scale was not fully appreciated or realised.[181] In November 2004, after a second inquest and after 50 years had elapsed, it was finally determined that Maddison had been unlawfully killed by nerve agent exposure.[182] For decades the public was kept in the dark about the extent of British nerve agent research and human experiments, which had reached this unfortunate climax in 1953.[183] This adherence to secrecy was all the more remarkable given that Churchill, in May 1953, when receiving the report of Maddison's death, had advised officials to 'tell the truth'.[184]

[175] For more details, see: Schmidt. *Secret Science*, 218–283.

[176] Tucker. *War of Nerves*, 147–148; Bolton, Talitha. 2008. Consent and the Construction of the Volunteer: Institutional Settings of Experimental Research on Human Beings in Britain during the Cold War. PhD diss., University of Kent, 65.

[177] Schmidt. *Secret Science*, 219–227.

[178] Ibid, 221.

[179] Ibid, 219. Some limited references were made on the nature of his death, but publicity did not gain as much traction as might be expected. See also: Liddell Hart Centre for Military Archives, H. C. Pincher, Press Cuttings, Vol.5 1953–1955, 'Poison Gas Tests Start Again', 12 October 1954. The author would also like to thank the Trustees of the Liddell Hart Centre for Military Archives for allowing the use of this material.

[180] Tucker. *War of Nerves*, 150.

[181] *Daily Mail*, 'Nerve Gas', 6 February 1960, 5.

[182] Evans, Michael. 2004. Nerve gas death of Porton Down guinea pig was unlawful killing. *The Times*, November 16, 4.

[183] Evans. *Gassed*, 168.

[184] TNA, PREM 11/3099, JS [full name unreadable]to Winston Churchill, 7 May 1953 [written comments by Churchill on the document].

Given the perceived scale of the Soviet threat, the lack of an effective deterrent, and with the urgency resulting from falling behind in the nerve agent arms race, these British experiments on human volunteers had been thought justified. This unethical drive for scientific advances came at a substantial cost, with Maddison's death marking the end of a particularly gloomy and questionable period in British defence policy, which had also been dominated by considerations and military support for changes in the first-use of lethal chemical weapons.

The period of 1951–1953 therefore marked both the zenith of postwar British nerve agent policy and signalled the start of a rapid decline. In 1951, crucial defence committees were pushing for a prominent role for nerve agents in British defence policy. This groundswell of support, based on the perceived military utility of nerve agent weapons, led to the CoS drastically attempting to change Britain's interpretation of the Geneva Protocol. This attempted change in first-use, however, met fierce resistance from both the Foreign Office and from senior politicians in the Cabinet Defence Committee. The CoS and military advocates of British nerve agent weapons were rebuffed. The political, economic, moral and reputational costs were deemed too significant. Only in an exceptional situation, and facing a grim fight for survival, would political approval be considered for a change in first-use policy. With no guarantee that nerve agent weapons would be used in a future war, British nerve agent policy suffered from uncertainty and cutbacks. The expansion of the Nancekuke facility to allow the production of 50 tons of sarin per week was halted, and serious questions emerged over the future of British CW policy. While economic constraints severely hampered policy, instead of scrapping offensive CW policy altogether, defence officials attempted to salvage the situation through a potential nerve agent deal with the United States. These protracted negotiations for an affordable nerve agent capability and deterrent, despite some early setbacks, would continue into the mid-1950s when British CW policy continued its drift and decline.[185]

[185] TNA, DEFE 13/265, 'Nerve Gas', Secretary of the Chiefs of Staff Committee to the First Sea Lord, CIGS, CAS, 1 October 1953; TNA, DEFE 5/50, 'Transportation of Nerve Gas from America', Ministry of Supply, 9 December 1953.

CHAPTER 4

Drift, Decline, and a Stubborn Prime Minister, 1954–1957

If we were to kill each other by the million with gases we might yet earn the gratitude of the historians and archaeologists of the future, who might say that at least the inhabitants of the western countries, in the second half of the 20th century, were civilised or cultured enough to kill each other but not to destroy everything their ancestors had made during thousands of years of civilisation.
Geoffrey de Freitas MP (Hansard, House of Commons, 'Civil Defence', Vol.529, cc1832–914, 5 July 1954)

After the Korean War, which drew to a close with the armistice agreement of July 1953, the Cold War briefly quietened.[1] The period of 1954–1955, despite coinciding with West Germany joining NATO and conflict in Indochina, represented a relative decline in bipolar Cold War tensions. This was also not just a trend in the Cold War, with overall levels of global

[1] Békés, Csaba. 2010. East Central Europe, 1953–1956. In *The Cold War, Volume I: Origins*, ed. Melvyn Leffler Odd and Arne Westad. Cambridge: Cambridge University Press, 334.

Policy drift is a recurring theme in CBW policy. The title for this chapter is inspired by the similar path of BW policy, as referred to in Balmer. *Britain and Biological Warfare*, 128.

W. King, *Nerve Agents in Postwar Britain*, Britain and the World,
https://doi.org/10.1007/978-3-030-70474-2_4

violence dropping for several years following 1954.[2] Nikita Khrushchev, the emerging Soviet successor after the death of Joseph Stalin in March 1953, initially appeared to embrace the new atmosphere. When he met President Dwight Eisenhower at Geneva in July 1955, he agreed to Austrian neutrality and aired ideas of disarmament.[3] However this climate of cooperation would soon prove short-lived.[4] Tensions would again re-ignite when Khrushchev forcefully suppressed the Hungarian uprising in 1956, and when Soviet Premier Nikolai Bulganin issued threats, mentioning nuclear weapons, to Britain regarding its actions in the Suez Crisis.[5] This environment of rising tensions, rhetoric, and confrontation, would take greater urgency with the 'Sputnik effect' in 1957, when Soviet breakthroughs in the area of ballistic missile technology added to fears in Britain and the United States of growing Soviet scientific and military capabilities in the thermonuclear age.[6]

Navigating the fall and rise in Cold War tensions was not an easy task for British Prime Ministers. Winston Churchill, keen to reinvent great power summitry after the death of Stalin, attempted to secure a more active, moderating role for Britain.[7] Suffering due to ill-health, in April 1955 he resigned. Anthony Eden, Churchill's heir-apparent, took on the mantle. Eden's premiership was unfortunately marked by the disaster of the Suez Crisis of 1956.[8] Branded as a 'traumatic clash' in Anglo-American relations, Suez underlined the limits of postwar British power, with pressure from the United States forcing a humiliating British retreat from its military action against Egypt.[9] In January 1957, following Eden's resignation, Harold Macmillan became the third Conservative Prime Minister in less than two years. Once in office, Macmillan sought to repair Anglo-American relations as a first priority. And, in May 1957, he witnessed

[2] Chamberlin, *The Cold War's Killing Fields*, 177.

[3] Mastny, Vojtech. 2010. Soviet foreign policy, 1953–1962. In *The Cold War, Volume I: Origins*, ed. Melvyn Leffler and Odd Arne Westad. Cambridge: Cambridge University Press, 315–319.

[4] Zubok, Vladislav and Pleshakov, Konstantin. 1996. *Inside the Kremlin's Cold War: From Stalin to Khrushchev*. Cambridge, MA: Harvard University Press, 280–281.

[5] Boyle, Peter. 2005. The Hungarian Revolution and the Suez Crisis. *History*, 90:4, 555.

[6] Mastny. Soviet foreign policy, 1953–1962, 320–323.

[7] Deighton, Anne. 2010. Britain and the Cold War. In *The Cold War, Volume I: Origins*, ed. Melvyn Leffler and Odd Arne Westad. Cambridge: Cambridge University Press, 130.

[8] Gaddis, John Lewis. 2005. *The Cold War*. London: Allen Lane, 126–127.

[9] Baylis, John. 1997. *Anglo-American Relations since 1939: The Enduring Alliance*. Manchester: Manchester University Press, 68; Reynolds. *Britannia Overruled*, 190–194.

Britain joining the thermonuclear club as the world's third member during the Grapple series of tests in the South Pacific, with yields around 100 times greater than the bombs used on Hiroshima and Nagasaki.[10]

In parallel with the growing dominance of nuclear weapons and deterrence, the economic situation also took on renewed importance during this period, especially in defence policy.[11] Defence expenditure was to witness a fall from 9.4% of GDP in 1954 to 6.8% in 1958.[12] Macmillan appointed Duncan Sandys as Minister of Defence, with a mandate to substantially reduce defence expenditure and manpower.[13] Sandys' famous Defence White Paper, issued in April 1957, announced the end of national service, embodied strategic and economic arguments to justify reductions in defence expenditure, and emphasised the pre-eminent role of the nuclear deterrent.[14] The British development of thermonuclear weapons during the Sandys era of defence policy was thought crucial. The nuclear arms race had begun to reach new heights. The United States had first tested a thermonuclear device in 1952, and the Soviet Union in 1955. For British policy-makers, this dawn of the thermonuclear age was a double-edged sword. Britain, as a close ally of the United States and member of NATO, was a nearby target for Soviet bombers. Yet, thermonuclear weapons would also diminish the advantage of physically larger countries, which had been a clear benefit the Soviet Union possessed over Britain.[15] Along with this substantial thermonuclear development, with its horrifying level of destructive capability, increasing focus and emphasis was also placed on the deterrent role of nuclear weapons in British defence policy. This prioritisation of nuclear weapons often came at the cost of other areas of defence. Thermonuclear weapons were deemed integral for deterrence, for prestige, and viewed as a more economical means of improving security in comparison to conventional forces.[16]

British CW policy, after the failure of the 1952 GSP, with substantial changes in the Cold War environment, and increasingly eclipsed by this

[10] Reynolds. *Britannia Overruled*, 171; Fry. *The Politics of Decline*, 158–160.

[11] Baylis. *Ambiguity and Deterrence*, 241.

[12] Baylis. *British Defence Policy*, 139.

[13] Self, Robert C. 2010. *British Foreign and Defence Policy Since 1945: Challenges and Dilemmas in a Changing World*. Basingstoke; Palgrave Macmillan, 62.

[14] Peden, G. C. 2009. *Arms, Economics and British Strategy: From Dreadnoughts to Hydrogen Bombs*. Cambridge: Cambridge University Press, 273.

[15] Clark and Wheeler. *The British Origins of Nuclear Strategy*, 210, 410.

[16] Baylis. *British Defence Policy*, 52.

ever-growing nuclear shadow, experienced a period of drift and decline. This chapter will first explore the legacy and fallout resulting from the 1952 GSP, and then it will analyse the crucial years of 1956–1957. Decisions made during this period, under Eden and Macmillan, held substantial and lasting ramifications for British CW policy in the Cold War.

Drift and Decline

By February 1954, any possible deal for acquiring nerve agents from the United States had fallen through. In 1953, despite the setbacks of the 1952 GSP, hopes had been pinned on acquiring a nerve agent capability from the United States. Even Churchill had offered to intervene to help in the endeavour. Energetic action by British officials proved not enough. Discussions had become bogged down in 'legal difficulties' and over the issue of payment.[17] Norman Brook, the Cabinet Secretary, informed Churchill that there were still no financial gains to be found with the United States deal, as the total cost of purchasing and transporting the nerve agents would be comparable to Britain domestically producing them.[18] British defence officials, asserting a more independent line, also increasingly thought it unwise to rely wholly on the United States for a CW capability. In line with earlier fears, this was a particular concern since United States nerve agent production was behind schedule, rendering it unlikely that they would have spare supplies for Britain to purchase in the immediate future.[19]

In light of this setback, the Cabinet Defence Committee initially and remarkably, under the strong urgings of the Minister of Supply Duncan Sandys, once again conditionally approved the domestic production of nerve agents in Britain.[20] It also confirmed that the nerve agent deal,

[17] TNA, PREM 11/3099, 'Policy for Chemical Warfare', Brownjohn to the Prime Minister, 8 February 1954; TNA, PREM 11/3099, 'Policy for Chemical Warfare', Norman Brook to the Prime Minister, 8 February 1954; NARA II, RG 218, Central Decimal File 1954–1956, Box 126, Secretaries of the United States Joint Chiefs of Staff to the United States Joint Chiefs of Staff, 15 February 1954, enclosed memorandum.

[18] TNA, PREM 11/3099, 'Policy for Chemical Warfare', Norman Brook to the Prime Minister, 8 February 1954.

[19] TNA, PREM 11/3099, 'Policy for Chemical Warfare', Brownjohn to the Prime Minister, 8 February 1954.

[20] TNA, CAB 131/14, 'Policy for Chemical Warfare', Memorandum by the Minister of Supply, 25 January 1954; TNA, CAB 131/14, Minutes of Meeting, Cabinet Defence Committee, 9 February 1954.

entailing reliance on the United States, would be 'abandoned'.[21] This approval was not, however, concrete. As the condition, like with the 1952 GSP, was that nerve agent weapons could still only be approved if they were deemed essential to defence.[22] In 1954, the CoS and the Ministry of Supply still both believed that the ability to retaliate with nerve agent weapons was important, and with the United States deal falling through, they revived plans for the expansion of the sarin plant at Nancekuke.[23] Use of nerve agent weapons was still thought effective against large numbers of Soviet troops, and so they were still believed to be an 'essential part of our military equipment'. [24] The immediate short-term consequence of the potential United States deal falling through was therefore that domestic production was once again considered. Proponents of nerve agent weapons remained keen to possess a nerve agent capability for both its military utility and deterrent value, even if first-use was off the table.

The expansion of the Nancekuke facility would cost millions of pounds, an outlay which still needed to be strongly justified in a period of economic cutbacks. This was all the more important given that it coincided with the Chancellor of the Exchequer, Rab Butler, warning that the overall cost of defence was 'exceedingly heavy'.[25] Britain was facing the challenge of meeting global defence commitments whilst remaining active in the thermonuclear arms race.[26] As in 1952, chemical weapons again needed to be shown as essential to defence in order to validate significant expenditure. Given these formidable pressures, and despite initially approving production, the nerve agent requirement was again put under the spotlight in mid-1954. But, this time, in addition to defence officials, Cabinet members and the Foreign Office, MPs in the House of Commons also began to engage with CW policy.

Throughout the early Cold War period, defence interests were often largely protected and shielded from more public forms of scrutiny.[27] A key

[21] Ibid.

[22] Ibid.

[23] TNA, PREM 11/3099, 'Policy for Chemical Warfare', Norman Brook to the Prime Minister, 8 February 1954.

[24] TNA, CAB 131/14, 'Policy for Chemical Warfare', Memorandum by the Minister of Supply, 25 January 1954; TNA, CAB 131/14, Minutes of Meeting, Cabinet Defence Committee, 9 February 1954.

[25] TNA, PREM 11/617, 'Defence', Rab Butler to the Prime Minister, 9 September 1954.

[26] Baylis. *Ambiguity and Deterrence*, 180; Baylis. *British Defence Policy*, 52.

[27] Hill. *Peace and Power in Cold War Britain*, 60–61.

event, and precursor to the substantial publicity characterising the coverage of scientific topics in the late 1950s and 1960s, was the Lucky Dragon incident. The Lucky Dragon was a Japanese fishing vessel, exposed to radioactive fallout following nuclear weapons tests by the United States at Bikini Atoll in March 1954. Of the crew, 23 suffered from radiation sickness, and one later died.[28] The incident was reported on internationally and, as noted by Alison Kraft, shortly after the incident key scientists such as Bertrand Russell and Joseph Rotblat intervened to publicly discuss the effects of nuclear weapons testing and radiation.[29] The BBC, attempting to run a scientific programme on nuclear weapons and fallout, was rebuffed by strong government pressure.[30] Increasingly, though, information was reaching the public domain. And, while Britain's initial decision to produce an atomic bomb was kept secret, cracks were beginning to show in the formerly tough facade.

It was in this context that, in a break from past practice, in July 1954, CW policy attracted political attention during a House of Commons debate on civil defence. Addressing the CW field via civil defence was undoubtedly in part enabled by the deliberate loosening of secrecy which had occurred almost two years before, when defence officials released information on the nerve agents to better inform medical and civil defence officials.[31] Remarkably though, this rare parliamentary coverage was almost positive. At the very least it pointed towards a degree of acceptance by those MPs addressing the controversial field. Labour MP Geoffrey de Freitas started the debate when he warned that we have 'to be careful, in dealing with this problem [civil defence], not to concentrate entirely on the hydrogen bomb type of warfare…I have not heard anyone refer to the problem of gas'.[32] This was not alarmist; rather, it was a drive against complacency and the lack of attention afforded to chemical weapons. Bizarrely for a Labour MP in the postwar period, de Freitas was, in a round-about fashion, reinforcing views over nerve agent weapons and Government CW policy. He accepted the horrors of chemical weapons but did not see how they could 'be any worse than blowing people to pieces with high

[28] Kraft, Alison. 2018. Dissenting Scientists in Early Cold War Britain: The "Fallout" Controversy and the Origins of Pugwash, 1954–1957. *Journal of Cold War Studies*, 20:1, 58.

[29] Ibid., 68.

[30] Hill. *Peace and Power in Cold War Britain*, 66–67.

[31] See: *Daily Mail*, 'New Gas Destroys Nerves', 8 August 1952, 3.

[32] Hansard, House of Commons, 'Civil Defence', Vol.529, cc1832–914, 5 July 1954.

explosives'.[33] While these comments were not reported on in any meaningful way in the press, they still reflected some information on CW trickling into the public domain. The comments also marked a stark change from prior practice, with previous Commons debates and interventions tending to emphasise the novelty, horrors and exceptional nature of chemical weapons.[34] The exact reasons and motivations as to why de Freitas made his observations are unclear, but in addition to coinciding with existing military thinking, his attempt at normalising chemical weapons could also have played a part in civil defence preparations through downplaying fears.

In Parliament, de Freitas was not alone in addressing and drawing attention to chemical weapons. Although he clearly went the furthest, especially when he stated that CW use might 'earn [us] the gratitude of the historians and archaeologists of the future'. His views were to a degree echoed by a select few MPs, who were keen on drawing further attention to the CW field.[35] Labour MP Austen Albu revealed to the Commons that he was hopeful because, similarly to the Second World War, both sides possessed chemical weapons, and thus they might not be used in the Cold War.[36] This line of thinking, which touched upon CW deterrence through the fear of reprisals, was to a degree representative of earlier political thinking seen with Clement Attlee and other politicians in the 1940s, and these limited disclosures also spanned party lines. Conservative MP Brigadier Otho Prior-Palmer, supporting attempts to address chemical weapons, drew lessons from CW deterrence to aid in further comprehending nuclear deterrence. He observed that:

> I think there is a great likelihood of it [nuclear warfare] being used, but there is just a faint possibility that, because of its appalling effect, rather in the same way that gas was not used in the last war, both sides may hold their hands.[37]

Select disclosures and this rare parliamentary attention reveals that the horrific nature of CW was occasionally, and surprisingly, seen as a form of

[33] Ibid.

[34] For example, see: Hansard, House of Commons, 'Defence', Vol.478, cc467–645467, 26 July 1950.

[35] Hansard, House of Commons, 'Civil Defence', Vol.529, cc1832–914, 5 July 1954.

[36] Ibid.

[37] Ibid.

reassurance. The thinking being that the nature of chemical weapons was so reviled that both sides would hesitate before inflicting the other to such an abhorrent form of warfare, and before exposing their own population to retaliatory attacks. In some form the taboo nature of CW was being expressed and acknowledged. The legacy of CW deterrence during the Second World War was also still clearly holding sway in political considerations of CW deterrence in the Cold War, with Prior-Palmer using wartime CW experiences as a glimmer of hope for successful nuclear deterrence in the Cold War.

While some political views were shifting and represented in multiple forms in the postwar period, military assessments of the tactical advantages and role of nerve agents had seen little change. Additional research and trials had merely reinforced the perceived military benefits of nerve agent use. One example was the assessment that the British Centurion tank was 'rather less vulnerable' to nerve agent weapons than the Soviet T.34.[38] It was still believed that Britain and its allies maintained the upper hand in CW use. In late 1954, the DRPC surmised that 'a nerve gas attack will always have considerable success'.[39] For even if casualties were low, fear of nerve agent exposure would lead to troops wearing gas masks for extended periods, which would severely hamper their manoeuvrability and fighting effectiveness.[40]

Continuing interest in the military utility of nerve agents was also fuelled by NATO's interest in formulating an approach to using chemical weapons on the battlefield.[41] During 1954, moves to involve the North Atlantic Council, the principal decision-making body of NATO, in any decisions made by the US Supreme Allied Commander Europe to release chemical weapons prompted the realisation that this would also entail the tripartite countries sharing information with other NATO countries.[42] Included within this were details on their advanced CW developments,

[38] TNA, DEFE 10/33, 'Review of the R and D Programmes', Note by the Secretaries of the Defence Research Policy Committee, 27 October 1954, Attached report.

[39] TNA, DEFE 10/33, 'Review of the R and D Programmes', Note by the Secretaries of the Defence Research Policy Committee, 18 November 1954, Attached report.

[40] Ibid.

[41] TNA, DEFE 10/448, 'SHAPE's Interim Requirements for Toxic Chemical Munitions', Note by the Joint Secretaries of the Chemical Warfare Sub-Committee, 14 July 1954, Attached draft report; NARA II, RG 218, Central Decimal File 1954–1956, Box 126, 'SACEUR's Interim Requirements for Toxic Chemical Munitions', Edwin Carns, 5 October 1954.

[42] TNA, DEFE 5/53, 'SHAPE's Interim Requirements for Toxic Chemical Munitions', Chemical Warfare Sub-Committee, 23 July 1954.

their level of CW preparedness and their research findings.[43] This was information which had previously been kept highly secret within the tight-knit group of Britain, the United States and Canada. British CW experts strongly opposed the recommendations. Despite being put forward by the United States, British officials were worried that the move would create more opportunities for the leakage of secret information.[44] It was also feared that Britain's 'unpreparedness' for CW, its limited ability to retaliate and the hollow threat of its CW deterrent would become known to the Soviet Union from NATO sources.[45] Unsurprisingly, British officials therefore strongly resisted United States proposals for more comprehensive NATO CW discussions. The resulting compromise was that while NATO forces would be provided with defensive equipment, including detectors, respirators, and decontamination gear, individual NATO countries would not be obliged to supply chemical weapons to NATO or disclose their capabilities to other NATO countries.[46] This brief disagreement added to the military desire in Britain for a nerve agent weapon. Otherwise British vulnerability could be uncovered and exploited by the Soviet Union, and British prestige could be adversely affected through allied discovery of its unpreparedness.

Despite recognition of the military utility of nerve agents, in November 1954, the DRPC still remained cautious over the sticking point of first-use and the ramifications this held for nerve agent weapons, particularly after the setback of the 1952 GSP. For the DRPC, even with Defence Committee approval, there needed to be some form of assurance that Britain could use nerve agent weapons in war for them to warrant funding and development. As such, the sub-committee advised the CoS that:

43 Ibid.

44 NARA II, RG 218, Central Decimal File 1954–1956, Box 126, 'SACEUR's Interim Requirements for Toxic Chemical Munitions', Office of the United Kingdom Representative to the Chairman of the United States Joint chiefs of Staff, 31 August 1954.

45 TNA, DEFE 5/53, 'SHAPE's Interim Requirements for Toxic Chemical Munitions', Chemical Warfare Sub-Committee, 23 July 1954.

46 TNA, DEFE 5/55, 'SHAPE's Interim Requirements for Toxic Chemical Munitions', Chemical Warfare Sub-Committee, 26 November 1954; NARA II, RG 218, Central Decimal File 1954–1956, Box 126, 'Tripartite Staff Conference on SACEUR's Interim Requirements for Toxic Chemical Munitions', Secretaries of the Joint Chiefs of Staff, 24 February 1955, enclosed memorandum.

> If, but only if, the Services can rely on being able to use chemical weapons from the beginning of a war [is] there ... a strong justification for continued development of production methods and weapon designs for the use of chemical agents. If such a change in policy is not agreed, consideration should be given to reducing the programme to the minimum necessary to ensure adequate defensive measures...[47]

Driven by the 'immense importance of nerve gas weapons' against 'overwhelming numbers of land forces', the DRPC advised the CoS to yet again review British first-use policy. But this time, it was made clear that a negative outcome would incur significant and lasting cutbacks. Nerve agent policy, oscillating between acquiring a nerve agent capability and an empty deterrent, while also balancing secrecy and security, was thus placed under another review.[48] The muddling approach to CW policy continued. But, it was hoped that the new review would provide nerve agent weapons with a solution for the Cold War, either production or reductions. Without this confirmation regarding use, and with no guarantee that Soviet forces would breach the Geneva Protocol (by using chemical weapons first) and allow Britain the option of retaliation, the DRPC warned there was little hope of justifying the mass-production of nerve agent weapons. If nerve agent weapons had no guaranteed option of use, or if they could not be included in defence planning, then they 'would inevitably duplicate other weapons on whose use the Services know they can rely'.[49] To warrant a place in defence planning chemical weapons thus required a confirmed role in a future global war, without one the field would suffer from significant curtailments.

British nerve agent policy was once again in a precarious state. It was again at a tipping point. The influential DRPC had raised serious questions over the role, place and possession of nerve agent weapons in British defence policy. The use of chemical weapons, which was the primary area of contention between Foreign Office officials, Conservative ministers and defence officials in 1952, again became the fulcrum on which the future of a British nerve capability rested. In 1952 defence officials had pushed for a first-use capability, and little consideration was given to a potentially negative outcome. Whereas on this occasion, the outcome, regardless of

[47] TNA, DEFE 10/33, 'Review of the R and D Programmes', Note by the Secretaries of the Defence Research Policy Committee, 18 November 1954, Attached report.

[48] Ibid.

[49] Ibid.

whether it was supportive or detrimental, was to have a more immediate impact agreed on beforehand.

Concurrently, while the CoS was planning the crucial review following warnings from the DRPC and concerns emerging over the greater harmonisation of CW policy within NATO, British politicians again found themselves publicly addressing CW policy. On this occasion, the Government found itself facing some difficult questions on sensitive areas of CW policy in the House of Commons, which was fuelled by Soviet actions. Coinciding with sensitive policy considerations on acquisition and this renewed political attention, defence officials remained alarmed that CW policy was being formulated in the dark, still hindered by a concerning absence of verifiable intelligence on the Soviet CW threat.

In early 1955, and reflective of the growing public interest in secret science, reports on CW were again emerging in British newspapers. This time, these accounts initially pertained to Soviet protests and parliamentary questions surrounding chemical weapons and the issue of West German re-armament and the Paris Agreements of October 1954. In January 1955, after reviewing the text of the Paris Agreements, Soviet officials alleged that it approved the production and stockpiling of chemical weapons by Western European powers.[50] Soviet officials charged Britain, and the other signatories, of subverting the 1925 Geneva Protocol on CBW. In its scolding appraisal, Soviet officials reminded Britain of the 'outstanding role' that the 1925 Geneva Protocol had played during the Second World War. Even proven war criminals of the 'Hitlerite Government' had abided by its principles.[51] Commitments to international treaties, reference to atrocities conducted by the Nazi regime and of chemical weapons use being beyond the pale for even them, again highlighted the exceptional nature of chemical weapons. The slightest allusion or reference to their possession drew immediate condemnation, as well as a quick publicity victory for Soviet officials keen to press their Cold War

[50] House of Commons Parliamentary Papers, 'Correspondence between Her Majesty's Government in the United Kingdom and the Soviet Government regarding Bacteriological and Chemical Warfare', Cmd.9384, Moscow, January 13/February 3, 1955. Also see: *The Manchester Guardian*, 'Russia Sees A New Danger: Chemical Warfare', 14 January 1955, 14; *The Manchester Guardian*, 'British Stock Of Poison Gas: No Breach of Agreement', 4 February 1955, 9.

[51] House of Commons Parliamentary Papers, 'Correspondence between Her Majesty's Government in the United Kingdom and the Soviet Government regarding Bacteriological and Chemical Warfare', Cmd.9384, Moscow, January 13/February 3, 1955.

rivals. Although undoubtedly manipulated for political point-scoring, with Soviet accusations misrepresenting the Paris Agreements, it did lead to some awkward questions for the British Government. In February 1955, Churchill found himself facing questions on the stockpiling of chemical weapons in West Germany, and on whether Britain was now tied in and 'the hostage of ex-Nazi military adventurers'.[52]

Even with Churchill shutting down the discussion with minimalist replies, in March 1955 there were further parliamentary questions fuelled by Soviet accusations and by reservations over West German rearmament. Harold Macmillan, then Minister of Defence, faced queries over British CBW first-use policy, and British commitments to CBW deterrence and retaliation.[53] These doubts and concerns forced Macmillan to clarify that while Britain reserved the means to retaliate with chemical and biological weapons, it was bound by the Geneva Protocol of 1925 which 'forbids the use of such weapons in war'.[54] These questions from MPs, fuelled by Soviet public accusations, reveals a degree of concern and fear of German rearmament and the controversial issue of CBW. The combination of the two was alarming. Though there were further questions over British adherence to the Geneva Protocol, the story and Soviet public remonstrations soon eased. Shutting down the discussion with limited responses has seemingly worked, with coverage of CW policy again returning to its shroud of secrecy.[55]

The direct questioning of government policy was itself revealing. Increasing access to information on the nerve agents, and snippets on the potential course of British policy, coincided with the re-emergence of peace and protest movements. As noted by Richard Taylor, in July 1955 the Golders Green Joint Committee for the Abolition of Nuclear Weapons was formed. The Committee represented a key early intervention by members of the public into the debate regarding controversial weapons. Greater access to information was crucial, with members spurred on by revelations of increased nuclear testing.[56] Though it would take some time before

[52] Hansard, House of Commons, 'Chemical and Bacteriological Weapons', Vol.536, cc2060–2061, 10 February 1955.

[53] Hansard, House of Commons, 'Bacteriological and Chemical Weapons', Vol.538, c26W, 8 March 1955.

[54] Ibid.

[55] For example, see: Hansard, House of Commons, 'Bacteriological and Chemical Weapons', Vol.539, cc368–369, 30 March 1955.

[56] Taylor, Richard. 1988. *Against the Bomb: The British Peace Movement 1958–1965*. Oxford: Clarendon Press, 6–7.

anti-CBW movements truly gained momentum, these were important early signs of more active and engaged members of the public when it came to controversial weapons.

While Soviet statements and accusations had prompted some difficult questions for British politicians domestically, in 1955 British intelligence still had little verifiable information as to what was occurring in the Soviet Union in terms of CW policy or preparedness. In assessments of the Soviet CW threat intelligence officials continued to struggle to uncover any tangible evidence that could inform policy.[57] In an attempt to counter this bleak situation, British intelligence liaised heavily with CW scientists in an attempt to gain further insights in spotting a nerve agent facility. This step would, however, only further highlight the difficulties involved in attaining accurate CW intelligence in the Cold War. British scientists at Porton Down gloomily reported that 'there is very little that can be said to be typical of a CW manufacturing plant'.[58] The equipment used in producing nerve agents is similar to any chemical industry. And, although ventilation at the site would be noticeably different, only a trained chemical engineer or plant operative would be able to spot these differences.[59] Requirements for electricity, water and refrigeration, although significant, would also not be outstanding. The production of intermediate chemicals such as phosphorus trichloride could be carried out off-site. Even the storage of nerve agents would likely be in underground bunkers, out of sight of prying eyes.[60] The difficulties of recognising and assessing nerve agent production were therefore extraordinary. This was compounded when taking into account the security-conscious nature of the Soviet Union. The JIC thus concluded that a large Soviet CW installation could exist, without there being any significant chance of it ever being discovered by British intelligence.[61]

In light of these intelligence gathering problems, the JIC admitted that 'we have no evidence to show whether large-scale production of any chemical warfare agent is or is not in progress in the USSR'.[62] Perceptions

[57] TNA, CAB 158/20, 'Russian Research and Development', Joint Intelligence Committee, 26 April 1955, Annex B.

[58] TNA, DEFE 41/157, Chemical Defence Research Department to H. S. Young, Deputy Director of Scientific Intelligence, 24 August 1955, Attached report.

[59] Ibid.

[60] Ibid.

[61] TNA, CAB 158/20, 'Russian Research and Development', Joint Intelligence Committee, 26 April 1955, Annex B.

[62] TNA, DEFE 5/58, 'Chemical Warfare Policy', Note by the War Office, 14 April 1955, Appendix.

of the Soviet CW threat were therefore based on rough estimates and the scouring of Soviet scientific publications. For planning purposes, it was assumed that there was parity in the level of development.[63] Similarly to Britain, intelligence officials in the United States also relied heavily on published scientific research to form intelligence assessments. More broadly, they found that the level of research in fields related to CW indicated that there was a 'scientific capability for the development of new or improved chemical agents'.[64] Ultimately, neither British nor United States intelligence agencies had any direct or verifiable information on Soviet chemical weapons development, production or intent.

The Anglo-American partners did however agree on a rapid reappraisal of Soviet defensive CW preparations. This was partly fuelled by the acquisition of a Soviet civilian respirator, dubbed the 'G-4' model.[65] Prior intelligence assessments had emphasised a relative Soviet weakness in civil and military defensive measures, but after 1953 this view had begun to shift. The respirator offered a high degree of protection against nerve agents and it showed rapid Soviet progression in the field.[66] This development in defensive preparations added to growing concerns over the Soviet threat. As, despite a lack of verifiable intelligence on Soviet nerve agent production, there were still some indicators available which revealed the seriousness with which Soviet officials were approaching the CW field.

It was in this period of poor intelligence but relative quiet after Soviet public accusations, that, acting on the advice of the DRPC, the first-use of nerve agent weapons was explored with renewed vigour. Based on the guidance of the DRPC, the CoS had launched a further evaluation of the value of chemical weapons in the thermonuclear age. The study,

[63] Ibid., From publications, a site at Kazan' was identified as carrying out fundamental research relevant to CW. Based on Soviet publications in chemistry, it was further uncovered that Ivan Knunyants, recipient of a 1948 Stalin prize, was involved in the Soviet CW program. For further details, see: TNA, CAB 158/20, 'Russian Research and Development', Joint Intelligence Committee, 26 April 1955, Annex B.

[64] NIE, 11-4-54, 'Soviet Capabilities and Probable Courses of Action Through Mid-1959', National Intelligence Estimate, 14 September 1954; NIE, 11-3-55, 'Soviet Capabilities and Probable Courses of Action Through 1960', National Intelligence Estimate, 17 May 1955.

[65] TNA, DEFE 10/174, 'Report on Russian Research and Development', Ministry of Defence, [exact date unspecified] February 1954.

[66] Ibid.; TNA, CAB 158/20, 'Russian Research and Development', Joint Intelligence Committee, 26 April 1955. Although designed for civilian purposes, British intelligence predicted that the same respirator, or another displaying the equivalent technical sophistication, would be employed by Soviet forces.

completed in April 1955, was thought a necessary precursor to a second attempt to secure political approval for a potential change in first-use policy, or, if this could not be achieved, then to side-line offensive nerve agent capabilities.[67] The first-use issue had therefore not gone away; it had lingered on.

In April 1955, an updated evaluation by the War Office provided a robust account of the benefits of chemical weapons in the thermonuclear age, and it strongly supported the need for a change in first-use policy. The War Office contended that nerve agent weapons would cost significantly less than conventional or nuclear weapons and that even in the thermonuclear age, Britain could not afford to forego such a 'powerful and comparatively cheap ancillary weapon'.[68] The War Office also controversially directly attacked existing British policy and the Geneva Protocol, lambasting a retaliation-only policy as 'neither logical nor militarily sound'.[69] Tactically, such a strong commitment to retaliation prevented defence officials from benefitting from the element of surprise in initiating CW.[70] In order to counter the decline and uncertainty surrounding CW policy, to fully utilise the military utility of nerve agent weapons, and to consolidate a role for nerve agent weapons, War Office officials thus thought it vital to again seek to re-define Britain's interpretation of the 1925 Geneva Protocol. Abiding by the Geneva Protocol was branded a 'grave handicap'.[71] With both the DRPC and the War Office pushing for this reconsideration of British first-use policy, alongside fears over the Soviet CW threat and with increasing pressure from the United States at the NATO level, the CoS acquiesced. The question of first-use was again pushed to the forefront of British nerve agent policy debates in mid-1955.[72]

[67] TNA, DEFE 5/58, 'Chemical Warfare Policy', War Office, 14 April 1955.

[68] Ibid. New developments in nerve agents, the V-series, also held substantial promise for the field, potentially entering the next level of effectiveness and military utility. The true military appreciation and ramifications of the V-agent discovery was however not fully recognised, at least in terms of policy, until 1956.

[69] TNA, DEFE 5/58, 'Chemical Warfare Policy', War Office, 14 April 1955.

[70] Ibid.

[71] Ibid.

[72] The DRPC had taken on an even more important role from mid-1955, with the disbanding of the CWSC. Also taking on a slightly more active/consultative role was the Chemical Defence Advisory Board, which provided advice on more technical aspects of research. For the precise division of work previously conducted by the CWSC, see: TNA, DEFE 5/58, 'Chemical Warfare and Biological Warfare Sub-Committees', Chairman of the Chiefs of Staff Committee, 2 June 1955, Annex. For the terms of reference of the Chemical

As in 1952, when the CoS was previously contemplating such a drastic change, it was again the Foreign Office which provided strong opposition to military officials and intervened to block any controversial change in first-use policy. The Foreign Office, representing legal and political arguments, once again found itself combatting defence officials keen to possess and potentially use nerve agent weapons. The Foreign Office informed the CoS that 'CW would only contribute to the deterrent if the Geneva Protocol were publicly denounced… in our view such a denunciation is out of the question'.[73] Political and diplomatic ramifications of publicly denouncing the protocol were thought to far outweigh the tactical benefits of nerve agent weapons, even if these weapons could be used at the outset of hostilities. Foreign Office officials were also sceptical as to whether a public denouncement of the Geneva Protocol would even render first-use viable. They argued that the Geneva Protocol was primarily intended as an expression of existing understandings in international law. The Protocol was representative of existing international norms and values, and so simply changing Britain's interpretation of it would still not render first-use viable. The internationally accepted understandings, underpinning and supporting the Protocol, would remain. Britain, in essence, was tied to a broader international framework that extended beyond the Protocol itself, with the Protocol reflective of widely held views and beliefs.[74] The alternative option of a secret shift in first-use policy was also dismissed, as such a change in policy would not be publicised, leaving British intent and preparedness unreported. This lack of publicity was thought to undermine military arguments that a re-interpretation of the Protocol would add to the deterrent; without a publicity aspect to a change in first-use policy, there was no justification that it would better deter the outbreak of CW.[75] In the eyes of the Foreign Office, unilaterally reinterpreting the Geneva Protocol was simply unacceptable. Arguments over deterrence and military utility could not override legal or political considerations.

Defence Advisory Board, see: University of East Anglia, Solly Zuckerman Collection, MOS (3)/3, Constitution, Terms of Reference and Membership of Council, Its Boards and Committees, Advisory Council on Scientific Research and Technical Development, Ministry of Supply, June 1956.

[73] TNA, DEFE 5/58, 'Chemical Warfare Policy', P.H. Dean, 7 May 1955.

[74] Ibid.

[75] Ibid.

Facing substantial pressure from the Foreign Office, the CoS nevertheless requested that the Joint Planning Staff examine the advisability of changing CW first-use policy. Trapped between an unyielding Foreign Office and defence officials adamant as to the military benefits of nerve agent weapons, the CoS went for the middle ground. The crucial deciding assessment was to be for the consideration of the Cabinet Defence Committee, who would have the final say. Most importantly, it was to be written with Foreign Office consultation.[76] The study thus represents a strange military and diplomatic hybrid. It provides a unique insight into British nerve agent policy in the Cold War. By addressing first-use policy, the report also effectively decided whether CW policy was to be deemed an essential item of defence and warrant significant funding, or if it was to be without a major role and potentially drastically reduced, as the DRPC had forewarned. This time, the first-use discussion would be largely decided and hold lasting and serious ramifications.

In July 1955, the Directors of Plans observed in their resulting report that in comparison to the superiority of thermonuclear weapons, CW 'pales into relative insignificance'.[77] However, in a tactical role, they believed that chemical weapons continued to hold many advantages. Chemical weapons could be used effectively against Soviet tanks and troops, and provide a means of temporarily neutralising an area without material destruction. Furthermore, there were surprise and shock benefits to be derived from first-use. The country initiating CW could gain a significant military advantage and probably already be prepared defensively for subsequent retaliation. The political implications of such first-use would be markedly less beneficial, but this was given minimal consideration. Such a scenario of beneficial first-use was only thought likely to occur in a global war, with CW use in a limited war deemed highly improbable. In a global war, where chemical weapons might be used, it was also doubtful as to whether other countries would abide by the terms of the 1925 Geneva Protocol. The Soviet Union was thought 'unlikely to adhere to its conditions', which would allow British retaliatory use, and the same assessment was made for the Chinese Government.[78] But, perhaps of greatest concern, it was also feared that there was a significant chance that the United States would initiate CW.

The sheer scale of United States production, research, and development in the CW field raised British concerns that the United States may be

[76] TNA, DEFE 6/29, 'Chemical Warfare Policy', Directors of Plans, 20 July 1955, Annex.
[77] Ibid.
[78] Ibid.

gearing up for the potential use of nerve agent weapons.[79] Adding to this fear was the perceived military utility of nerve agent weapons, which could offer substantial tactical benefits against numerically superior forces, and the United States not being a signatory of the 1925 Geneva Protocol. As with justifications made by the CoS in 1952, and despite the slightly tenuous link between preparation and actual use, the Directors of Plans discerned that any Soviet retaliation or use of chemical weapons against NATO forces would release Britain from the confines of the Geneva Protocol, allowing the use of nerve agent weapons by British forces. The challenge was attempting to juggle British assessments of potential first-use, with there being great uncertainty as to potential Soviet, Chinese and United States actions during a global war. This led to the Directors of Plans ruefully observing that 'we might…find ourselves confronted with the initiation of chemical warfare by our major ally or by the enemy'.[80]

The Directors of Plans also crucially intervened on the subject of CW deterrence and publicity, finding that:

> A decision now to initiate the use of chemical weapons in war would have no deterrent effect in the Cold War, since it would have to be kept secret. Such a decision would also have to be taken in the light of the effect of the probable enemy retaliation on the population of this country, and the enormous provision that would then inevitably have to be made for it.[81]

This went against existing military thinking and interpretations of CW deterrence, which rested upon a public announcement regarding an intent to use chemical weapons alongside preparedness to retaliate. A commitment to secrecy seriously undermined one of the key military arguments for acquisition, that of deterrence. It also went against the legacy and ingrained interpretation of successful CW deterrence present during and after the Second World War. This aversion to publicity, and of any change entailing adherence to secrecy, consolidated the view that even if first-use policy were changed then there would be no positive effect on CW deterrence and thus no clear justification for doing so. While the moral dilemmas of the first-use of lethal nerve agent weapons were included with reference to the shocking price in terms of civilian casualties, the same

[79] Ibid.
[80] Ibid.
[81] Ibid.

arguments could also be made for soldiers. Defending against a large-scale CW attack would be a substantial task and one which would likely cause numerous casualties and take a considerable psychological and moral toll.

Despite acknowledging the arguments for chemical weapons, and even with the uncertainty surrounding the intent of other powers, the Directors of Plans ultimately sided with the Foreign Office. Given the overwhelmingly negative factors involved in a change in first-use policy, the Directors of Plans recommended to Ministers and to the CoS that Britain should only plan to use these weapons in retaliation.[82] The military benefits of using chemical weapons again failed to provide sufficient justification for repudiating political and diplomatic concerns and Britain's commitments under the Geneva Protocol.[83]

After this second major review of first-use policy in less than three years, the CoS withdrew any attempted revision to first-use policy.[84] A second rebuttal confirmed that first-use policy would not be changing. This second failure was to prove all the more damaging. It would trigger substantial ramifications in the future of British CW policy by creating great uncertainty over the role and place of nerve agent weapons in defence policy. As forewarned by the DRPC, without a clear and confirmed role in defence planning the decline of offensive CW policy appeared inevitable. In a harsh economic climate and with the dominance and ascendency of thermonuclear weapons, expenditure on such an uncertain area of defence policy was infeasible.

Soon after this second failed attempt to alter British interpretations of the Geneva Protocol, the DRPC, as promised, returned to the troublesome issue of British nerve agent policy in October 1955. This time, with a lack of accurate intelligence and with no change in first-use policy, the DRPC withdrew its support for offensive nerve agent weapons

[82] Ibid.

[83] TNA, DEFE 5/58, 'Chemical Warfare Policy', P.H. Dean, 7 May 1955; TNA, DEFE 6/29, 'Chemical Warfare Policy', Director of Plans, 20 July 1955, Annex.

[84] The CoS deferred and downplayed any potential change, focusing instead on the Long Term Defence Programme review, as well as the condition that CW would likely on be used in a global war. See: TNA, DEFE 4/78, Minutes of Meeting, Chiefs of Staff Committee, 22 July 1955. For later confirmation, see: TNA, DEFE 10/34, 'Research and Development Programmes', Chairman of the Defence Research Policy Committee, 8 October 1955, Attached report.

altogether.[85] From the late 1940s, defence officials had supported the acquisition of nerve agent weapons, but they were now seemingly out of reach. Carrying through with its prior warning over the impact of failing to change first-use policy, the DRPC proposed that plans for the large-scale production of nerve agents, and the development of anti-tank and anti-personnel nerve agent weapons, all be cancelled.[86] The DRPC was not alone in its more pessimistic outlook. The Service Chiefs were now united in recognising the decline of offensive CW policy, which had centred on nerve agent weapons. The War Office and the Air Ministry also acknowledged that cuts to CW policy were now necessary, with economic pressures cited as the cause.[87] It is however worth noting that these reductions in expenditure were only seriously considered after attempts to alter first-use policy had been exhausted, the nerve agent deal with the United States had fallen through, and efforts to consolidate and justify a role and place for nerve agent weapons had failed.

While uncertainty and decline dictated British CW policy from within, oddly enough public accounts were emerging that argued that there was, in fact, a potential role and value to possessing chemical weapons, especially in terms of deterrence. In a similar vein to early political discussions in Parliament, this assessment was erring on the supportive side, showing a mixture of views regarding this extremely controversial form of warfare. In September 1955, the *Daily Telegraph* had reported on the publishing of a book by T. H. O'Brien on civil defence during the Second World War. In the book, O'Brien repeated the claim that Germany did not use chemical weapons due to fear of Allied reprisals.[88] Shortly after, in November, an article in *The Economist* also alluded to a potential role for chemical weapons, with the emergence of concepts of 'graduated deterrence'.[89] In the article, it was argued that chemical weapons could operate and be used in the middle tier of graduated deterrence, after conventional weapons but before nuclear weapons.[90] Unfortunately for the few remaining advocates

[85] For a summary, see: TNA, DEFE 10/34, 'Research and Development Programmes', Chairman of the Defence Research Policy Committee, 8 October 1955, Attached report.

[86] Ibid.

[87] TNA, DEFE 13/265, War Office to the Minister of Defence, 11 June 1956; TNA, DEFE 13/265, Air Ministry to the Minister of Defence, 6 June 1956.

[88] *The Daily Telegraph*, 'German Fear of Gas Reprisals', 19 September 1955, 9.

[89] *The Economist*, 'Graduated Deterrence?', Issue 5854, 5 November 1955, 457. For a response to the article, see: A.W. Buzzard, 'Graduated Deterrence', *The Economist*, Issue 5856, 19 November 1955, 647.

[90] *The Economist*, 'Graduated Deterrence?', Issue 5854, 5 November 1955, 457.

of nerve agent weapons, these arguments and assessments gained little purchase inside the defence establishment. Even major new developments in the CW field failed to alter the DRPC's adverse forecast.[91]

A substantial advance in the CW field came in the form of the V-agents. Although unable to alter the immediate direction and decline of British nerve agent policy in the mid-1950s, they would have a legacy far beyond the Cold War. The V-agents, apparently named V for venomous as a result of their skin-penetrating characteristics, are a series of chemical compounds of immense lethality. Researchers at Plant Protection Limited (PPL), a subsidiary of Imperial Chemical Industries (ICI), discovered them in the early 1950s.[92] The discovery of the V-agents was part of a long-running search for new CW agents, during which government officials co-opted colleagues from the British chemical industry. Government officials requested that any particularly lethal toxic compounds, which were not viable as commercial products, be shared with the Ministry of Supply. In a similar vein to the German discovery of tabun, British scientists at PPL had been searching for an effective insecticide. Instead, they discovered what would later be branded as Amiton. Too lethal for civilian use as an insecticide, and revealing the close and secret ties between industry and scientific defence research, Amiton was shared with Porton Down.[93] Soon after, scientists at Porton Down began actively researching what was then designated C11, the structure of which was either identical or extremely closely related to Amiton. Labelled with the military code VG, by May 1954 this V-agent development had been shared with the United States.[94]

The initial assessment by Porton officials was that the V-agents might further enhance the military advantages of CW use, as unlike sarin which lasted around one day once released, some of the V-agents could remain

[91] Graduated deterrence seems to have gathered less traction in Britain than in the United States. With defence officials unable to overcome the plethora of obstacles involved in acquisition. As will be seen in the 1960s, even after graduated deterrence increasingly gained prominence in the United States, it did not represent the same boon for CW advocates in Britain.

[92] For further details, see: McLeish and Balmer. Development of the V-Series Nerve Agents, 273–287.

[93] Ibid., 278.

[94] Ibid., 279–280.

lethal and persistent for just under a week.[95] Reflective of their designated name, the V-agents were also known for their high percutaneous toxicity, as they could take effect quickly through the skin.[96] Aiding in assessments of its potential military application was also the consistency of V-agents, with agents such as VX having a thick, oil-like quality, meaning that they can adhere to surfaces and that they are subsequently extremely difficult to remove.[97] By early 1956, after tests and trials, British officials fully appreciated that the V-agents represented a substantial upgrade in CW capabilities. The persistent nature of the V-agents led to the conclusion that:

> The plight of a man whose person, clothing, equipment and surroundings are contaminated is certainly unenviable. He could possibly save himself from the effects of the first shell by careful decontamination of his skin, stripping off his clothing, avoidance of surroundings and administration of atropine, but he is then in a sorry state as a fighting soldier and unprepared for the next shell.[98]

The V-agents were a weapon to kill quickly. They were thought lethal in 1–10 minutes, and the best of the V-agents was believed to be five to ten times more poisonous than the best-known German nerve agents, such as sarin.[99] Additionally, V-agents such as VX were found to have little or no smell, and they could be produced at 'attractively low expenditure rates'.[100] In April 1956, the Ministry of Supply informed the DRPC that the V-agents represented 'another big stride forward in chemical warfare research'.[101] Remarkably though, the significant V-agent discovery yielded few immediate changes in policy. It alone could not reverse the tide resulting from the recent failures to change first-use policy. In the

[95] TNA, DEFE 10/33, 'Review of the R and D Programmes', Secretaries of the Defence Research Policy Committee, 18 November 1954, Attached report; Sidell et. al. Nerve Agents, 167; Balmer. *Secrecy and Science*, 118–119.

[96] McLeish and Balmer. Development of the V-Series Nerve Agents, 279–280.

[97] For more information on VX and VX testing, see: Schmidt. *Secret Science*, 265–279.

[98] TNA, DEFE 10/281, 'V-Gases', W. Makinson, 24 February 1956, Attached report.

[99] TNA, DEFE 7/700, 'Interim Appreciation of the Offensive Potential of V-Gases', Ministry of Supply, 25 April 1956; Robinson. *The Rise of CB Weapons*, 85.

[100] TNA, DEFE 7/700, 'Interim Appreciation of the Offensive Potential of V-Gases', Ministry of Supply, 25 April 1956.

[101] Ibid.

thermonuclear age, and even with this substantial development, nerve agent weapons still lacked a clearly defined role as an essential item of defence.

The decline of offensive CW policy culminated in June 1956, when the Minister of Defence, Walter Monckton, recommended substantial changes to British CW policy. Based on the DRPC's original analysis, Monckton advised that Britain dispose of CW stocks, cancel programmes for the mass-production of nerve agents and halt the development of chemical weapons.[102] Monckton's initial recommendation overlooked British stockpiles of sulphur mustard, but the Chancellor of the Exchequer, Harold Macmillan, soon reminded him that Britain still possessed 6000 tons of the agent, along with considerable facilities for its production.[103] In addressing his oversight, Monckton thus included sulphur mustard into his sweeping cutbacks. He pushed for the abandonment of large-scale nerve agent production, weapons development, and the disposal of existing stocks of sulphur mustard and leftover tabun.[104] Such drastic changes had not only political support, but also senior military support. The First Sea Lord even cited the potential moral benefits of not possessing nerve agent weapons, and the War Office and the Air Ministry both accepted the need for cuts to CW expenditure.[105] The moral argument seemed to again be tied to the taboo nature of chemical weapons, an exceptional weapon but also one that had no justifiable place in defence policy. Whether this had roots in perceptions of a more 'honourable' form of warfare, was geared towards concerns over adverse publicity, or the abhorrent nature of chemical weapons, is unclear. While Monckton's views on CW policy received support from key military officials and senior Cabinet members,

[102] TNA, DEFE 7/700, Minister of Defence to the Prime Minister, 15 June 1956.

[103] TNA, PREM 11/3099, Harold Macmillan to Walter Monckton, 5 June 1956.

[104] TNA, DEFE 7/700, Minister of Defence to the Prime Minister, 15 June 1956; Walker, John R. 2012. *Britain and Disarmament: The UK and Nuclear, Biological and Chemical Weapons Arms Control and Programmes 1956–1975*. Farnham, Surrey: Ashgate, 8.

[105] TNA, DEFE 7/700, R.N. Smith to P.J. Stephens, 6 June 1956; TNA, DEFE 7/700, Minister of Defence to the Prime Minister, 15 June 1956; TNA, AIR 19/1097, 'Chemical Warfare Policy', DCAS to the Secretary of State, 30 May 1956; TNA, PREM 11/3099, Harold Macmillan to Walter Monckton, 5 June 1956. Although to be taken with a degree caution given the Navy's minimal interest in the CW field, the First Sea Lord did represent the beneficial arguments and the moral perks resulting from unilateral disarmament.

there remained one significant and a somewhat surprising obstacle to his drastic reductions in CW policy, the Prime Minister, Sir Anthony Eden.[106]

The Last Man Standing

Even with military and political support, Monckton faced stiff opposition to his proposals from Eden in June 1956. Disagreements over CW policy were taking place around one-month before Egyptian leader Gamal Abdel nationalised the Suez Canal and at a time of tense relations between Britain and Egypt. Eden's opposition to reductions in the CW field was despite his keen focus on defence cuts after succeeding Churchill in April 1955, fuelled by his concerns over increasing inflation, rising import costs and growing defence expenditure.[107] To combat this negative slide, Eden believed in restoring the vitality of the economy. Britain must 'now cut its coat according to the cloth'.[108] This tough stance on defence expenditure did not, however, cover CW policy. Eden was resistant to Monckton's proposals. While the Prime Minister conceded that it might be best to dispose of sulphur mustard given the economic and military circumstances, he strongly opposed any plans to reduce facilities for the production of nerve agents.[109]

The comparatively small amount of expenditure required to acquire a nerve agent capability was likely one factor for Eden's resistance, but of even greater importance was his commitment to CW deterrence. When questioning the rationale behind the proposed cuts in June 1956, Eden recalled 'how valuable chemical warfare preparations have been to us in the past as a deterrent'.[110] Here he was referring to the role of chemical weapons during the Second World War, where the threat of retaliation alongside British preparedness to wage CW was thought to have

[106] TNA, DEFE 7/700, R.N. Smith to P.J. Stephens, 6 June 1956; TNA, DEFE 7/700, Minister of Defence to the Prime Minister, 15 June 1956; TNA, DEFE 7/700, Prime Minister's Personal Minute to the Minister of Defence, 20 June 1956.

[107] James, Robert Rhodes. 1986. *Anthony Eden*. New York, NY: McGraw-Hill, 416; Self. *British Foreign and Defence Policy Since* 1945, 161.

[108] Eden, Anthony. 1960. *The Memoirs of Sir Anthony Eden: Full Circle*. London: Cassell & Company, 213–216, 369–372; Baylis. *Ambiguity and Deterrence*, 209.

[109] TNA, DEFE 7/700, Prime Minister's Personal Minute to the Minister of Defence, 20 June 1956.

[110] Ibid.

successfully deterred German first-use.[111] Interestingly his take on the success of CW deterrence during the war also aligned very closely to the views of his former War Cabinet colleague, Clement Attlee, who had approved the British acquisition and development of nerve agent weapons in 1950. It is, though, difficult to trace the particular evolution of Eden's attitude towards CW deterrence, especially since he was not present at the crucial 1952 Cabinet Defence Committee meeting on first-use policy. In his reflections on his First World War experiences, though, CW can be seen to have had an impact. In 1916, he had overseen the placing of gas canisters on the Western Front, and encountered a near-miss with a German gas attack, noting that a gas alarm had saved him from a 'grave danger'.[112] Eden's experiences in the First World War and the Second World War, as well as the impact of the nerve agent discovery and the Cold War climate, were all influential in shaping his position on CW deterrence in 1956.

Eden, perhaps inadvertently, was also supporting and reflecting prior military arguments by emphasising preparedness to wage CW as a critical component of successful deterrence. Yet he was doing so at a time when the military support for this form of CW deterrent had drifted away. Had the request for nerve agent weapons reached the Defence Committee earlier, with Eden present, and before defence officials and given up hope, a very different direction could well have been decided upon. The DRPC and the Service Chiefs, though, had accepted that the CW deterrence requirement alone, without the confirmed possibility of use, was not sufficient to justify the production or possession of nerve agent weapons. Some defence officials were, in fact, extremely frustrated by the Prime Minister's involvement, noting that Eden was attempting to redirect policy 'as soon as we try to save money!'[113]

Irrespective of Eden's scepticism and resistance, Monckton pressed ahead. He sought approval from the Defence Committee in July 1956.[114] In a persuasive memorandum to his ministerial colleagues, which was based on the judgements of the DRPC and the earlier report by the Directors of

[111] For some examples of this thinking see: TNA, PREM 3/89, H. L. Ismay to Churchill, 24 May 1944; TNA, CAB 131/9, 'Chemical Warfare Policy', Chiefs of Staff Committee, 4 September 1950; TNA, DEFE 10/447, 'Annual Review 1950', Joint Secretary of the Chemical Warfare Sub-Committee, 10 August 1951, Appendix A, Section III.

[112] Eden, Anthony. 1976. *Another World, 1897–1917*. London: Allen Lane, 81, 132.

[113] TNA, DEFE 7/700, Prime Minister's Personal Minute to the Minister of Defence, 20 June 1956.

[114] TNA, CAB 131/17, 'Chemical Warfare Policy', Memorandum by the Minister of Defence, 4 July 1956

Plans, Monckton pushed for a significant shift in offensive CW policy.[115] His memorandum sought approval for the abandonment of large-scale nerve agent production, the disposal of existing CW stocks and the abandonment of offensive weapons development. The two main exceptions, and thus spared from these sweeping cuts, were broader CW research and defensive CW research, both of which would continue.[116] In justifying this substantial reduction in CW policy, Monckton stated that under the Geneva Protocol Britain could only ever retaliate with chemical weapons, as first-use was not an accepted policy.[117] The limitations of a retaliatory-only policy brought into question the financial expenditure necessary to acquire a nerve agent capability, and whether the weapons were 'absolutely essential' to defence.[118] Monckton also believed that for CW deterrence, Britain did not need to possess a nerve agent capability, and he stated that:

> Our American allies are devoting a considerable effort to the development of nerve gas weapons and this will provide a powerful deterrent against the initiation by Russia of chemical warfare.[119]

While acknowledging the tactical benefits of nerve agents, Monckton also included justifications which were aimed at placating advocates of nerve agent weapons. For example, he outlined how by continuing to work on CW research knowledge of the field would be kept alive. Stemming from this, Monckton used the development of the lethal V-agents, the next generation of nerve agent weapons, to support his policy shift. After revealing that the V-agents provided a far greater hazard than nerve agents such as sarin, he stated that the future production of these new more effective and persistent nerve agents would not be adversely affected by abandoning plans for sarin production. This was perhaps a slightly more tenuous point, given that cutbacks to sarin production would certainly have an impact on the future production and

[115] Ibid.

[116] While the term "defensive" is slightly ambiguous, here it is taken to mean predominantly a focus on defences against CW attack, which while requiring a degree of information on offensive techniques is very limited in its scope and is mainly confined to defensive equipment, detection methods and defensive training.

[117] TNA, CAB 131/17, 'Chemical Warfare Policy', Memorandum by the Minister of Defence, 4 July 1956.

[118] Ibid.

[119] Ibid.

development of VX, particularly when it came to drawing on past experiences, expertise, and lessons learnt from the large-scale nerve agent production. His reference to the V-agents also implied that nerve agents such as sarin were already outdated by more recent developments in the field and that additional discoveries might yet be made.[120] Monckton essentially used the discovery of the V-agents, a hugely significant development in the CW field, as an argument to reduce aspects of existing offensive CW policy. Brandishing the stick of drastic cuts, he also offered the carrot of future V-agent production. To further undermine the importance of sarin-based weapons, Monckton informed the Committee that:

> I feel myself that our possession of nuclear weapons and the massive American nuclear armoury together with their chemical warfare potential justify us in our present economic circumstances in abandoning our own capacity to wage offensive chemical warfare.[121]

The rest of the Defence Committee agreed with Monckton's policy changes, yet Eden remained unimpressed. Eden frustratingly wrote 'I am not convinced by this. We always seem to make small cuts at the deterrent and none at the "Admirals"… Something should be kept going'.[122] Factors such as prestige and a reluctance to be further reliant on the United States were not directly mentioned by the Prime Minister, but his resistance does reveal a strong desire for an independent British capability and CW deterrent. While Eden's comments were reflective of his disgruntlement with the direction of wider defence policy, they also reveal the importance he attached to Britain possessing a nerve agent capability. The Prime Minister thus continued to strongly resist cutbacks to nerve agent weapons policy and development.[123]

Eden further believed that a nerve agent capability would not just deter the outbreak of CW, but form a part of the deterrent as a whole, with British possession of nerve agent weapons discouraging the Soviet Union

[120] Ibid.; Walker. *Britain and Disarmament*, 8.

[121] TNA, CAB 131/17, 'Chemical Warfare Policy', Memorandum by the Minister of Defence, 4 July 1956.

[122] TNA, PREM 11/3099, 'Chemical Warfare Policy', Memorandum by the Minister of Defence, 4 July 1956. (Eden's comment is hand written on the first page of the report)

[123] Ibid.; In broader defence policy, Eden believed that economies in defence were being carried out in an unbalanced way, with the Navy not cutting back adequately on defence expenditure and the burden falling too heavily on items involved in deterring war.

from any attack on Western Europe.[124] In this interpretation of CW deterrence, Eden's views differed greatly from British defence officials. In contrast to Eden, defence officials firmly believed that CW preparedness and the ability to retaliate would primarily deter Soviet use of chemical weapons during a war, rather than war itself. Increasingly in defence circles, nuclear weapons were seen as not just a deterrent for conventional war or nuclear war, but also as a deterrent against Soviet use of any form of WMD.[125] Sir Frederick Brundrett, the Chief Scientific Advisor to the Ministry of Defence, was a key supporter of this, informing Monckton that 'in a world in which we rely on a nuclear deterrent…the only sensible answer to an assault on this country by chemical weapons would be nuclear retaliation'.[126] This was in itself something which had not garnered significant attention in British CW planning—what was the threshold for use and different eventualities, especially in terms of other forms of warfare. If nerve agents were used against Britain, it seems unlikely that officials would have responded with a nuclear strike. As officials in the DRPC, CWSC, and CoS had all recognised before, even a large-scale chemical weapons attack would pale into comparison against one with a nuclear weapon. Brundrett's observation would, if followed, likely represent a tremendous escalation in terms of the scale of damage and lives lost.

Eden's interpretation of CW deterrence also collided with the views and judgement of his Cabinet Secretary, Norman Brook, who informed the Prime Minister that the possession of chemical weapons was 'unlikely to be a factor in preventing a major war'.[127] Brook further counselled Eden that with the need for defence cuts, 'very grave risks' had to be taken.[128] The Prime Minister's views were thus under criticism from two key angles—that of economics and deterrence.

Facing this significant opposition, Eden gave ground. However, while he conceded to the proposed policy changes, in July 1956, he erected one

[124] For an excerpt of this discussion see: TNA, PREM 11/3099, 'Chemical Warfare Policy', Norman Brook to Anthony Eden, 9 July 1956. For a more general picture of Eden's active engagement in CW policy throughout this period, see file: TNA, PREM 11/3099.

[125] TNA, DEFE 13/265, Minute Sheet for the Minister of Defence, 12 November 1956; Tucker. *War of Nerves*, 154–155; Spiers. *A History of Chemical and Biological Weapons*, 61.

[126] TNA, DEFE 7/700, Brundrett to the Minister of Defence, 12 November 1956.

[127] TNA, PREM 11/3099, 'Chemical Warfare Policy', Norman Brook to Anthony Eden, 9 July 1956.

[128] Ibid.

final and considerable hurdle.[129] The Prime Minister wanted assurances from the United States before Britain unilaterally abandoned chemical weapons, as after the policy change Britain would be largely dependent on the United States for CW deterrence and completely dependent on its close ally for a CW retaliatory capability.[130] Although initially, the United States was merely to be informed of Britain's policy change, Eden soon escalated this after the meeting to securing guarantees from the United States over CW retaliation on behalf of British forces, and of continued United States nerve agent production.[131] Eden had thus deemed it prudent to secure guarantees; if Britain abandoned its CW capabilities it should ensure that there would be at least one power in NATO with the ability to retaliate against Soviet first-use.[132]

In late July, Monckton, wary of Eden's interventions and his desire for strong United States commitments, attempted to push through changes in CW policy without United States consultation. Eden firmly resisted. He was adamant that cutbacks to CW policy would not be finalised until talks with the United States.[133] The Prime Minister then became even more involved in CW policy when he judged that these talks should be carried out at the political level, rather than the military level. Again going against the advice of Monckton and senior defence officials, including Brundrett, Eden decided that Britain's shift in CW policy warranted a personal message from himself to the President of the United States.[134] In pushing for such a communication, Eden caused significant friction with his senior advisors and with his Minister of Defence. One disgruntled official noted that Eden's decision was a 'very great mistake', and that the Prime Minister's course of action was due to him having a very different view on

[129] TNA, CAB 131/17, Minutes of Meeting, Cabinet Defence Committee, 10 July 1956.

[130] Ibid.; TNA, DEFE 13/265, Anthony Eden to Walter Monckton, 24 July 1956.

[131] For example, see the shift from: TNA, CAB 131/17, Minutes of Meeting, Cabinet Defence Committee, 10 July 1956 to TNA, DEFE 13/265, Anthony Eden to Walter Monckton, 24 July 1956.

[132] TNA, DEFE 13/265, Anthony Eden to Walter Monckton, 24 July 1956.

[133] Ibid. For further details on this running dispute, see communications in July 1956 in the files: TNA, PREM 11/3099 and TNA, DEFE 7/700.

[134] TNA, DEFE 7/700, Brundrett to Forward, 13 July 1956; TNA, DEFE 7/700, 'Chemical Warfare Policy', Monckton to the Prime Minister, 20 July 1956; TNA, DEFE 13/265, Anthony Eden to Walter Monckton, 24 July 1956. For a more strongly worded rebuttal by Monckton, see the drafted but not sent communication: TNA, DEFE 7/700, 'Chemical Warfare Policy', draft communication from Monkton to the Prime Minister, [undated] July 1956.

CW policy from everyone else.[135] Monckton, in particular, bemoaned Eden's decision. He argued that there was no need to bring such British defence policy changes to the attention of the highest authority in the United States. Here there was also perhaps a sense of frustration or even embarrassment, with an admission to the President that due to economic constraints Britain could no longer afford to keep up in terms of nerve agent capabilities. This coming after Britain had expended considerable effort in producing and developing a nuclear weapons capability, in part to maintain a seat at the top table. In view of this, the Minister of Defence, supported by Reginald Maudling, the Minister of Supply, and Norman Brook, sought a more nuanced approach to Britain's disposal of offensive chemical weapons.[136] Monckton believed that 'it would be better to represent them not so much as a major change in policy but as a re-arrangement of our resources in the light of present conditions'.[137] He also concluded that it should be the role of the Joint Staff Mission in Washington, not the Prime Minister, to seek guarantees from the United States and to convey and downplay Britain's decline in offensive CW policy.[138]

Despite this strong opposition, Eden again steadfastly resisted any further encroachment on his stance, insisting that it should be he who communicated directly to President Eisenhower.[139] Whereas Monckton wanted to downplay Britain's substantial shift in CW policy, Eden thought that securing United States involvement and a confirmed CW deterrent for Britain trumped all else.[140] Willingly revealing British deficiencies was thought an acceptable price to pay for the Prime Minister. A key reason for Eden's commitment and desire to be heavily involved was rooted in his strong views on CW deterrence and the importance he attached to it, but another was actually a desire to maintain national prestige. Eden cautioned that if the United States were informed of the policy change by someone other than himself, then 'the effect of such action would surely be to diminish still further the regard which the United States Service

[135] TNA, DEFE 7/700, Forward to Powell, 30 August 1956.

[136] TNA, PREM 11/3099, 'Chemical Warfare Policy', Walter Monckton to the Prime Minister, 20 July 1956. (Norman Brook's agreement is attached to the front).

[137] Ibid.

[138] Ibid.

[139] TNA, DEFE 7/700, Prime Minister's Personal Minute to the Minister of Defence, 24 July 1956.

[140] Ibid.

authorities have for our defence effort'.[141] Clearly concerned over the potential damage that Britain's withdrawal from offensive CW capabilities would cause to Anglo-American relations, Eden continued to insist on taking the lead. While Monckton and defence officials thus wanted to downplay the policy reductions, Eden wanted to be direct and open with the United States. Eager to limit United States criticism, Eden was thus adamant that he play the personal card with Eisenhower, to both secure United States CW commitments and to reduce the potential damage to Anglo-American relations. Given Eden's steadfast commitment to communicating directly with Eisenhower, and in order to avoid further confrontation with the entrenched views of the Prime Minister, Monckton and defence officials relented.[142]

Before Eden could inform Eisenhower of Britain's CW policy reductions, or request assurances, other events came to dominate political considerations. From mid-1956, the upcoming Presidential election in the United States overrode all considerations of contacting Eisenhower. The agreed consensus was to not disturb the President with matters of CW policy at such a sensitive time.[143] Behind the scenes, defence officials continued in their attempts to persuade the Prime Minister to accept the cutbacks to CW policy without assurances from the United States.[144] Eden, however, yet again proved uncompromising. He continued to delay any CW policy change. In the meantime, his government was consumed by the final phases of the Suez Crisis which reached a climax with the Anglo-French invasion of Egypt in November 1956.[145] Following the Suez fail-

[141] Ibid.

[142] TNA, DEFE 7/700, Walter Monckton to the Prime Minister, 3 September 1956. For an earlier, heavily revised draft see: TNA, DEFE 7/700, Powell to Forward, 29 August 1956, Attached draft letter.

[143] TNA, DEFE 7/700, Forward to Powell, 8 October 1956; TNA, DEFE 7/700, Minute sheet for Powell, 10 October 1956.

[144] Ibid.; TNA, WO 32/21950/2, 'Chemical Warfare Policy', Lt. Col. Lodes, 3 December 1956.

[145] Although this author has found no evidence of this, in his controversial book 'Spycatcher', Peter Wright claimed that Eden's interest in chemical weapons took on an even greater importance—alleging that MI6, with Eden's approval, formulated a plan to assassinate Nasser with chemical weapons. According to Wright, MI6 planned to introduce nerve gas into the ventilation system of Nasser's military headquarters. For details, see: Wright, Peter. 1987. *Spycatcher: The Candid Autobiography of a Senior Intelligence Officer*. London: Viking, 159–161; Shpiro, Shlomo. 2009. Poisoned Chalice: Intelligence Use of Chemical and Biological Weapons. *International Journal of Intelligence and CounterIntelligence*, 22:1,

ure, and the rift in Anglo-American relations that was opened through a lack of transparency, Eden unsurprisingly chose to further delay communicating with President Eisenhower on CW policy, and ignored all requests for political approval to implement the changes agreed in the July 1956 Defence Committee meeting. By the end of the year, despite defence officials claiming that the 'dust of Suez has settled', British CW policy continued in limbo, with the Defence Committee decision still awaiting final confirmation and implementation.[146] British defence officials found themselves without political guidance, and Eden was unable to attend to this aspect of policy as his premiership collapsed.[147]

This policy stalemate continued until, in poor health and politically weakened by Suez, Eden resigned on 9 January 1957.[148] As Prime Minister, Eden had fought for a British nerve agent capability and deterrent. He attached great importance to a nerve agent capability, so much so that he was willing to go against Monckton, who was representing the assessments of the DRPC, the Directors of Plans and the CoS. While Eden did eventually accept the arguments in favour of cutting British CW commitments, he hindered and delayed the change until the very end. Eden had thus remained firmly committed to Britain maintaining access to some form of CW deterrent, whether belonging to Britain or the United States, and with his resignation, British CW policy lost its most influential advocate.

In the short-term, Eden's actions had directly clashed with the emerging consensus in military circles, that without assurances over first-use, nerve agent weapons were not a vital area of defence. As Brundrett summarised in March 1957:

> Unfortunately, the late Prime Minister, against our advice, ruled that the notification to the American Government must be done by him direct to the President. As events turned out, it was never considered an appropriate

9–11. Also for other British considerations of chemical weapons use for assassinations around this time, see: Aldrich, Richard. 2001. *The Hidden Hand: Britain, America and Cold War Secret Intelligence*. London: John Murray, 480.

[146] TNA, DEFE 7/700, Philip de Zulueta to J.M. Gibbon, 12 November 1956; TNA, DEFE 7/700, 'Chemical Warfare Policy', Brundrett to the Minister of Defence, 19 March 1957.

[147] TNA, DEFE 7/700, Philip de Zulueta to J.M. Gibbon, 12 November 1956.

[148] James. *Anthony Eden*, 596–597.

moment to make the communication. Consequently nothing has been done and we go on, therefore, as we were.[149]

British CW policy, primarily as a result of Eden's intervention, had been in stasis for over eight months. However, after Eden's resignation, with a lack of senior political support, the increasing dominance of thermonuclear weapons, economic pressures, and the failure to change first-use policy in both 1952 and 1955, British offensive CW policy suffered the severe setback which had long been on the horizon.

Finalising decisions on CW policy coincided with the arrival of Duncan Sandys as the newly appointed Minister of Defence, and with Harold Macmillan becoming Prime Minister in January 1957.[150] Sandys was a key advocate of reducing defence expenditure, and with little military support leftover, implementing the Defence Committee's suggested reductions to offensive CW policy was a mere formality.[151] This was despite his earlier support, in early 1954 as the Minister of Supply, for nerve agent production in Britain. Without Eden blocking the path, changes and cutbacks to CW policy were rapidly implemented in mid-1957, with the winding down of offensive weapons development programmes, the deletion of programmes for the mass-production of nerve agents, and the disposal of stockpiled chemical weapons.[152]

Although the Defence Committee's decision and its implementation were detrimental to offensive CW policy, for advocates of nerve agent weapons it was not all doom and gloom. After forecasting the impact of the 1956 policy decision, investment into CW policy for the 1957–1958 financial year was predicted to be reduced by around 30%, from an annual amount of around £1.3 million, with broader CW research and research

[149] TNA, DEFE 7/700, 'Chemical Warfare Policy', Brundrett to the Minister of Defence, 19 March 1957.

[150] Baylis. *Ambiguity and Deterrence*, 241.

[151] Rees, Wyn. 1989. The 1957 Sandys white paper: New priorities in British defence policy? *Journal of Strategic Studies*, 12:2, 215–229; Self. *British Foreign and Defence Policy*, 162; French, David. 2013. Duncan Sandys and the Projection of British Power after Suez. *Diplomacy & Statecraft*, 24:1, 41–58.

[152] TNA, DEFE 7/700, 'Chemical Warfare Policy', Brundrett to the Minister of Defence, 19 March 1957. For a later summary/review, see: TNA, DEFE 7/700, 'Report on Biological Warfare and Chemical Warfare', Defence Research Policy Committee, 15 April 1958, Annex B.

into defensive measures continuing.[153] Though a substantial percentage, this in itself did not mark the end of British interest or activities in the nerve agent field. The decision to follow through with the policy reductions should therefore be viewed, and to a degree moderated, with the caveats that extensive research continued, Britain still had access to cutting edge research through increasingly important tripartite cooperation and could still decide to produce the recently discovered V-agents.

The V-agents were thought of as a potential weapon to cover a large area with lethal effect, and perhaps most importantly, they could do so with only a 'small expenditure'.[154] It was estimated that a V-agent plant for mass-production would initially cost around £4–5 million.[155] This area for further research and potential production thus represented another avenue for offensive weapons, which was not directly ruled out by the Defence Committee decision of 1956. As Monckton had implied, this technological development was in itself a reason to dispose of outdated wartime stocks, and it had been spun as a possible ray of hope for advocates of an offensive CW capability.

Continued British access to cutting edge research, and its ability to later produce V-agent weapons, remained greatly aided by tripartite cooperation with Canada and the United States. This collaboration in CW research and development had continued to flourish despite wider fluctuations in Anglo-American relations, and even after Britain's decision to dispose of chemical weapons and to cut weapons research and development.[156] Through this intensive cooperation, Britain retained access to the advanced and extremely well-funded CW research effort in the United States. Officials in the United States, after British scientists had supplied them

[153] TNA, DEFE 7/700, Brundrett to the Minister of Defence, 12 November 1956. Also see: TNA, DEFE 7/700, Brundrett to the Minister of Defence, 12 November 1956, Attached minute by the Ministry of Supply.

[154] TNA, DEFE 7/700, 'Working Party on BW and CW', Defence Research Policy Staff, 21 February 1957, Annex.

[155] Ibid.

[156] Cooperation was not all plain sailing, especially in April 1956, as clashes over NATO defensive CW preparedness caused some disgruntlement. The War Office noted that American defensive preparedness is 'designed to cover every contingency whereas British equipment is based on a more practical realism'. Concerns were predominantly focused on the perceived 'lavish scale' of United States CW defensive preparations. For details, see: TNA, DEFE 5/66, 'SHAPE's Chemical Warfare Requirements for Individual and Unit Protection and Detection Equipment', War Office, 3 April 1956. Also see: Fredericks. United States Chemical Warfare Policy, III-2, 14–16.

with information on the V-agents, had already begun a crash course in V-agent research.[157] In April 1956, United States officials had even established a specialist V-agent Team, which was tasked with testing and evaluating V-agent delivery systems.[158] And, after extensive research leading to the United States military creating around 50 different types of V-series agents, they settled upon VX for large-scale production in February 1957.[159]

The substantial gulf in scale between the Anglo-American partners was thus growing even larger in 1957. Britain was implementing policy reductions, and terminating plans for the mass-production and development of nerve agent weapons, whereas the United States had reached its peacetime capacity for sarin production and its programme was a comparative behemoth.[160] Britain did, however, contribute to this much more extensive United States research effort through tripartite cooperation, and through bilateral cooperation in the form of the United States-United Kingdom Mutual Weapons Development Program. This bilateral programme, originating from US Congressional approval for the Mutual Security Act of 1953, covered the costs of six CW-related projects.[161] These projects were run on a conjoint cost-sharing basis and included research on V-agent chemistry, methods for V-agent detection and methods for V-agent production.[162] As a core part of the country's contribution to tripartite and bilateral cooperation, British scientists still used the Nancekuke facility, despite it not reaching mass-production, to produce small amounts of VX to test industrial production techniques.[163]

This remarkable level of cooperation, unbeknownst to Eden and other senior politicians, had also rendered aspects of the debate over informing the United States of Britain's potential policy change slightly redundant. For example, during the Eleventh Tripartite Meeting in 1956, Porton officials had already revealed the virtual abandonment by the Government

[157] Tucker. *War of Nerves*, 154–158.

[158] Ibid., 160.

[159] McLeish and Balmer. Development of the V-Series Nerve Agents, 279–280.

[160] Tucker. *War of Nerves*, 156.

[161] Carter and Pearson. North Atlantic Chemical and Biological Research Collaboration, 87–90.

[162] Ibid.

[163] Tucker. *War of Nerves*, 165.

of all offensive CW capabilities.[164] Even without a formal letter from the Prime Minister to the President, United States and Canadian experts were thus already well aware of the CW policy situation in Britain.[165] During the Twelfth Tripartite Meeting in 1957, the three countries also agreed to a further division of labour in V-agent research, which continued to provide Britain with invaluable information. At the meeting, it was decided that the United States would take on the bulk of the work by developing land and air munitions for VX, Britain would evaluate the military potential of VX, and Canada would determine the hazards from contaminated terrain.[166] Tripartite cooperation facilitated the division of resources and the sharing of significant research findings, which after the 1956 Defence Committee decision was vital for British research. This advanced cooperation facilitated Britain staying abreast of CW developments and remaining at the forefront of research. This very scale and scope of the substantial United States CW research effort also provided British officials with the appealing option and incentive of willingly increasing dependence, as while the country could continue reductions in defence expenditure it might still retain access to cutting edge research.

This continuing access to highly advanced research into the V-agents was all the more critical given evolving perceptions of the Soviet CW threat in 1957. Although exact figures on Soviet capabilities were unknown, United States intelligence estimated that the Soviet Union had the materials and skills available to produce a staggering 40,000 to 60,000 tonnes of CW agents.[167] Soviet reserves of CW agents were also thought sufficient to allow the sustained use of nerve agents for several months in a global war.[168] In these assessments of Soviet offensive CW capabilities, Anglo-American intelligence agencies increasingly drew on the United States programme to determine Soviet progress, a practice that

[164] Carter and Pearson. North Atlantic Chemical and Biological Research Collaboration, 87–90.

[165] Ibid., 87. This of course does not mitigate the important aspect of United States guarantees, which were not secured at this lower-level. It does however highlight how advanced cooperation was at the time, that such a sensitive policy decision was shared.

[166] Tucker. *War of Nerves*, 164.

[167] NIE 11-4-56, 'Soviet Capabilities and Probable Courses of Action Through 1961', National Intelligence Estimate, 2 August 1956. The JIC concluded that Britain had no direct knowledge of Soviet CW research, development, or its level of production. See: TNA, CAB 158/28, 'Russian Research and Development up to the end of 1956', Joint Intelligence Committee, 23 May 1957, Annex B.

[168] Ibid.

given the severe difficulties in attaining intelligence was one of the few courses of action left.[169] In line with this process of mirroring, when the V-agents were discovered it was assumed that the Soviets would probably already be aware of this latest development, or that they already possessed something similar.[170] These fears were further fuelled by the fact that ICI had published details surrounding the V-agents in the magazine *Chemical Review*.[171] The JIC thus believed that 'Soviet workers cannot have failed to notice some closely related work which has been published in the United Kingdom'.[172] In reality, however, Soviet scientists had only developed R-33 by this stage, otherwise known as Substance-33, which was not as efficient as VX and too unstable for long-term storage.[173] The potential advantage that the United States and Britain possessed with the V-agents was therefore not recognised. The perceived balance of power in the CW field was actually increasingly thought to be in favour of the Soviet Union.

The belief in a growing Soviet threat, and of expanding Soviet capabilities, was again further exacerbated by intelligence gains on Soviet CW defensive measures, one of the few areas where CW intelligence continued to yield results. In 1957, Soviet forces were now thought to be well prepared to defend against CW attacks, with United States National Intelligence Estimates concluding that Soviet defensive preparations for

[169] The Anglo-American also partners placed slightly different emphasis on areas of CW intelligence. For example, a regular theme in mid-1950s United States intelligence reports on CW was the inclusion of psychogenic drugs. Psychogenic drugs, like LSD, can make individuals indifferent to their surroundings and can induce apprehension and confusion. In comparison, British intelligence did not give much weight to this area; it was rarely mentioned in JIC assessments of the Soviet CW threat. For examples, see: NIE 11-56, 'Soviet Gross Capabilities for Attacks on the United States and Key Overseas Installations and Forces Through Mid-1959', National Intelligence Estimate, 6 March 1956; NIE 11-4-56, 'Soviet Capabilities and Probable Courses of Action Through 1961', National Intelligence Estimate, 2 August 1956.

[170] TNA, CAB 158/24, 'Russian Research and Development up to the end of 1955', Joint Intelligence Committee, 11 May 1956.

[171] TNA, DEFE 10/281, Minutes of Meeting, Defence Research Policy Staff, 1 March 1956.

[172] TNA, CAB 158/28, 'Russian Research and Development up to the end of 1956', Joint Intelligence Committee, 23 May 1957, Annex B.

[173] This observation is based on secondary sources on the Soviet CW programme. For more information on R-33, see: Tucker. *War of Nerves*, 181–186; Mirzayanov, Vil. 2009. *State Secrets: An Insider's Chronicle of the Russian Chemical Weapons Program*. Denver, CO: Outskirts Press, 121–123; Kaszeta. *A History of Nerve Agents*, 122–126.

CW might, in fact, be superior to those of major Western nations.[174] A shift in appreciations was also evident in intelligence assessments of Soviet CW doctrine. As United States intelligence surmised that extensive programmes were underway to train both military and civilian personnel in defensive techniques.[175] British intelligence also reached a similar conclusion. They observed that the 'Soviet Army possesses a chemical arm which is specially trained in both the offensive and defensive aspects of this type of warfare'.[176] This growing appreciation of the Soviet threat came at a time when Britain had significantly decreased its own CW capabilities, and it was increasingly reliant on the United States for CW deterrence and retaliation. These intelligence forecasts, however, changed little in terms of the direction of British CW policy, which was predominantly shaped by the issue of first-use, economic restraints, the strength of the United States CW programme, and by the resignation of Eden.

It was Anthony Eden, amongst British Prime Ministers, who took the most active interest in British CW policy during the 1950s. Eden, keen on a British nerve agent capability for deterrence and retaliation, faced substantial opposition. This opposition included his Minister of Defence and defence advisors, and even after giving ground, he delayed and resisted any actual change to CW policy by insisting he personally notify President Eisenhower. Unfortunately for the few remaining advocates of nerve agent weapons development, Eden's resignation condemned offensive CW policy to cancellation and confined British CW policy to research and defensive measures. Without Eden's support, and with no guarantee of use, nerve agent weapons failed to justify a place as an essential item of defence in an age dominated by thermonuclear weapons and economic cutbacks. In particular, it was the disagreement over, and failure to change, CW first-use policy which had the most significant bearing on military support for nerve agent weapons. Without a guarantee of use, funding and production were simply not viable options even with the CW deterrent requirement.

[174] NIE 11-4-57, 'Main Trends in Soviet Capabilities and Policies 1957–1962', National Intelligence Estimate, 12 November 1957.

[175] Ibid.

[176] TNA, CAB 158/28, 'Russian Research and Development up to the end of 1956', Joint Intelligence Committee, 23 May 1957, Annex B; TNA, CAB 158/28, 'Soviet Research and Development 1956–1957', Joint Intelligence Committee, 17 June 1957.

CHAPTER 5

Another Reverse Course: From Dependence to Acquisition, 1958–1963

We rely on the United States for CW retaliation, although rather curiously we appear not to have appraised them of this reliance.
Brief for the Vice-Chief of the Air Staff (TNA, AIR 20/9440, 'Chemical, Biological and Radiological Warfare', Brief for the Vice-Chief of the Air Staff, 4 August 1960)

From 1958, Cold War relations showed clear, alarming, and unmistakable signs of deterioration. The earlier hopes of a potential détente had faded. In November of that year, Nikita Khrushchev issued an ultimatum, with a six-month deadline, calling for the withdrawal of allied military forces, including British, from West Berlin.[1] The ultimatum served to dramatically increase tensions between East and West, as the United States, Britain, and their allies made clear that they would rather fight than surrender their position or recognise the legitimacy of the East German state.[2] In June 1961 Khrushchev revived his ultimatum and calls to create a free city in Berlin when he met the new American President, John F. Kennedy, in Vienna. This fresh crisis was only averted when the Communist authori-

[1] Gaddis. *The Cold War*, 113–115.

[2] Troyanovsky, Oleg. 2000. The Making of Soviet Foreign Policy. In *Nikita Khrushchev*, ed. William Taubman, Sergei Khrushchev, and Abbott Gleason. London: Yale University Press, 219–221.

W. King, *Nerve Agents in Postwar Britain*, Britain and the World, https://doi.org/10.1007/978-3-030-70474-2_5

ties opted to instead construct a wall to isolate West Berlin in August 1961.[3] Following this, superpower tensions would reach their climax during the Cuban Missile Crisis in October 1962, when the world was brought to the very brink of nuclear war.[4]

In this age of superpower confrontation, dominated by the figures of Kennedy and Khrushchev, Prime Minister Harold Macmillan attempted to chart a path for Britain.[5] Macmillan, as with many British Prime Ministers in the postwar period, was gravely concerned by the country's economic woes and by the substantial costs of defence.[6] He judged that one clear avenue and a potential remedy for these economic woes was for Britain to strengthen further the Anglo-American relationship. He believed that Britain could 'play Greece to America's Rome'.[7] This desire for greater cooperation with the United States was reflected throughout British defence policy, but particularly in nuclear weapons cooperation. Where, after the signing of bilateral agreements in July 1958 and May 1959, Anglo-American cooperation on nuclear weapons research reached new levels.[8] These agreements facilitated the transfer of highly secret nuclear technology. Even seemingly independent British nuclear capabilities, like Blue Streak, benefitted greatly from information supplied by the United States.[9] Given the scale and depth of information being sent across

[3] Westad. *The Cold War*, 294–297. For a brief overview of Britain and Macmillan's role, see: Sandford, Christopher. 2018. *Union Jack: John F. Kennedy's Special Relationship with Great Britain*. United Kingdom: The History Press, 185–188.

[4] Paterson, Thomas. 1989. *Kennedy's Quest for Victory: American Foreign Policy, 1961–1963*. Oxford: Oxford University Press, 22–24; Taubman, William, Khrushchev, Sergei, and Gleason, Abbott. Introduction. In *Nikita Khrushchev*, ed. William Taubman, Sergei Khrushchev, and Abbott Gleason. London: Yale University Press, 5; Hershberg, James. 2010. The Cuban missile crisis. In *The Cold War, Volume 2: Crises and Détente*, ed. Melvyn Leffler and Odd Arne Westad. Cambridge: Cambridge University Press, 65.

[5] Nunnerley, David. 1972. *President Kennedy & Britain*. London: The Bodley Head, 49–56.

[6] Mahan, Erin. 2002. *Kennedy, De Gaulle and Western Europe*. New York, NY: Palgrave Macmillan, 5.

[7] Baylis. *Anglo-American relations since 1939*, 84; Jones. *UK Strategic Nuclear Deterrent, Volume I*, 50.

[8] Ball, S. J. 1995. Military Nuclear Relations Between the United States and Great Britain Under the Terms of the McMahon Act, 1946–1958. *The Historical Journal*, 38:2, 452; Dumbrell, John. 2006. *A special relationship: Anglo-American relations from the Cold War to Iraq*. Basingstoke: Palgrave Macmillan, 48.

[9] Gott, Richard. 1963. The Evolution of the Independent British Deterrent. *Royal Institute of International Affairs*, 39:2, 247; Jones. *UK Strategic Deterrent, Volume I*, 49.

the Atlantic, it is no surprise Macmillan thought that he had acquired the 'great prize'.[10] In public, and to maintain appearances, Macmillan stressed Britain's independent nuclear capabilities, but behind closed doors, he was putting substantial emphasis on further interdependence with the United States.[11]

Reflective of Macmillan's drive for greater Anglo-American collaboration, highly secret plans were also underway to further integrate Anglo-American cooperation in the CW field. After the implementation of the 1956 Defence Committee decision to abandon offensive weapons development and mass-production, attention shifted to consider whether revitalised Anglo-American cooperation could pave the way for even greater British reductions in CW expenditure. While discussions over Anglo-American cooperation in nuclear weapons oscillated between independence and interdependence, in CW, the only options appeared to be even greater interdependence, or complete dependence on the United States for CW deterrence, retaliation and research. British CW policy, after the resignation of Eden, was in a parlous and precarious state in 1958. The difficulties in formulating CW policy were compounded not just by a complex period of renewed superpower rivalries, the dominance of nuclear weapons, and necessary economic cutbacks, but also by emerging public and press engagement with, and criticism of, government policy and science. The age of reverence was giving way to one of criticism and scepticism. After exploring discussions of greater interdependence and reliance on the United States, this chapter will analyse the influence and impact of a spate of negative publicity in the CBW field in 1959–1960 and the emergence of an unprecedented level of postwar political and public scrutiny over British CW policy. This chapter will then end by revealing the crucial influence Sir Solly Zuckerman exerted on British CW policy, leading to a drastic change in direction in 1962.

[10] Dumbrell. *A Special Relationship*, 48; Pine, Melissa. 2009. Transatlantic Nuclear Cooperation: The British Perspective, 1945–1991. In *The British Way in Cold Warfare: Intelligence, Diplomacy and the Bomb, 1945–1975*, ed. Matthew Grant. London: Continuum Books, 109; Young. *Britain and the World*, 171–172; Reynolds. *Britannia Overruled*, 200.

[11] Dockrill. *British Defence since 1945*, 80; Jones. *UK Strategic Nuclear Deterrent, Volume I*, 50; Rees. The 1957 Sandys white paper, 227.

Growing Dependence

After the decision to abandon both CW stocks and the mass-production of nerve agents in 1956–1957, Britain was completely dependent on the CW deterrent and retaliatory capability of the United States. British CW capabilities and potential production were 'negligible', and the country was not able to produce nerve agents on a significant scale. Nerve agents were possessed, but only in an extremely small quantity for laboratory testing and experiments.[12] In terms of sarin, while Britain still maintained the one-ton-a-week pilot plant at Nancekuke, its production had been halted in 1956.[13] The site, from 1954, had only produced around 20 tons of sarin, and even if production restarted it was unable to meet military requirements which were thousands of tons for the initial stages of war.[14] Conservative estimates had still placed the initial British war requirement at 2500 tons. Ramping up the production of nerve agents domestically would have taken years. It would also have necessitated the restarting of production at Nancekuke, as well as its expansion. Other developments, such as in tear gases, held promise as incapacitating agents for military use. But, most of Britain's domestic production was already 'booked for [military and private] customers', rendering mass-production for military stockpiling highly unlikely in the short-term.[15]

In stark contrast, by 1958 British defence officials observed that the United States CW programme was running 'full bore'.[16] The United States had stockpiled thousands of tons of sarin, along with blister agents such as sulphur mustard (a chemical weapons agent under some debate as to its uses in the nerve agent age).[17] Advanced delivery methods for lethal

[12] TNA, DEFE 5/84, 'Biological and Chemical Warfare', Note by the Secretary of the Chiefs of Staff Committee, 19 August 1958, Annex.

[13] TNA, DEFE 7/700 Ministry of Supply to C. Wright, 6 July 1956.

[14] Ibid.; TNA, DEFE 5/84, 'Biological and Chemical Warfare', Note by the Secretary of the Chiefs of Staff Committee, 19 August 1958, Annex; Tucker. *War of Nerves*, 120–121. For a summary of the work and research conducted at Nancekuke, see: The National Archives Web Archive. 2010. 'Nancekuke Remediation Project', Defence: about defence, December 8.

[15] TNA, DEFE 5/84, 'Biological and Chemical Warfare', Note by the Secretary of the Chiefs of Staff Committee, 19 August 1958, Annex.

[16] TNA, WO 32/21950/2, Untitled Minute, Lt. Col. Saunders, 8 October 1958.

[17] Estimated cumulative output in the United States by July 1955 was hoped to be around 24,000 tons, with the aim of it reaching around 26,000 tons for the outbreak of war, see: United States NARA II, RG 218, Central Decimal File 1951–1953, Box 153, [title redacted], Memorandum for the Secretary of Defense, 14 April 1953. For scale of production see:

CW agents had been researched and developed. This included mortar shells, artillery shells, aerial bombs and the M55 rocket.[18] In all areas of CW policy, in production, stockpiling, weapons and research, the United States CW programme dwarfed that of Britain's. From 1958, the United States clearly dominated the West's ability to retaliate with, and deter Soviet use of, chemical weapons.[19] British defence officials consoled themselves, stating that while 'the West must continue to possess an offensive capability in BW and CW', this did 'not require the United Kingdom itself to possess such a capability'.[20] This willing reliance on the United States for a CW retaliatory and deterrent capability opened up a further avenue of dependence, when, from late 1957 onwards, attention turned to developing and deepening Anglo-American CW cooperation yet further.

While British CW research had continued and survived the cutbacks of 1956–1957, given the scale of the United States CW programme, the failure to change first-use policy, and Britain's economic woes, senior British defence officials contemplated even scrapping this last vestige of British activity in the CW field.[21] Defence officials thus considered the option of eliminating all CW research at Porton Down, including on the nerve agents, either through 'reliance on the United States or as an acceptable military risk'.[22]

The proposed policy shift towards even greater reliance on the United States had the support of the Air Staff. Edmund Hudleston, the Vice-Chief of the Air Staff, believed that as the United States was devoting considerable resources to CW research, Britain might be able to eliminate

TNA, CAB 21/3912, 'Policy for Chemical Warfare', Minister of Defence, 6 March 1953; Tucker. *War of Nerves*, 136. (United States production would be mothballed after initial wartime requirements were met).

[18] Ibid., 134–137.

[19] Even though France also had a strong CW programme, as with Britain, it could not equal the scale of the United States CW effort. Spiers. *Chemical Warfare*, 136.

[20] TNA, DEFE 5/84, 'Biological and Chemical Warfare', Note by the Secretary of the Chiefs of Staff Committee, 19 August 1958, Annex.

[21] TNA, AIR 8/1936, 'Biological and Chemical Warfare', Brief for the Vice-Chief of the Air Staff, 7 August 1958; TNA, DEFE 4/110, Minutes of Meeting, Chiefs of Staff Committee, 14 August 1958; TNA, DEFE 7/1395, T. M. Crowley to Mr. Chilver, 27 August 1958.

[22] TNA, AIR 8/1936, 'Biological and Chemical Warfare', Brief for the Vice-Chief of the Air Staff, 7 August 1958; TNA, DEFE 7/1395, T. M. Crowley to Mr. Chilver, 27 August 1958.

its own.[23] This was a drastic change from the Air Staff's prior assessments of CW research, which just nine months before had branded such work a 'comparatively cheap insurance'.[24] In August 1958, though, the Air Ministry had changed its tune. It was now keen to secure British access to the substantial United States CW programme through the framework of interdependence. Hudleston also suggested that British CW scientists, who would be without a role after the elimination of expenditure on British CW research, could be sent to work on the United States CW programme.[25] This remarkable level of dependence on the United States could offer the dual benefits of strengthening collaboration, whilst reducing British defence expenditure. Painting an extremely optimistic picture of British reliance, being utterly dependent on the United States was framed as a positive move for Britain. Just two years after Prime Minister Anthony Eden's staunch defence of a nerve agent capability, the prospects for the CW field had declined and seemingly collapsed.

These strong policy recommendations, for almost complete dependence on the United States, were also supported by the DRPC. The DRPC increasingly referred to the CW capabilities and research of the 'Western Powers', rather than focusing on the British effort in isolation.[26] This interpretation of CW requirements, and the placement of them within this much wider framework of an overall Western effort, was used to justify increasing British dependence on the United States. In July 1959, after exploring the options and requirement for British CW research, the CoS agreed with the Air Ministry and with the DRPC, concluding that while 'the West as a whole' must continue research in the CW field, there was no obligation for British CW research.[27] By framing CW requirements as those needed by the West as a whole, British CW research was

[23] TNA, AIR 8/1936, 'Biological and Chemical Warfare', Brief for the Vice-Chief of the Air Staff, 7 August 1958; TNA, DEFE 4/110, Minutes of Meeting, Chiefs of Staff Committee, 14 August 1958.

[24] TNA, AIR 20/9440, 'Biological and Chemical Warfare', Brief for the Chief of the Air Staff, 22 October 1957.

[25] TNA, DEFE 4/110, Minutes of Meeting, Chiefs of Staff Committee, 14 August 1958.

[26] TNA, DEFE 5/84, 'Biological and Chemical Warfare', Note by the Secretary of the Chiefs of Staff Committee, 19 August 1958, Annex.

[27] TNA, DEFE 4/119, Minutes of Meeting, Chiefs of Staff Committee, 14 July 1959. A slightly earlier adherence to this 'West as a whole' argumentation can also be found in: TNA, DEFE 41/156, 'Biological and Chemical Warfare', Joint Planning Staff, 30 July 1958; Balmer, Brian. 1997. The drift of biological weapons policy in the UK 1945–1965. *Journal of Strategic Studies*, 20:4, 133–134.

seen as an unnecessary expenditure and a duplication. At its core, this was effectively another argument in favour of greater dependence on the United States. When referring to the 'Western Powers' senior British defence officials were, in essence, referring to the United States programme, which dwarfed those of all the other Western powers.[28] The CoS, supporting and reflecting the opinions of defence officials, thus seriously considered the option of eliminating nearly all British CW expenditure and research through a 'framework of interdependence' with the United States.[29]

Shortly after Britain had disposed of CW stocks, the development of offensive weapons and the scrapping of large-scale production, this drastic step would have deprived Britain of nearly all CW research, both offensive and defensive. The CoS had gone far beyond considerations of interdependence. It had demonstrated a potential willingness to be completely dependent on the United States for all areas of CW policy.[30] The possibility of complete dependence in military circles had therefore spread from CW deterrence and retaliation, which was already a reality, to CW research as well; putting the very existence of British CW research in doubt.

This potentially extraordinary shift in CW policy was however conditional. In a remarkably similar vein to Eden in 1956, the CoS stipulated that this level of dependence would only take place if Britain attained assurances from the United States.[31] Even though it was willing to explore complete dependence, the CoS still appreciated in some form the importance of CW deterrence and nerve agent weapons. However, given the economic climate and with no guarantee of use, they could not justify a British nerve agent capability. Confirmation that Britain would continue to have access to the United States effort, and that the West would still possess a CW deterrent, was therefore vital to determining whether the country could realistically remove its own CW expenditure. Bizarrely though, after enquiries from the Foreign Office in August 1958, it was realised that the United States had never at the senior level officially been

[28] A notable exception here would be France, which was proceeding with CW research, but the scale was not comparable to the United States.

[29] TNA, DEFE 4/110, Minutes of Meeting, Chiefs of Staff Committee, 14 August 1958; TNA, DEFE 5/84, 'Biological and Chemical Warfare', Note by the Secretary of the Chiefs of Staff Committee, 19 August 1958, Annex; TNA, DEFE 4/119, Minutes of Meeting, Chiefs of Staff Committee, 14 July 1959.

[30] TNA, DEFE 4/110, Minutes of Meeting, Chiefs of Staff Committee, 14 August 1958.

[31] Ibid.

informed of Britain's drastic CW policy reductions in 1956–1957.[32] Eden had insisted that he inform President Eisenhower directly, but this had never transpired. This was also despite the policy changes entailing British reliance on the United States for CW retaliation on behalf of British forces, and for CW deterrence. Senior United States officials were thus seemingly oblivious of the fact that Britain had chosen to become completely dependent on their CW capabilities. They were also unaware that in 1958 British officials were contemplating complete dependence on the United States in CW research as well.

Although surprising, not informing senior United States officials of British reliance was, strangely enough, in part a deliberate policy. In July 1960 defence officials observed that politically, there was no benefit to informing senior United States officials of Britain's dependency, which was thought a weakness.[33] At the technical and scientific levels of cooperation, United States officials were already well aware of Britain's position through close bilateral and trilateral cooperation.[34] Updates on the status of Britain's CW policy had been shared through the yearly Tripartite Conferences.[35] In a sense the officials who needed to know already did, and if more senior US officials were informed there could be negative political consequences. British officials also feared that if more senior United States officials found out about Britain's unpreparedness for CW, they would insist on storing chemical weapons and the means of delivery in Britain.[36] Had such a situation arisen and been uncovered, of the United States storing lethal nerve agent weapons on British soil, then the public and political backlash would have proved severe for any political leader.

In evading this feared outcome, this strange state of play and adherence to secrecy had continued into the mid-1960s, several years after British policy changes. In August 1960, Hudleston was briefed by his staff that

[32] For examples of defence officials trying to discover whether the United States had been informed, see: TNA, DEFE 7/1395, 'Biological and Chemical Warfare', J. G. Boyd to D. J. P. Lee, 20 August 1958; TNA, DEFE 7/1395, 'Biological and Chemical Warfare', Secretary of the Chiefs of Staff Committee to Mr. E. C. Williams, 25 August 1958; TNA, AIR 20/9440, 'Chemical, Biological and Radiological Warfare', Brief for the Vice-Chief of the Air Staff, 4 August 1960.

[33] TNA, WO 32/21379, E. Haddon to the War Office. 21 July 1960.

[34] Ibid.

[35] Ibid.; TNA, DEFE 10/355, Minutes of Meeting, Defence Research Policy Committee, 17 November 1959.

[36] TNA, WO 32/21379, E. Haddon to the War Office. 21 July 1960.

'we rely on the United States for CW retaliation, although rather curiously we appear not to have appraised them of this reliance'.[37] This remarkable situation also shows a substantial divide between the technical and political levels of CW cooperation between the Anglo-American partners. At the technical level, United States officials were keen to continue close collaboration and they were already long aware of the reductions to British CW policy. Yet at the senior political level, United States officials were deliberately not informed of a substantial shift in British defence policy, despite the policy reductions leading to complete British reliance on the United States for both nerve agent capabilities and for CW deterrence. Another reason for this secrecy and the withholding of information was likely that of prestige and/or embarrassment. There was a realisation that for years Britain had been completely reliant on the United States, but that senior United States officials had not been informed of this British reliance and dependence, nor fully consulted on Britain's substantial policy reductions in 1956–1957. With complete dependence on the United States requiring guarantees, necessitating that senior United States officials actually be informed of British dependence, consideration of full dependence on the United States for CW research, development and deterrence stuttered. British secrecy, and the stumbling and muddled nature of CW policy, thus itself served as a substantial obstacle to continued cutbacks.

Coinciding with this significant and largely self-constructed hurdle of informing and securing guarantees from the United States, serious concerns were emerging over the potential damage that British dependence would actually cause Anglo-American and tripartite cooperation in practice. Throughout 1959, and after its earlier scepticism, the DRPC began to fully appreciate the immense damage that British withdrawal from the CW field would cause to bilateral relations.[38] Reminiscent of Crawford's arguments some years prior, in which the former Chairman of the CWSC in his desperate attempt to reverse the decline had lambasted the direction of CW policy, the DRPC observed that:

[37]TNA, AIR 20/9440, 'Chemical, Biological and Radiological Warfare', Brief for the Vice-Chief of the Air Staff, 4 August 1960.

[38]TNA, DEFE 10/357, 'Biological & Chemical Warfare', Note by the Chairman of the Defence Research Policy Committee, 20 July 1959, Annex. These concerns were also expressed slightly earlier, see for example: TNA, DEFE 10/356, 'Biological & Chemical Warfare', Defence Research Policy Committee, 28 January 1959.

> The tripartite integration which has been achieved in these fields is a model and information flows more freely between the three countries in these subjects [CBW] than in almost any other. In short, interdependence in these fields was a reality ten years before the word was coined.[39]

The DRPC, supported by some officials involved in CW research keen to avoid almost complete cutbacks, judged that this remarkable level of secret cooperation would be threatened if Britain drastically cut back its engagement and involvement in the CW field.[40] The DRPC warned that if Britain brought nothing to the table and had nothing to offer, then there could be 'no doubt' that the United States would cease or severely limit bilateral cooperation. This would be the worst-case scenario. Cut adrift from the United States programme, and without a meaningful one of its own, Britain would have been isolated and unprepared. Furthermore, the DRPC concluded that the potential savings from cutting CW policy were not worth the political and scientific consequences.[41] The yearly cost of around £1 million for CW research was deemed a small subscription to pay, especially in light of all the benefits accrued from both domestic research and trilateral cooperation.[42] Warnings over the potential damage to Anglo-American relations and the awkward situation of withholding Britain's CW dependence from senior United States officials thus combined and undermined arguments for complete dependence on the United States. By mid-1960, arguments in favour of keeping some British CW research going prevailed. An extremely awkward conversation was avoided.

[39] Ibid.

[40] Schmidt. *Secret Science*, 316.

[41] TNA, DEFE 10/357, 'Biological and Chemical Warfare', Note by the Chairman of the Defence Research Policy Committee, 20 July 1959, Annex.

[42] TNA, CAB 21/4505, 'Biological & Chemical Warfare', Note by the Defence Research Policy Staff, 28 January 1959; TNA, DEFE 10/357, 'Biological and Chemical Warfare', Note by the Chairman of the Defence Research Policy Committee, 20 July 1959, Annex. War Office officials had also cautioned that to become completely dependent on the United States would do 'great damage' to the relationship. See: TNA, WO 32/21950/2, Untitled Minute, Lt. Col. Saunders, 8 October 1958.

Unwelcome Publicity

While resisting complete dependence on the United States for all aspects of CW policy, occurring concurrently with these secret policy discussions was an unprecedented postwar tide of publicity striking the CW field in 1959. There had been minimal warning of this impending publicity in 1958, with coverage of the CW field relatively quiet. Brief mentions were made regarding the deterrent value of CW, and the almost reassuring lessons that could be drawn from its non-use during the Second World War. For example, in March 1958, Lord Freyberg confidently announced to the House of Lords that 'if chemical warfare was a deterrent, how very much greater will the deterrent effect be of nuclear war!'[43] To some extent, even in the late 1950s, CW non-use during the Second World War continued to be seen as vindication for nuclear deterrence, with successful preparation and an effective retaliatory capability thought crucial to preventing the outbreak of that form of warfare. Many observers at the time also believed that the more horrific and devastating the weapon, then the lower the chance that it would actually be used. Eric Reading, writing for the *Daily Telegraph*, elucidated this when he wrote that Britain should go on producing bigger and better nuclear weapons, as 'they need never be used—any more than poison gas was used in the last war'.[44]

Although publicity and coverage of CW was relatively quiet in 1958, in 1959, this changed dramatically.[45] British defence officials, when considering greater dependence on the United States, were already cautious of the impact of publicity in the CW field, fearing that it would 'have a deplorable effect both at home and abroad'.[46] Deeming it likely that if information on British CW research went public, then the situation would easily get out of control. Journalists and opposition MPs would not heed the justification or explanations that nerve agent weapons would only be used

[43] Hansard, House of Lords, 'Defence', 5 March 1968, Vol.207, cc1092–196.

[44] Eric Reading et. al., 'Peace Slogans', *The Daily Telegraph*, 10 April 1958, 8. For another example of this line of thinking, see: Peter Simply, 'What Use?', *The Daily Telegraph*, 16 July 1958, 10. Simply observes that 'is the use of poison gas implicit in its possession? In the last war both sides possessed it, neither used it'.

[45] Other sources were emerging in 1958 on CW, especially with Chinese accusations that the United States was supplying poison gas to Chinese nationalists on Quemoy. For example, see: *The Daily Telegraph*, 'Poison Gas "Lie" by Peking', 5 November 1958, 14.

[46] TNA, DEFE 7/2140, Ministry of Defence to British Joint Staff Mission Washington DC, 17 June 1958.

in retaliation, or that research was focused on defensive aspects.[47] Unfortunately for defence officials, their secret defence planning met with a much larger shift, one characterised by changes in public perceptions, critical engagement by the press, and an overall change in attitudes towards the coverage of politics and science.

Formerly ambivalent and positive coverage of government policy and science was beginning to give way to a far more critical approach.[48] As observed by Tony Shaw, by the late 1950s, signs of dissent and strong scepticism had begun to emerge. Comedies, melodramas, and science-fiction were reflecting a growing sense of unease with the spiralling arms race and the role of the state.[49] A willingness to criticise government policy, rather than to hold it in awe and simply accept it, was growing. It was greatly aided by an increasing flow of information.[50] Early signs of this could be seen with the BBC's 'Tonight' programme and ITV's 'Under Fire', where differing perspectives were shown to the audience.[51]

The changing attitudes and availability of information would have a particular resonance for single-issue movements. Christopher Hill notes that the emergence and growth of these single-issue movements, such as against nuclear weapons testing, were inextricably tied to the sourcing and reworking of information in the press, on the radio and television.[52] These changes were connected to a rise of protest movements, characterised by the strong performance of the Campaign for Nuclear Disarmament (CND). Opinion polls revealed that 22% of people questioned supported the CND's unilateralist policy in February 1958, which rose to 30% in June 1959.[53] There would be some overlap with the momentum established by the CND, when the Committee of 100, which split from the CND in September 1960, started to engage with CBW. Willing to take more drastic measures to receive active press coverage of their cause, the Committee of 100 carried the slogan 'fill the jails' in 1960–1961.[54] Later

[47] Ibid. As explored in more depth by Balmer, this also reveals the strikingly different approaches to secrecy by different actors. See: Balmer, Brian. 2010. Keeping Nothing Secret: United Kingdom Chemical Warfare Policy in the 1960s. *Journal of Strategic Studies*, 33:6, 871–893.

[48] Gregory and Miller. *Science in Public*, 45.

[49] Shaw. *British Cinema and the Cold War*, 193.

[50] Hill. *Peace and Power*, 45–50.

[51] Ibid.

[52] Ibid., 49.

[53] Taylor. *Against the Bomb*, 48.

[54] Hill. *Peace and Power*, 211.

on, in 1963, two members of the Committee of 100 would force their way into Porton Down, leading to raids on the headquarters of the Committee of 100 by Special Branch.[55] While the emergence of strong single-issue movements and the greater availability of information would have a lasting effect on British CW policy, in 1959, we see a key turning point and important early marker of change.

In January 1959, publicity on British BW policy dragged British CW policy into the spotlight. Labour MPs, including Emrys Hughes and Emanuel Shinwell, enquired about the nature and role of British BW research.[56] At this early stage, and attempting to stifle any negative coverage, the Government provided minimal responses to questions, outlining how the perceived scale of poisons developed in Britain was 'greatly exaggerated' and that the BW programme was 'almost entirely defensive'.[57] This stance of extremely limited disclosures when facing questions on CBW was remarkably similar to the one taken by Churchill in 1955. These attempts to moderate some of the emerging negative coverage fell short. By mid-1959, questions over British BW research had expanded to animal testing, which included coverage of controversial CW research.[58] During these debates, Conservative MP Eric Johnson questioned his own Government. On the 15 June 1959, he informed the House that 3000 animals had been killed in just six months of CBW experiments.[59] Following this revelation, on 29 June, Johnson asked in Parliament for details on British CW research, with the MP requesting information on highly sensitive information such as on staff numbers, costs, when British CW research was started, and what it was intended to provide defence against.[60] A set of previously unthinkable questions on a highly sensitive issue. While initially thrust into the public spotlight through its close

[55] Schmidt. *Secret Science*, 384.

[56] Hansard, House of Commons, 'Microbiological Research Centre and Chemical Defence Experimental Establishment', 26 January 1959, Vol.598, cc693-5; *The Manchester Guardian*, 'Poison Threat "Exaggerated"', 27 January 1959, 2; Hammond, Peter and Carter, Gradon. 2002. *From Biological Warfare to Healthcare: Porton Down, 1940–2000*. Hampshire: Palgrave, 222.

[57] Ibid.

[58] *The Manchester Guardian*, 'Experiments and Animal Deaths', 16 June 1959, 2; Schmidt. *Secret Science*, 381.

[59] Hansard, House of Commons, 'Chemical Defence Experimental Establishment (Animals)', 15 June 1959, Vol.607, cc29–31.

[60] Hansard, House of Commons, 'Chemical and Microbiological Research (Animals)', 29 June 1959, Vol.608, cc29–33.

association with BW policy, CW policy soon became a focal point for public and parliamentary attention. Parliamentary scrutiny proved unsatisfied, unrelenting, and it continued to gain momentum throughout 1959.

With British CW research receiving such scant coverage and operating under the highest levels of secrecy in the postwar period, one official had to look back almost 30 years to find a comparable scenario to draw lessons from.[61] After some digging, it was ruefully observed that 'it is often said that history repeats itself…in the years 1930 and 1931 the CW organisation suffered attacks in Parliament and adverse articles in the press not dissimilar from those we have had to deal with in the last few months'.[62] For defence officials accustomed to the quiet of secrecy, this was only the beginning.

Fuelling further parliamentary questions, and much to the consternation of British defence officials, MPs and newspapers began reporting on growing CW publicity and commentary in the United States. In July 1959, a United States press release on Soviet CW capabilities led Sydney Silverman MP to ask the Prime Minister, Harold Macmillan, whether he had discussed chemical and biological weapons disarmament with Khrushchev and whether Khrushchev would abide by the 1925 Geneva Protocol.[63] During this time the Pugwash Conferences on Science and World Affairs, established by Bertrand Russel and Joseph Rotblat, discussed CBW during their 5th Conference in August. Parliamentary questions on British CBW research and disarmament continued. [64] Due to this ongoing interest, Macmillan felt obliged to clarify British policy. In December 1959, the Prime Minister informed the Commons that:

> Our position is perfectly clear towards both chemical and bacteriological weapons. We are pledged or bound not to use them except in retaliation, but I remember, for instance, that in the Second World War it was necessary for us to prepare methods of retaliation in chemical warfare, and perhaps the fact that we were known to have prepared them had some effect on their not being used.[65]

[61] TNA, WO 188/2772, E. Haddon (DCDRD) to DGSR, AS, CIO and DPBR, 17 August 1959.

[62] Ibid.

[63] Hansard, House of Commons, 'Prime Minister (Visit To Soviet Union)', 2 July 1959, Vol.608, cc59–60W.

[64] *Pugwash Newsletter*. 2007. Special Edition commemorating 50 years of Pugwash Editor: Jeffrey Boutwell. 44:22, 26.

[65] Hansard, House of Commons, 'Porton Microbiological Research Station', 8 December 1959, Vol.615, cc224–5.

Although scrapping offensive weapons development in 1957, Macmillan chose not to mention this. Instead he hinted that Britain may well still be preparing a retaliatory CW capability in some form. There was a clear degree of ambiguity. In the face of strong questions from Labour MP Emrys Hughes over this 'new kind of horrible deterrent', Macmillan remained steadfast. He stated that defence and retaliatory capabilities had 'certainly served us well in the last war'.[66] Macmillan was reflecting and representing one core interpretation of CW deterrence and wartime experiences, in a similar vein to Attlee and Eden before him, and Hughes another. Macmillan's views coincided with those MPs who, throughout the postwar period, had argued in favour of CW deterrence and its positive reinforcement of nuclear deterrence. Hughes, by contrast, was a part of a growing movement against chemical and biological weapons, which was increasingly prevalent within the Labour Party and which emphasised the immoral and controversial nature of chemical and biological weapons. The latter grouping, able to mobilise emerging information on the CBW field, was growing in influence and membership.

British CW policy was to receive yet more attention when it was thrust centre stage by journalists and MPs in early 1960. Even after the best efforts of British officials and Macmillan, the shroud of secrecy around British CW policy was beginning to show signs of potentially unravelling. A key reason for this was again CW policy in the United States. United States officials were keen to make a public statement on its mass-production of VX, which they thought necessary to aid and inform civil defence and to pre-empt domestic criticism.[67] For British officials however, this proposed statement caused great alarm. The statement included revealing where the V-agents were discovered. It was feared that Britain's discovery of the lethal V-agents and its continued involvement in CW research would perforate across the Atlantic, and into British newspapers and Parliament. This potential dilemma had been on the radar for British defence officials since 1958, yet in early 1960 it was gaining traction and urgency. United States officials were increasingly adamant that a statement needed to be made.[68]

A clear sign of the impending trouble came on 22 February 1960. Macmillan was pessimistically informed that our 'position has been

[66] Ibid.

[67] TNA, DEFE 7/2140, Ministry of Defence to British Joint Staff Mission Washington DC, 17 June 1958; TNA, PREM 11/3099, 'V-agents', Brief for the Prime Minister, 22 February 1960.

[68] Ibid.

complicated by the fact that Mr Chapman-Pincher has now got hold of this story and may reveal it at any moment'.[69] Henry Chapman Pincher was an investigative journalist working for the *Daily Express* who was attracted to the most secret areas of British defence policy and intelligence.[70] While reflective of a trend in press coverage of science and government policy, which was moving towards more robust criticism, Pincher took this to another level. In February 1960, he had approached British defence officials for comment on an article he was writing, and in doing so, revealed that he knew that British scientists had discovered the V-agents and that the details had been shared with the United States.[71] Surprisingly Chapman Pincher had actually had the nerve agent story for over three months before he approached officials. He had simply been 'keeping it on ice'.[72]

On 24 February 1960, after recognising the impending and looming danger of negative publicity, Macmillan took the step of forewarning the Cabinet Defence Committee.[73] Ominously though, and on the very same day that the Defence Committee met, the *Daily Mail* ran with an article titled 'New war gases "worse than H-bombs"'.[74] The article, based on a speech made by Dr Cecil Coggins in the United States, alleged that the Soviet Union possessed enough stocks of nerve agents to 'wipe out the

[69] Ibid.

[70] Pincher, Chapman. 1978. *Inside Story: A Documentary of the Pursuit of Power*. London: Sidwick & Jackson, 11–13; Andrew, Christopher. 2009. *The Defence of the Realm: The Authorized History of MI5*. London: Allen Lane, 210. For the massive breadth and contribution of Chapman-Pincher's stories in highly secret areas, see: Liddell Hart Centre for Military Archives, H. C. Pincher, Press Cuttings, Volumes 1–12. For the final released story, see: Chapman Pincher, 'American V-Gas Starts Rumpus', *Daily Express*, 7 March 1960, 1.

[71] TNA, PREM 11/3099, 'V-agents', Brief for the Prime Minister, 22 February 1960.

[72] TNA, CAB 21/4505, L.G. Lohan to Mr. Reeve, 25 February 1960. Chapman Pincher was to later prove an even bigger thorn in Macmillan's side. Just a year after he discovered that Soviet spy George Blake had worked for MI6, with the revelation causing Macmillan to despair 'can nothing be done to suppress or get rid of Mr. Chapman Pincher'. Andrew. *The Defence of the Realm*, 489. Later in his career, Chapman Pincher would go on to reveal highly secret information on the work and methods of GCHQ; Levy, David. 2015. The Traitor Hunter. *International Journal of Intelligence and Counter Intelligence*, 28:2, 405; Aldrich, Richard and Cormac, Rory. 2016. *The Black Door: Spies, Secret Intelligence and British Prime Ministers*. London: William Collins, 210, 226.

[73] TNA, CAB 131/23, Cabinet Defence Committee, Minutes of Meeting, 24 February 1960.

[74] Jeffrey Blyth, 'New war Gases "worse than H-bombs"', *Daily Mail*, 24 February 1960, 9.

entire population of 1000 cities the size of Manchester and Liverpool'.[75] Though shocking, the story contained no information on Britain's CW policy or on the sharing of the V-agent discovery with the United States, but it was a sign of things to come. In attempting to organise a coherent response, during the Defence Committee meeting members were urged to avoid referring to the British V-agent development and to say 'as little as possible' about the United States CW programme.[76] In order to limit the flow of information from across the Atlantic, the Defence Committee also decided to try exert some influence over what United States officials would say in their public statement on VX production. Cabinet members were keen on pressing United States officials to focus on the defensive aspects of their CW programme and to emphasise that VX production was a necessary retaliatory capability and a deterrent, in a similar vein to prior British justifications.[77]

Macmillan was particularly concerned by this planned United States public statement. He made clear his 'misgivings' and proposed contacting President Eisenhower personally, to try to head-off the unwelcome publicity. As with Churchill and Eden before him, Macmillan thus became yet another Prime Minister in the postwar period who thought CW policy important enough and sensitive enough to warrant his personal involvement.[78] However, even with these doubts and Prime Ministerial misgivings, United States authorities remained set on issuing a public statement. With some form of publicity seemingly inevitable, Macmillan was advised that 'what, if anything, we do now, is a question of tactics'.[79]

One such tactic British officials used was to try and continue to delay any official United States press release on VX production, thus buying time. The Foreign Office, under political pressure, managed to secure a temporary delay on the public statement until early March 1960.[80] This was achieved by arguing that both the V-agents, and the level of tripartite cooperation, were exceptionally sensitive areas of British CW policy, of which the British public was largely unaware.[81] In justifying its requests for

[75] Ibid.

[76] TNA, CAB 131/23, Cabinet Defence Committee, Minutes of Meeting, 24 February 1960.

[77] Ibid.

[78] Ibid.

[79] TNA, PREM 11/3099, 'V-agents', Brief for the Prime Minister, 29 February 1960.

[80] Ibid.

[81] TNA, PREM 11/3099, British Foreign Office to Washington, 4 March 1960.

a delay, the Foreign Office was also keen to point out that a United States statement would seriously affect Britain on political grounds, and that it would also impact East/West relations and 'be especially damaging to us in the uncommitted countries'.[82] This line of argumentation reveals a strong apprehension over the moral and ethical qualms of nerve agent research and the reputational damage any revelations would cause both at home and abroad. As a result of these feared ramifications, Britain had 'not shown the same alacrity as the United States Chemical Corps in telling the world about new [chemical warfare] agents'.[83]

Even after applying diplomatic pressure and attempting to manage publicity in the United States, British officials could do little when, on 7 March 1960, Chapman Pincher finally decided to publish his story on the front page of the *Daily Express*.[84] Titled 'American V-gas Starts Rumpus' Chapman Pincher divulged that the highly lethal V-agents were discovered by British scientists and supplied to the United States, which was now mass-producing them.[85] Allegedly he received this significant breakthrough from someone working at the British company ICI.[86] Pincher proclaimed that these V-agents were of 'unprecedented power', 1000 times more lethal than German nerve agents such as sarin, and 'a powerful extra deterrent to the H-bomb'.[87] Grossly inflating the threat, the release of this information highlights the dangers of CW publicity, with the potential for it to create fear and shock through misunderstandings or exaggerations, whilst also making these weapons seem more appealing in terms of their military effectiveness. Interestingly though, Pincher did touch upon the divide between the Anglo-American partners over the publicity and role of the V-agents. He revealed that an 'American proposal that poison gas weapons…be brandished as a new Western deterrent is causing consternation in Whitehall'.[88]

This investigative journalism by Chapman Pincher opened the flood gates for unprecedented scrutiny of postwar British CW policy by MPs

[82] Ibid.

[83] TNA, WO 32/21950/1, British Joint Services Mission Washington DC to W. N. Hewson, 21 April 1960, Attached letter.

[84] Chapman Pincher, 'American V-Gas Starts Rumpus', *Daily Express*, 7 March 1960, 1.

[85] Ibid.

[86] It is also possible that Chapman-Pincher acquired this information from a scientist in PPL, a subsidiary of ICI, which discovered Amiton. See: McLeish and Balmer. Development of the V-Series Nerve Agents, 277–280.

[87] Chapman Pincher, 'American V-Gas Starts Rumpus', *Daily Express*, 7 March 1960, 1.

[88] Ibid.

and journalists. Capitalising on the rare appearance of information on CW policy, swiftly after the press release MPs asked difficult and direct questions. The majority of the MPs to pounce upon Pincher's revelations were, in a similar vein to their earlier critiques, predominantly from the Labour Party. On 21 March 1960, Marcus Lipton raised concerns over CW research and human experiments, and Barbara Castle warned of the 'disquiet among public opinion' over British research.[89] Two days later, on 23 March, Konni Zilliacus queried what information, about 'a powerful nerve gas development', had been shared with the United States, and Arthur Henderson questioned whether the dangers of CBW were greater now than in 1944.[90] On 30 March 1960 MPs asked further questions; Frank Allaun asked what information on new methods of CW was supplied to the United States, Silverman enquired about the impact of the United States not being a signatory of the Geneva Protocol, and William Warbey questioned whether the Minister of Defence knew of United States plans for the mass-production of V-agents.[91] Press coverage and parliamentary scrutiny were not only holding Government policy to account but also exerting considerable pressure in a highly sensitive area.[92]

Under this barrage of questions, government officials attempted to stick to what was agreed in the Defence Committee: minimal and tactical responses to any and all questions. Amidst the scale of public and parliamentary interest, though, they were forced into providing some limited concessions. The Minister of Defence, Harold Watkinson, conceded that

[89] Hansard, House of Commons, 'Brain-washing and Nerve Gas', 21 March 1960, Vol.620, cc30–32. In addition to Chapman Pincher, observations by Professor Alexander Kennedy also stoked the flames and spurred greater publicity and attention in the CW field by writing about brainwashing through psychoactive chemicals. This was conflated with CW activities, and the two became meshed together in some parliamentary questions. For example, see: *The Observer*, 'Spies Brainwashed in Britain: Wartime Method Adapted to Help Old People', 28 February 1960, 17; Hansard, House of Commons, 'Brain-washing and Nerve Gas', 21 March 1960, Vol.620, cc30–32.

[90] Hansard, House of Commons, 'Chemical Warfare (Exchange of Information)', 23 March 1960, Vol.620, c33W; Hansard, House of Commons, 'Biological and Chemical Weapons', 23 March 1960, Vol.620, cc497–8.

[91] Hansard, House of Commons, 'Chemical and Biological Warfare', 30 March 1960, Vol.620, cc1322–4. Unfortunately, there are too many names to list them all and as such this is not a complete list, but a select sample.

[92] For some considerations of the constraints and constraining factors on CW policy and chemical weapons' acquisition, see: Meselson, Matthew. 1972. Gas Warfare and the Geneva Protocol of 1925. *Bulletin of the Atomic Scientists*, February, 33–37.

there was indeed an agreement in place for the exchange of CW information between Britain and the United States and that this held no restrictions on how either country used the information.[93] In another debate, when asked about British chemical weapons production, the Secretary of State for War, Christopher Soames, remarkably conceded that Britain only produced chemical weapons in laboratory quantities.[94] This in itself was a substantial admission, and one which was necessitated by the level of public and parliamentary scrutiny, as well as by the desire of British officials to distance themselves from the large-scale production activities of the United States. It is hard to imagine that Soames was adhering to his guidance, as this admission willingly revealed a striking British vulnerability in the CW field. But, it is testament to the significant political pressure placed on senior government officials. On more complicated questions, government representatives simply evaded and provided extremely limited replies.[95] Frustrated by the relative lack of information, on one occasion Sydney Silverman called a point of order in the Commons, to which the Speaker simply replied, 'I cannot make the Minister answer if he does not want to'.[96]

Further revelations, coinciding with the Chapman Pincher article and parliamentary questions, expanded the debate to human experiments and the disposal of chemical weapons. This pressure pushed the War Office to publicly confirm that Britain had indeed been testing nerve agents on volunteers from the armed forces. But, as with other partial disclosures, information on deaths and the true extent remained secret. The public justification for these human experiments, as reported in *The Observer*, was that they were vital to discovering suitable antidotes, antidotes which had 'saved many lives'.[97] In March 1960 *The Times* had also reported on the British scuttling of ships filled with CW agents, which could cause a 'chain reaction explosion that would…let loose enough gas to poison the whole

[93] Hansard, House of Commons, 'Chemical and Biological Warfare', 30 March 1960, Vol.620, cc1322–4.

[94] Hansard, House of Commons, 'Nerve Gases', 18 March 1960, Vol.619, cc141–2W.

[95] Hansard, House of Commons, 'Biological and Chemical Weapons', 23 March 1960, Vol.620, cc497–8; Hansard, House of Commons, 'Chemical and Biological Warfare', 30 March 1960, Vol.620, cc1322–4.

[96] Hansard, House of Commons, 'Biological and Chemical Weapons', 23 March 1960, Vol.620, cc497–8.

[97] Abraham Marcus, 'War Gas Scientists Save Lives Now', *The Observer*, 20 March 1960, 2.

of the Western Baltic and much of the northern coast of Germany'.[98] It was alleged that as a result of poor planning, Britain had inadequately disposed of chemical weapons at the end of the Second World War. Toxic chemicals were now reported to be leaking into the Baltic Sea. In order to remedy this, divers spent three months bringing these toxic shells to the surface, which were then placed in metal drums, coated in cement, and simply re-dumped at sea at a deeper location.[99] Soames, and government officials, attempted to reassure MPs and the public that there was in fact 'no danger to anyone' from these chemical weapons.[100]

Under pressure, Soames made yet further admissions when he conceded that from 1955 around 25,000 tons of chemical weapons had been disposed of in the Atlantic.[101] Remarkably he also conceded that the disposal of chemical weapons included 17,000 tons of German 'gas bombs', which was a striking revelation.[102] Soames had essentially confirmed British vulnerability in the CW field, with the dual admission that Britain had disposed of its captured German tabun bombs, and by previously stating that Britain possessed only a laboratory scale production capacity for new nerve agents. Yet even after making these concessions, the magnitude of which seemed to go largely unnoticed, the Government was unwilling to fully capitalise on the fact that it had essentially unilaterally disarmed in the CW field. Rather than seek to benefit from the disposal of chemical weapons and abandonment of offensive chemical weapons development, just enough information was revealed to quieten hostile questioning, but this stopped far short of outwardly and deliberately capitalising on a situation which could have offered numerous political benefits. Policy, publicity and secrecy had thus created a strange situation. Tactical concessions amounting to the admittance of non-possession, but with adherence to some form of secrecy still preventing outright, clear, and direct admission.

Three key reasons can be identified that help explain why the Conservative Government proved unwilling to benefit from what was in

[98] *The Times*, 'Dangerous Cargo', 10 March 1960, 9.

[99] Ibid. The story, while somewhat exaggerating the threat and potentially of questionable accuracy, contributed to growing fear and concerns over British activities in the CW field.

[100] Hansard, House of Commons, 'Gas Bombs (Dumping)', 31 March 1960, Vol.620, c164W; *The Manchester Guardian*, '"No danger" from bombs dumped in the sea', 1 April 1960, 2.

[101] Ibid.

[102] Ibid.

effect unilateral chemical weapons disarmament in 1960. Firstly, there was the Government's ongoing commitment to some degree of secrecy and security considerations, with officials unwilling to contemplate directly and explicitly revealing British vulnerability in the CW field. This was especially so for defence officials. Secondly, there was the Anglo-American dimension, with any British decision to unilaterally renounce chemical weapons and take the moral high ground, at the same time as the United States was publicly announcing its own mass-production of VX, undoubtedly causing substantial friction. Such a step would also have made it abundantly clear to senior officials in the United States the extent of British reliance for CW deterrence and retaliation. In addition, it would have entailed the United States not just bearing the entire economic burden, but also all the political and moral costs of nerve agent production as well—leaving Britain to benefit from United States retaliatory capabilities and deterrent while reaping the potential political benefits of unilateral disarmament. Thirdly, despite secretly disposing of chemical weapons and offensive weapons development, military officials did not support the indefinite removal of the nerve agent weapon option. The 1956–1957 reductions were viewed by some as a temporary measure given the economic climate, the issues over first-use and the dominance of thermonuclear weapons in defence planning. Indeed, one key argument put forward by Monkton in 1956–1957 was the promising development of the V-agents. The earlier shift was therefore not a resounding moral reorientation or a drastic and permanent revision of the perceived military value of nerve agent weapons and CW deterrence. Ultimately, despite effectively unilaterally disarming and essentially revealing that the country possessed no military CW capabilities, British officials were unable to fully capitalise on the potentially politically beneficial situation.

Before the United States press release and Chapman Pincher's article, the majority of MPs were oblivious of trilateral cooperation in the CW field, and of the British discovery of the V-agents. After these stories emerged however, backbench Labour MPs and British newspapers succeeded in pressing a reluctant Government into confirming the existence of an agreement with the United States, acknowledging the British discovery of the V-agents, and revealing Britain's experiments on volunteers. Even after securing these rare admissions, on the whole MPs generally were provided with limited and tactical responses designed to leave the shroud surrounding CW policy somewhat intact. This was very much in line with what the Defence Committee had agreed in February 1960, before the rush of publicity. As Cabinet Secretary Norman Brook was informed in the immediate aftermath of the Chapman Pincher article, 'it looks as though we shall be able to hold the position...by saying virtually nothing'.[103] Limited and tactical scraps of information and non-answers, despite concessions by Soames, thus played a substantial part in the Government's approach to discouraging further criticism. This absence of further information starved informed criticism, and with but a few exceptions, questions over British CW policy gradually subsided from April 1960.[104]

The British Government's handling of this adverse publicity reveals the very different approaches towards CW publicity taken by Britain and the United States. Britain took a more cautious and secretive path, designed to mitigate and reduce publicity, whereas United States officials were not afraid to proactively engage and combat the negative image of chemical weapons. In the United States, stoking media attention was occasionally a tactic to aid CW policy and to secure greater funding. As astute writers in *The Times* noted in March 1960, 'there is constant and strong pressure from certain groups within the Pentagon, notably the Army Chemical Corps, for more funds and more publicity for chemical and biological weapons'.[105] The newspaper further reported that the United States Chemical Corps had hired civilian experts in public relations to try and portray chemical weapons as a more 'humane' weapon.[106] This publicity coincided with, and was to an extent part of, 'Operation Blue Skies', which was an attempt by the United States Chemical Corps to attain greater funding for CBW policy.[107] The United States Chemical Corps and influential figures interested in securing greater CW funding were more than willing to use the press as a tool in accomplishing this aim. This was in stark contrast to British officials. As Macmillan was counselled, 'why should we run the risk of political criticism (both domestic and

[103] TNA, CAB 21/4505, F.A. Bishop to Sir Norman Brook, 8 March 1960.

[104] In July 1960 Allaun attempted to gain information on West German chemical weapons production in parliament. However due to the sheer paucity of information, Allaun was enquiring about West German nerve agent production that did not exist. See: Hansard, House of Commons, 'Nerve Gases', 18 July 1960, Vol.627, c26.

[105] *The Times*, 'Pressure in United States for Chemical Weapons', 8 March 1960, 12.

[106] Ibid.

[107] *The New York Times*, 'Pentagon Spurs Chemical Arms: Versatility of Gas Warfare Held Underrated by United States—More Funds Urged', 9 August 1959, 3; Adams, Valerie. 1989. *Chemical Warfare, Chemical Disarmament: Beyond Gethsemane*. Basingstoke: Macmillan, 127–128, 146–147.

international) by giving in to the insidious pressure of the American publicity machine?'.[108]

British officials had little desire to draw attention to the country's CW capabilities. While Britain was withholding its reliance on the United States for CW deterrence and retaliation, British intelligence was also concerned by the publicity drive by the United States Chemical Corps. The JIC was fearful that by drawing attention to the CW field, and to the Soviet Union's chemical weapons stockpile, the United States Chemical Corps risked escalating the chemical weapons arms race by prompting an increase in Soviet CW preparations.[109] This United States publicity drive could trigger significant alarm as it attempted to normalise chemical weapons and make their use more acceptable to the public. Any attempt to alter public perceptions of CW could, therefore, potentially be seen by rival powers as laying the foundations for the future initiation of chemical weapons use through normalising them. And, at the very least, it further reinforced the narrative regarding the necessity of acquiring nerve agent weapons, whether that be for deterrence or use.

For many connected to British defence policy, publicity was problematic and unwelcome. Despite political conclusions rendering the period not too damaging, War Office officials noted that the coverage of CW policy had been 'adverse'. In September 1960, the CoS warned that CW policy 'should be treated as one of great sensitivity in view of the serious consequences of any publicity'.[110] This secrecy was paramount given that Britain was about to undertake another substantial review of CW policy, during which controversial arguments over the development and possession of nerve agent weapons, and their role in conflict, would again resurface.

The Zuckerman Effect

In September 1960, after the period of intense negative publicity and with the dismissal of complete reliance on the United States, defence officials returned to the CW field. This time, in stark contrast to prior reviews, there was one notable difference. For joining the CoS was the new Chief Scientific Advisor to the Ministry of Defence and Chairman of the DRPC,

[108]TNA, PREM 11/3099, 'V-agents', Brief for the Prime Minister, 23 February 1960.

[109]TNA, CAB 158/39, 'Soviet Biological and Chemical Warfare', Joint Intelligence Committee, 11 March 1960.

[110]TNA, DEFE 4/129, Minutes of Meeting, Chiefs of Staff Committee, 27 September 1960, Confidential Annex and Attached Report.

Sir Solly Zuckerman.[111] Those more favourable to Zuckerman's character and role would describe him as a one-man think-tank and of having a significant influence on policy.[112] However, he has also been accused of having a disproportionate influence on the military and strategic policies of the Government, and of being ambitious, arrogant, and of having a propensity for making enemies.[113] At the time, Chapman Pincher even observed that several career civil servants resented Zuckerman's appointment so deeply that they resigned in protest.[114] Zuckerman was also believed to have routinely 'delighted in undermining' and challenging the views of his own officials, which was particularly the case in sensitive areas such as nuclear weapons policy.[115]

In 1960, though, Zuckerman emerged as a hugely significant figure at a crucial time in British CW policy. Despite later branding the DRPC a 'gentlemanly forum in which the Services competed with each other for a share of the trivial amount of new money', he wasted little time in acting upon a changing mood in military circles in regards to nerve agent weapons. Defence officials, searching for new options and developments, increasingly questioned previous assessments and policy decisions.[116]

Zuckerman, supported by the DRPC and the War Office, now informed the CoS that previous conclusions and assessments of CW were 'no longer valid'.[117] A substantial factor for this change in approach were 'great advances' in the CW field, with developments in dissemination methods and with the toxicity of the V-agents now deemed even higher than previ-

[111] TNA, DEFE 4/129, Minutes of Meeting, Chiefs of Staff Committee, 27 September 1960, Confidential Annex. Zuckerman also had a particularly close working partnership with Admiral Louis Mountbatten, Chairman of the CoS, leading to them being referred to as the 'Zuckbatten Axis' in British defence policy. See: Peyton, John. 2001. *Solly Zuckerman: a scientist out of the ordinary*. London: John Murray, 140–142.

[112] Peyton. *Solly Zuckerman*, 134, 140–142.

[113] Ibid., 140–141.

[114] Liddell Hart Centre for Military Archives, H. C. Pincher, Press Cuttings, Vol.11 1960–1961, 'The ever-spreading Sir Solly', *Daily Telegraph*, 27 July 1960.

[115] Jones. *UK Strategic Deterrent, Volume I*, 183–185,

[116] TNA, DEFE 4/129, Minutes of Meeting, Chiefs of Staff Committee, 27 September 1960, Confidential Annex; Zuckerman, Solly. 1988. *Monkeys, Men, and Missiles: An autobiography 1946–88*. London: W.W. Norton & Company, 197; Peyton. *Solly Zuckerman*, 138. The DRPC was particularly concerned that Britain was losing the technical and mass-production experience. See: TNA, CAB 131/24, 'The Defence Research and Development Programme', Minister of Defence, 20 December 1960, Annex.

[117] TNA, DEFE 4/129, Minutes of Meeting, Chiefs of Staff Committee, 27 September 1960, Confidential Annex and Attached Report.

ously thought. Zuckerman also relayed the point that developments in incapacitating agents had opened up another avenue for CW, with these relatively new weapons offering the means of nullifying a threat without the ensuing level of casualties often associated with chemical weapons.[118] The War Office, in particular, looked very favourably upon them. They believed that incapacitating agents might actually 'revolutionise the *attitude* to the use of chemical agents'.[119] With incapacitating agents, defence officials hoped that there might be some wiggle room as to whether these weapons were included under the 1925 Geneva Protocol and in terms of how they were perceived. These new capabilities and advances in the CW field were thought all the more imperative given the potential risks and opportunities associated with nuclear sufficiency, with both superpowers soon thought to possess enough nuclear weapons to inflict such a scale of devastation on the other that they would be deterred from using nuclear weapons altogether. Zuckerman thought that this scenario might rule out the nuclear option, leaving room for other weapons in British defence policy, such as the nerve agents.[120]

After receiving Zuckerman's advice, the CoS agreed, believing that a far-reaching review of CW policy was now essential 'to establish the facts'.[121] Members of the CoS also cautioned that in addition to the dangers and potentialities brought about by nuclear sufficiency, there was a risk of becoming too 'mesmerised' by the concept of a nuclear exchange which was to the detriment of other areas of defence, such as the possibilities of CBW.[122] These new advances in, and appreciations of, the CW field coincided with simmering tensions in the Cold War. The superpower confrontation over Berlin was still ongoing. This amalgamation of events added a degree of urgency and potency to this re-appraisal of the utility and role of chemical weapons in British defence policy.

A core part of this substantial review of CW policy was an updated intelligence assessment of the Soviet CW threat.[123] The early 1960s, con-

[118] Ibid.

[119] Ibid.

[120] TNA, DEFE 4/129, Minutes of Meeting, Chiefs of Staff Committee, 27 September 1960, Confidential Annex; Balmer. *Britain and Biological Warfare*, 176.

[121] TNA, DEFE 4/129, Minutes of Meeting, Chiefs of Staff Committee, 27 September 1960, Confidential Annex.

[122] Ibid.

[123] Ibid.

trary to prior difficulties, represented a golden era for British CW intelligence. By 1962 the JIC had acquired 'good evidence' that the Soviet Union was actively involved in CW research.[124] One key reason for this change in fortunes was the acquisition of sources on Soviet CW preparedness, which provided much-needed insights into Soviet capabilities and intent. From 1959, increasing sources had started to emerge.[125] Intelligence gains included a Soviet publication which stated that the ability to mass-produce chemical weapons and deliver them on a large scale had 'considerably increased the prospects of using war gases in modern warfare'.[126] In the same year, Anglo-American intelligence agencies attained an article by Major-General Drugov, who wrote that Soviet scientists regarded it 'as their patriotic duty to study the action of poisons and to develop countermeasures'.[127] In 1960, sources also included references to Soviet civil defence measures, which further emphasised the seriousness with which Soviet officials were taking CW.[128]

One of the most important intelligence gains in this period came from a jointly run MI6 and CIA informant in the Soviet Union, Soviet Army Colonel Oleg Penkovsky. During 1961 and 1962, Penkovsky furnished his handlers with some much-needed information on Soviet military capabilities.[129] While the value of Penkovsky's information has been debated and his credibility questioned, for British intelligence he provided some crucial insights into Soviet CW preparations, with prior intelligence assessments hindered by a scarcity of sources.[130] In a meeting in July 1961, Penkovsky disclosed that while he was not directly involved in CW, he did

[124] TNA CAB 158/45 Part I, 'Sino-Soviet Bloc War Potential 1962–1966', Joint Intelligence Committee, 16 February 1962.

[125] University of East Anglia, Solly Zuckerman Collection, MOS (3)/1, Minutes of Meeting, Advisory Council on Scientific Research and Technical Development, 26 November 1959.

[126] Adams. *Chemical Warfare, Chemical Disarmament*, 127–128.

[127] TNA, DEFE 44/204, 'Soviet Chemical Warfare', Scientific and Technical Intelligence Sub-Committee, [exact date unpublished] September 1960.

[128] See for example: The National Security Archive, Chemical and Biological Warfare, Box 2, 'Chemical Weapons of Foreign Armies and Antichemical Protection', A. Malsinskiy, September 1960.

[129] Aldrich and Cormac. *The Black Door*, 214.

[130] Scott, Len. 1999. Espionage and the Cold War: Oleg Penkovsky and the Cuban missile crisis. *Intelligence and National Security*, 14:3, 23–26.

have some knowledge of Soviet activities in the area.[131] He subsequently revealed limited details on Soviet delivery methods for nerve agents, which included spray tanks, artillery shells and warheads filled with CBW agents.[132] Penkovsky also informed MI6 and the CIA that there was a 7th Directorate in the Soviet General Staff working on CBW. He claimed that there was also a dedicated section within the Ministry of Defence working on CW, an experimental station near Moscow, and a testing ground at Kaluga, south-west of Moscow.[133] Additional intelligence came in December 1961, when Penkovsky supplied key insights from the top-secret Soviet journal *Military Thought*.[134] In one article addressing the deployment of Soviet soldiers in a global war, Lieutenant-General S. Andryushchenko warned Soviet Commanders to 'evaluate thoroughly and in the quickest possible time the…complex radiation, chemical, and bacteriological situation evolving in the entire area of army deployment'.[135]

On Soviet defensive capabilities and intent, Penkovsky provided his handlers with reports that Soviet planners believed they were, in fact, better prepared for the outbreak of CW than Western powers. This coincided with existing Anglo-American concerns.[136] The JIC already believed that Soviet troops would be well-equipped to mitigate against a CW attack, with respirators, protective suits, manually-operated gas detectors and

[131] Digital National Security Archive, Central Intelligence Agency. Meeting No. 21, Oleg Penkovskii transcript, 24 July 1961, 12.

[132] Ibid.

[133] Penkovsky, Oleg. 1965. *The Penkovsky Papers*. London: Collins, 165; Harris and Paxman. *A Higher Form of Killing*, 146–148. Without access to Soviet sources, it is difficult to verify the true accuracy of his claims, but they did provide some insights for CW policy planners.

[134] Digital National Security Archive, Central Intelligence Agency, 'Military Thought', Memorandum for the Director of the Defense Intelligence Agency, 29 December 1961, Attached translated report.

[135] Ibid.

[136] NIE, 11-4-58, 'Main Trends in Soviet Capabilities and Policies, 1958–1963', National Intelligence Estimate, 23 December 1958.

Digital National Security Archive, Central Intelligence Agency, 'Military Thought', Memorandum for the Director of the Defense Intelligence Agency, 29 December 1961, Attached translated report. For British assessments, see: TNA, CAB 158/27, 'Sino-Soviet Bloc War Potential 1957–1961', Joint Intelligence Committee, 15 February 1957; TNA CAB 158/45 Part I, 'Sino-Soviet Bloc War Potential 1962–1966', Joint Intelligence Committee, 16 February 1962.

atropine syrettes.[137] Advanced Soviet defensive measures, reinforced by published Soviet articles and Penkovsky, thus revealed a degree of preparedness for the outbreak of CW.[138] These Soviet defensive preparations seemed to confirm long-held suspicions regarding Soviet interest and intense activities in the CW field. Furthermore, Penkovsky provided a glimpse into Soviet intent. He alleged that not only was Khrushchev preparing for CW, but that alarmingly the political decision regarding first-use had already been made, with Soviet doctrine allowing individual Field Commanders to decide.[139] Such a decision, if true, made CW use far more likely in a continental war.

In light of these alarming reports, the Soviet CW threat was gradually revised upwards from 1960, placing even greater urgency on the review of British CW policy.[140] This more informed intelligence picture would add to, and coincide with, growing military support for a significant re-thinking of British CW policy, its direction, and the need for a British nerve agent deterrent.

In tandem with British intelligence gaining valuable sources on the Soviet CW programme, Zuckerman established an independent sub-committee to review policy in late 1960, even though the War Office had already begun the process.[141] The CoS had requested a single unified report to inform policy, yet with Zuckerman's intervention, there were now two concurrent reviews of British CW policy taking place at the same time. One review, the original, was to be conducted by military officials in the War Office. The other was an independent panel established by

[137] TNA, CAB 158/39, 'Sino-Soviet Bloc War Potential, 1960–1964', 1 March 1960; TNA CAB 158/45 Part I, 'Sino-Soviet Bloc War Potential 1962–1966', Joint Intelligence Committee, 16 February 1962. (Atropine is used to offset the effects of nerve agent exposure).

[138] TNA, CAB 158/39, 'Sino-Soviet Bloc War Potential, 1960–1964', 1 March 1960; TNA, CAB 158/39, 'Sino-Soviet Bloc War Potential, 1960–1964', 1 March 1960; TNA CAB 158/45 Part I, 'Sino-Soviet Bloc War Potential 1962–1966', Joint Intelligence Committee, 16 February 1962.

[139] Penkovsky. *The Penkovsky Papers*, 165–166.

[140] DEFE 10/355, Minutes of Meeting, Defence Research Policy Committee, 13 September 1960; TNA CAB 158/45 Part I, 'Sino-Soviet Bloc War Potential 1962–1966', Joint Intelligence Committee, 16 February 1962.

[141] TNA, DEFE 4/129, Minutes of Meeting, Chiefs of Staff Committee, 27 September 1960, Confidential Annex; TNA, DEFE 10/417, Minutes of Meeting, Defence Research Policy Committee, 12 April 1961; Balmer. The UK Biological Weapons Program, 75.

Zuckerman and chaired by Sir Alexander Todd.[142] Unsurprisingly, Zuckerman had chosen someone he could rely on for this task, for he and Todd had a long working relationship dating back to the Second World War.[143] The two also met for dinners and discussions, and later once becoming Master of Christ's College, Cambridge, Todd invited Zuckerman to become a Fellow Commoner.[144] As decided by Zuckerman, the compromise was that Todd's assessment was to look at the bigger picture for CBW policy, while the War Office review, with input from the Air Ministry and the Admiralty, was to take the form of operational assessments of offensive and defensive CBW.[145] Before Zuckerman's intervention in the form of the Todd Panel, the War Office review had been intended to provide the foundation for future considerations of British CBW policy. In one of his first actions when reviewing CBW policy, Zuckerman had thus overridden military officials by establishing his own committee. This new committee reported directly to him. From the outset, his strong character was already beginning to directly impact the course and direction of British nerve agent policy.

By late 1961, both substantial reviews were complete. Unsurprisingly, the Todd report closely reflected Zuckerman's own views and vice versa. Importantly, the Todd report acknowledged the changing intelligence picture, warning that British civilians and soldiers, both at home and abroad, were seriously vulnerable to Soviet CW attacks.[146] In response to this threat, and with developments in chemical weapons, Todd argued that Britain

[142] Carter and Balmer. Chemical and Biological Warfare and Defence, 298–299.

[143] Zuckerman had also taken over Todd's position on the Ministry of Supply Advisory Council on Scientific Research and Development. See: University of East Anglia, Solly Zuckerman Collection, MOS (3)/2, Eric Rideal to Solly Zuckerman, 15 February 1956. Todd had also previously served on the Chemical Defence Advisory Board, see: McLeish and Balmer. Development of the V-Series Nerve Agents, 275.

[144] University of East Anglia, Solly Zuckerman Collection, Communications with Lord Todd, Clerks of the Salters' Company to Solly Zuckerman, 5 January 1962; University of East Anglia, Solly Zuckerman Collection, Communications with Lord Todd, Lord Todd to Solly Zuckerman, 16 October 1965; University of East Anglia, Solly Zuckerman Collection, Communications with Lord Todd, Solly Zuckerman to Lord Todd, 6 December 1966.

[145] For an outline of the Tri-Service report on CW, see: TNA, WO 32/20166, 'Operational Assessment of Chemical Warfare', Note by the Chairman of the Defence Research Policy Committee, 11 July 1961, Attached Report.

[146] TNA, WO 32/20166, 'Biological and Chemical Warfare', Brief for DCIGS, MGO and Chief Scientist by the Army Council Secretariat, 7 November 1961; TNA, DEFE 4/149, Chiefs of Staff Committee Meeting, 1 November 1962, Appendix to Annex.

should now do everything up to the point of chemical weapons mass-production, namely offensive and defensive research, weapons development and weapons testing.[147] He placed much greater emphasis on offensive research, in contrast to existing CW policy. Previous policy decisions were blamed for creating 'uncertainty and diffusion'.[148] The War Office reports also left open this possibility of developing a nerve agent capability, and its assessments, like that of the Todd report, supported the expansion of British CW policy.[149] The War Office also went slightly further, by stating that an independent chemical weapons capability 'would materially increase the potential of our forces'.[150] This nod to nerve agent weapons development, by both the Todd report and the War Office, was a marked change from years of policy. Nerve agent weapons thus still had military advocates, who had resurfaced, with possession thought useful for deterring enemy first-use and in slowing down and hindering Soviet ground forces.

The two reviews, despite not being vastly different in terms of assessments, caused significant friction. The CoS required a single cohesive assessment to inform discussions on the future of British CBW policy, but now they had two. Initially, the Defence Research Policy Staff attempted to combine the findings of the Todd report and those of the War Office.[151] However, there was a clear split within the DRPC over how this should be done. Zuckerman again clashed with his own staff. He fought for the pre-eminence of Todd's assessment, while military personnel in the Defence Research Policy Staff favoured the War Office.[152] Zuckerman, undeterred, directly confronted his own staff, accusing them of favouring the views of the War Office. He made it abundantly clear that the future of British CW policy would be considered 'on the basis of the Todd report'.[153]

As a result of the disjointedness between the Chairman of the DRPC and his staff, the DRPC was unable to crystalise its views on CW policy and so was unable to provide any concrete policy recommendations to the

[147] TNA, WO 32/20166, 'Chemical and Biological Warfare', Note by the DRP Staff, 31 October 1961.

[148] TNA, WO 32/20166, 'Biological and Chemical Warfare', Brief for DCIGS, MGO and Chief Scientist by the Army Council Secretariat, 7 November 1961.

[149] TNA, WO 32/20166, 'Operational Assessment of Chemical Warfare', Note by the Chairman of the Defence Research Policy Committee, 11 July 1961, Attached Report.

[150] Ibid.

[151] TNA, WO 32/20166, Minutes of Meeting, Defence Research Policy Committee, 8 November 1961.

[152] TNA, WO 32/20166, 'Biological and Chemical Warfare', Brief for DCIGS, MGO and Chief Scientist by the Army Council Secretariat, 7 November 1961.

[153] Ibid.

CoS in 1961. Zuckerman, clearly unhappy with the work of his own staff, pushed for the establishment of another committee. Under the chairmanship of Dr Walter Cawood, this new committee would be responsible for producing a combined and unified assessment of CBW policy.[154] It would be responsible for providing a 'single, simplified short paper' for the CoS.[155] Cawood, as both Scientific Advisor to the War Office and answerable to Zuckerman, certainly had an unenviable position, which was further hampered by, as Cawood noted, some of the 'conflicting' aspects of the Todd and War Office reports.[156] With all these organisational divisions over the future of CBW policy, it was not until November 1962 that the CoS received a combined assessment on CBW policy, an assessment it had requested over two years before.[157] Given Zuckerman's hugely influential and forceful nature, it was perhaps to be expected that Cawood's review for the CoS leaned towards the Todd report. The committee, though, agreed with key aspects of both Todd's assessment and the War Office, in that it approved the drive for increased funding and for an expansion in CW research and weapons development.[158]

Coinciding with Zuckerman changing his mind over the issue, Cawood's review also remarkably pushed for the acquisition of a tactical nerve agent capability. Defence officials were thus once again actively supporting the acquisition of a nerve agent capability. Helpfully alongside the official report by Cawood, the CoS also received a special appendix written by Zuckerman.[159] In this appendix, Zuckerman informed the CoS that:

> We should give thought to the implications of developing an offensive lethal CW…capacity:-

[154] TNA, WO 32/20166, Minutes of Meeting, Defence Research Policy Committee, 8 November 1961.

[155] Balmer. *Britain and Biological Warfare*, 177.

[156] TNA, DEFE 10/417, Minutes of Meeting, Defence Research Policy Committee, 3 July 1962.

[157] TNA, DEFE 4/149, Minutes of Meeting, Chiefs of Staff Committee, 1 November 1962; TNA, DEFE 5/131, 'Chemical and Biological Warfare', Note by the Secretary of the Chiefs of Staff Committee, 6 November 1962. (The date of the report is later than the meeting, as it is the collection of documents from the meeting after CoS approval).

[158] TNA, DEFE 4/149, Minutes of Meeting, Chiefs of Staff Committee, 1 November 1962. The report did however diverge on Todd's recommendation for a new CBW policy making body.

[159] Ibid.; TNA, DEFE 5/131, 'Chemical and Biological Warfare', Note by the Secretary of the Chiefs of Staff Committee, 6 November 1962, Appendix to Annex.

(i) to make it clear to the enemy, as a deterrent, that we have the means of retaliation; and
(ii) for actual use in retaliation if this would be to our military advantage.[160]

On 1 November 1962, just days after the ending of the Cuban Missile Crisis, the CoS approved the recommendations of the Cawood report, and they heeded Zuckerman's guidance; Britain should once again acquire a lethal CW capability, which in essence referred to nerve agent weapons.[161]

In addition to the prolonged review and Zuckerman's influence, this CoS support for a nerve agent capability was undoubtedly spurred on by events in the Cuban Missile Crisis. The crisis had either shown the effectiveness of nuclear deterrence, and implicitly reinforced the need for a nerve agent deterrent, or it had highlighted the fragility of the superpower balance and the need for British preparedness. The CoS thus believed that while arguments for chemical weapons had typically struggled on political grounds, in the thermonuclear age these political constraints could yet again be questioned.[162] In a similar vein to arguments over graduated deterrence, the CoS concluded that chemical weapons might, in fact, delay or reduce the escalation to all-out nuclear war, or at the very least that they would provide a capability somewhere in between conventional forces and nuclear weapons.[163] The crisis had also alarmingly shown the risks of weapons proliferation in the Third World. Britain, like the United States in Cuba, potentially could be caught out by the proliferation of advanced weaponry in Third World countries. This would render British forces either vulnerable to the weapons themselves or to any threat of their use if left without a retaliatory capability. A nerve agent capability was now dubbed a vital addition to British capabilities.

With CoS backing, the request for a British nerve agent capability was passed up to the Secretary of State for Defence, Peter Thorneycroft, who based on this emerging consensus in defence committees also supported

[160] Ibid.

[161] TNA, DEFE 4/149, Minutes of Meeting, Chiefs of Staff Committee, 1 November 1962. This shift also coincided with a wider change in defence policy, towards a more flexible approach, within which the nerve agent weapons could play a part. For further details on this shift, see: Speller, Ian. 2008. Corbett, Liddell Hart and the "British Way in Warfare" in the 1960s. *Defence Studies*, 8:2, 227–239.

[162] TNA, DEFE 4/149, Minutes of Meeting, Chiefs of Staff Committee, 1 November 1962.

[163] Ibid.

the need for the acquisition of a lethal nerve agent capability.[164] In April 1963, almost seven years after Britain had decided to dispose of chemical weapons and abandon offensive weapons development, Thorneycroft advised his Defence Committee colleagues that 'we should equip ourselves with a small retaliatory capability with a lethal chemical agent'.[165] In presenting his case for the acquisition of nerve agent weapons, Thorneycroft was formulating policy in the shadow of the Cuban Missile Crisis, while representing the views of the CoS, the DRPC and Zuckerman. It was the coming together of these key supporters and events, alongside perceptions of the external threat and advances in the CW field, which drove Thorneycroft to seek this political approval. The approval of nerve agent acquisition thus continued to coincide with significant periods of hostility and tension. Nerve agent acquisition from Nazi Germany was approved in the closing stages of the Second World War (and for potential use against Imperial Japan). Clement Attlee approved domestic production during the early stages of the Korean War, and now following the Cuban Missile Crisis potential acquisition was again re-visited.

In justifying the necessity of nerve agent weapons, Thorneycroft gave the Defence Committee the impression that Britain was falling behind. He argued that while the United States and the Soviet Union had extensive research programmes and large stockpiles of chemical weapons, Britain had no chemical weapons and possessed inadequate defensive measures. The superpowers were pulling ahead. Britain was even falling behind NATO requirements, which from 1962 had requested that forces be equipped with a retaliatory CW capability.[166]

Thorneycroft, most disturbingly of all, reported that Britain was even at risk of CW attack in the Third World, likely by Soviet proxies armed with Soviet chemical weapons.[167] Seemingly referring to an array of countries in Africa, the Middle East, and Asia, Thorneycroft warned of the dangers of weapons proliferation and of British forces being exposed to chemical

[164] TNA, DEFE 11/660, 'Biological and Chemical Warfare Policy', J. K. Watkins to the Chief of the Defence Staff, First Sea Lord, CIGS and CAS, 14 March 1963; Balmer. *Britain and Biological Warfare*, 178.

[165] TNA, DEFE 11/660, 'Biological and Chemical Warfare Policy', Memorandum by the Minister of Defence, 16 April 1963. Thorneycroft further counselled that Britain not only needed lethal chemical weapons, but also 'extremely interesting' incapacitating agents.

[166] Ibid.; Balmer. *Britain and Biological Warfare*, 177; Balmer. Keeping Nothing Secret, 878.

[167] TNA, DEFE 11/660, 'Biological and Chemical Warfare Policy', Memorandum by the Minister of Defence, 16 April 1963; TNA, CAB 131/28, Minutes of Meeting, Cabinet Defence Committee, 3 May 1963.

weapons. This aspect to nerve agent considerations was gaining increasing traction, fuelled by the legacy of the Cuban Missile Crisis, which had displayed the alarming situation of Soviet-supplied weapons abroad. The increasing focus on the Third World also coincided with emerging reports and accusations that the United States was using chemical weapons in Vietnam, of Castro using sarin in Cuba against rebels hiding in the Escambray Mountains, and shortly after stories also started surfacing of Egypt using chemical weapons in Yemen.[168] Without a nerve agent capability, Thorneycroft warned that Britain would be at a significant, unacceptable, and potentially humiliating disadvantage in the Third World. British forces 'would have no power to retaliate'.[169] Thorneycroft was insinuating that Britain was falling behind not just the Soviet Union or NATO allies, but potentially globally to countries formerly under imperial rule. Interestingly, though, his approach fell short of revealing fears of scientific advances abroad, and the stated concern was of Soviet-supplied rather than domestically produced chemical weapons. Regardless of the origin, though, a humiliating disadvantage was deemed unacceptable. He therefore strongly advised that Britain acquire a lethal CW capability 'in view of the access which our potential limited war enemies might have to Soviet technology'.[170]

While acknowledging the military utility of possessing nerve agents and the retaliatory option they provided, Cabinet members raised concerns over the political consequences of producing and deploying nerve agent weapons in peacetime. Another variation of postwar political concerns and normative considerations clashing with military requirements. They believed that the:

> Manufacture of these agents would present political problems. If we manufactured them it would be desirable to keep the fact from becoming known, but this would not be possible if we told our NATO Allies what we were doing.[171]

As the feared political ramifications were thought to be substantial, the condition and price of a nerve agent capability was thus secrecy, even from

[168] *Daily Mail*, 'Nerve-gas attack kills Cuba rebels', 19 March 1962, 2; *The Sunday Telegraph*, 'United States Poison Gas Denial', 10 March 1963, 2; Utgoff, Victor A. 1990. *The Challenges of Chemical Weapons: An American Perspective*. London: Macmillan, 70–71; Terrill, W. Andrew. 1991. The chemical warfare legacy of the Yemen war. *Comparative Strategy*, 10:2, 111; Tezcür, Güneş Murat & Horschig, Doreen, A conditional norm, 374.

[169] TNA, CAB 131/28, Minutes of Meeting, Cabinet Defence Committee, 3 May 1963.

[170] Ibid.; Walker. *Britain and Disarmament*, 11.

[171] TNA, CAB 131/28, Minutes of Meeting, Cabinet Defence Committee, 3 May 1963.

NATO allies. Providing secrecy was adhered to, the Defence Committee supported Thorneycroft's recommendations. It was thus again agreed that Britain would again acquire a nerve agent capability.[172] As Prime Minister Harold Macmillan concluded in May 1963, British CW policy would now prioritise:

1. Research in order to keep up to date with techniques and with American information.
2. The development of [an] offensive capability as a deterrent against such agents being used against us.
3. Defensive measures.[173]

In this approval of nerve agent acquisition, intense political desire to avoid any negative publicity and political aversion to nerve agent weapons had essentially meshed and clashed with arguments over military utility. This strange blending of conditions and requirements led to a bizarre CW policy, which was almost impossible to implement. The May 1963 Cabinet Defence Committee decision had at its core two inherent contradictions, which either went unnoticed or were swept aside.[174]

Firstly, the Defence Committee had approved the development of a secret deterrent, which resulted in a very confusing policy. Macmillan wanted a CW capability to deter others from using chemical weapons against British forces, but he also wanted the British acquisition of chemical weapons kept secret.[175] A secret deterrent is slightly paradoxical, as how can a weapon be secret and still function as an active deterrent? The very nature of deterrence entails an enemy being somewhat aware of the weapon's existence, as otherwise an enemy cannot be deterred by it: an invisible deterrent has no obvious deterrence value. While secrecy in the domestic sphere was seen as a good thing, as it mitigated against potential criticisms of Government policy, in the international arena a secret weapon

[172] Ibid.; Tucker. *War of Nerves*, 185; Balmer. Keeping Nothing Secret, 879; Schmidt. *Secret Science*, 324.

[173] TNA, CAB 131/28, Minutes of Meeting, Cabinet Defence Committee, 3 May 1963.

[174] Secrecy leading to contradictions and confusion in policy is not purely limited to CW, for as Balmer notes, it was also the case for BW policy. See: Balmer. *Britain and Biological Warfare*, 186.

[175] TNA, CAB 131/28, Minutes of Meeting, Cabinet Defence Committee, 3 May 1963.

had a limited, possibly non-existent role as a deterrent.[176] This seemingly strange contradiction, which went unclarified, further reveals another example where political concerns directly clashed with military utility. This Defence Committee approved policy reveals a clear red-line for politicians, who, while appreciating the military necessity of a nerve agent capability, could not contemplate the political ramifications of publicly defending that same capability. This political fear of publicity was all the more potent given the experiences of limited revelations in 1959–1960.

The second contradiction in the Defence Committee's 1963 policy is again tied to the stringent desire for secrecy, and it further reveals the strength of political concerns surrounding any potential public backlash, as well as the influence of negative connotations and perceptions of nerve agent weapons. In addition to keeping a British deterrent secret from the domestic population and rival powers, the Defence Committee also stipulated that Britain's production and development of nerve agent weapons should be kept secret from NATO allies. This was a remarkable position. In practice British defence officials had initially taken it to mean even excluding the United States.[177] British officials were worried about informing close allies such as Australia, Canada and the United States of the decision to develop chemical weapons, and of it leaking. The more who knew, the higher the chance that the Macmillan Government's decision to acquire nerve agent weapons would be unveiled, which was thought extremely likely to draw negative political repercussions; secrecy was therefore of paramount importance.

After attempts to maintain this stringent level of secrecy, it was soon realised that keeping Britain's decision to develop lethal chemical weapons a secret from the United States was exceedingly impractical.[178] Remarkably close levels of collaboration in the CW field rendered the attempted secrecy almost impossible. As defence officials noted, the United States would undoubtedly have soon realised that Britain was keeping something secret.[179] In addition, if the country was to develop a CW capability

[176] Ibid.

[177] See for examples: TNA, WO 32/20126, MOD London to BDS Washington, 22 July 1963; TNA, DEFE 24/31, 'Restrictions on Discussion of Chemical Warfare', DEP, 3 December 1963, Attached draft report; TNA, DEFE 24/31, A. MacKintosh to W. Brown, 3 April 1964; TNA, WO 32/20126, 'WEPC Meeting on 21 April '64', AEP 4, 16 April 1964.

[178] TNA, WO 32/20126, G. Dodds to Bernard Burrows, 13 January 1964.

[179] TNA, DEFE 24/31, 'Restrictions on Discussion of Chemical Warfare', DEP, 3 December 1963, Attached draft report; TNA, DEFE 24/31, A. MacKintosh to W. Brown,

'efficiently and economically', it needed to make full use of United States experience and expertise in the field.[180] Withholding this policy change from the United States thus appears to be more an issue of miscommunication between politicians and defence officials over the level of secrecy, with the secrecy condition in practice referring to NATO but excluding the tripartite countries. The fear of political consequences and the strict secrecy measures created confusion even amongst British defence officials, and erring on the side of caution they initially informed no allies of the major policy change.

This attempt at secrecy, or the very least this miscommunication in terms of the tripartite angle, occurred even with the importance of the United States CW programme fully recognised in the Defence Committee's decision. As Macmillan instructed, keeping in touch with the advanced research taking place in the United States was of the highest priority.[181] The sheer scale and effort of the United States CW programme made continued access vital, especially if Britain wished to remain in the chemical weapons arms race and develop its own capability. By 1963, the United States Army had produced over 25,000 tons of sarin, and it had 4000 tons of it stockpiled at the Rhine Ordnance Depot in West Germany.[182] United States officials, wary of the Soviet threat, planned to increase its stockpile of chemical weapons in Europe to over 30,000 tons, with the weapons stored in West Germany, France and Sicily.[183] The United States now also possessed an operational VX mass-production facility, and it had developed a frightening array of CW capabilities. Its chemical weapons included mortar shells, artillery shells, land mines, M55 rockets and aerial bombs containing varying CW agents such as sulphur mustard, sarin and VX.[184]

3 April 1964; TNA, WO 32/20126, 'WEPC Meeting on 21 April '64', AEP 4, 16 April 1964.

[180] TNA, DEFE 24/31, 'Restrictions on Discussion of Chemical Warfare', DEP, 3 December 1963, Attached draft report.

[181] TNA, CAB 131/28, Minutes of Meeting, Cabinet Defence Committee, 3 May 1963.

[182] United States NARA II, RG 218, Central Decimal File 1961, Box 49, 'Negotiations for Storage Rights in United States ECOM (U)', Memorandum for the Secretary of Defense, 17 December 1962, Appendix A; United States NARA II, RG 218, Central Decimal File 1961, Box 49, 'Action to Improve CBR Capability (U)', Memorandum for the United States Commander in Chief, Europe, 31 December 1962; Tucker. *War of Nerves*, 166–168.

[183] Ibid.

[184] United States NARA II, RG 218, Central Decimal File 1961, Box 49, 'Negotiations for Storage Rights in United States ECOM (U)', Memorandum for the Secretary of Defense, 17 December 1962, Appendix A.

The colossal scale of the United States CW programme, while underlining just how far behind Britain was, continued to provide Britain with unrivalled access to significant advances and developments.[185]

While there were some inherent contradictions within the Defence Committee's policy changes, there were still substantial benefits for CW advocates. Approval had been granted for the research and development of nerve agent weapons, and the acquisition of a nerve agent capability and deterrent. Another important gain came with the securing of greater funding. British expenditure on CW research and development rose by around 40%. Annual funding for CW research at this time was only around £1.1 million, but with the additional funding, this was increased to just over £1.5 million.[186] This significant increase now meant that CW research was receiving around three times as much direct funding as BW research.[187]

From 1958 to 1963, the direction of British CW policy had therefore almost completely reversed course. In 1958, serious consideration had been given to abandoning all British CW research and of becoming almost entirely dependent on the United States, but in 1963, Macmillan's Government had approved nerve agent weapons development and production, providing it was kept secret. This strong desire for secrecy had clear ties to 1960, with Cabinet members facing hostile Labour backbenchers and a multitude of questions on a highly sensitive area of defence policy. The uncomfortable questioning, both in newspapers and in Parliament, seemingly confirmed the negative public and parliamentary image of CW policy, and the potential political backlash of any information leaking. While this experience was politically difficult, it was not enough to override or prevent the seismic shift in British CW policy which

[185] Cost of the V-agent facility referred to in: TNA, WO 32/21950/2, Untitled Minute, Lt. Col. Saunders, 8 October 1958; Tucker. *War of Nerves*, 166.

[186] Yearly expenditure on CW research at this time was around £1.1 million. See: TNA, CAB 131/24, 'The Defence Research and Development Programme', Minister of Defence, 20 December 1960, Annex. This was increased by just over £2.1 million across five years. See: TNA, CAB 131/28, Minutes of Meeting, Cabinet Defence Committee, 3 May 1963.

[187] For expenditure forecasts in the period, see: TNA, CAB 131/24, 'The Defence Research and Development Programme', Minister of Defence, 20 December 1960, Annex. For the increase in funding, see: TNA, DEFE 11/660, 'Biological and Chemical Warfare Policy', Memorandum by the Minister of Defence, 16 April 1963. For confirmation of the proposed increase, see: TNA, CAB 131/28, Minutes of Meeting, Cabinet Defence Committee, 3 May 1963.

occurred in the early 1960s. Developments in the CW field, fears of Third World proliferation, Cold War confrontations, a better appreciation of the Soviet threat, the need for a CW deterrent and the influence of Sir Solly Zuckerman all amalgamated to dramatically reverse the direction of policy from 1963.

CHAPTER 6

A 'Secret Deterrent' and a 'Campaign of Criticism', 1964–1969

"Deterrence" in nuclear terms means spending an enormous amount of money and making loud noises about it; in chemical and biological terms, there is no need to spend much, and there are strong political objections to boasting about what we do spend.
—Sir Joseph Percival William Mallalieu MP (TNA, DEFE 13/557, 'Chemical Warfare and Biological Warfare—The Future of MRE and CDEE, Porton', Sir Joseph Percival William Mallalieu, 22 August 1967)

The period of 1964–1969 was characterised by substantial changes and a significant shift in both the core dynamics of the Cold War and in domestic British politics. While the early-1960s had been fraught with international dangers and fears of escalation, the latter part of the decade witnessed a comparative de-escalation in superpower confrontations, with tensions eventually giving way to a period of détente.[1] This climate of greater cooperation was shaped and facilitated by the arrival of nuclear parity, the emergence of the Sino-Soviet split, the stabilisation of the Berlin problem, and reduced US global power and influence as a result of its bruising war in

[1] Reynolds, David. 2000. *One World Divisible: A Global History Since 1945*. London: Penguin Books, 322–326; Savranskaya, Svetlana and Taubman, William. 2010. Soviet foreign policy, 1962–1975. In *The Cold War, Volume 2: Crises and Détente*, ed. Melvyn Leffler and Odd Arne Westad. Cambridge: Cambridge University Press, 134.

W. King, *Nerve Agents in Postwar Britain*, Britain and the World,
https://doi.org/10.1007/978-3-030-70474-2_6

Vietnam. This de-escalation was reversed in the sphere of domestic politics, though, as this period witnessed unprecedented levels of public engagement in politics, mass-movements, and fierce criticism of government policy. These were movements and engaged individuals not afraid to challenge and question official policy, nor were they shy in drawing attention to sensitive issues in foreign and defence policy. In Britain, as in much of Western Europe and the United States, this public criticism and engagement would reach a crescendo in 1967–1968.

The 1960s were also a crucial juncture in the Anglo-American relationship. Britain did not send troops to assist United States forces in Vietnam.[2] A conflict where, despite rapidly increasing involvement from 1964, with the aftermath of the substantial Tet Offensive the Johnson administration pursued a negotiated settlement.[3] Taking up office in January 1969, President Richard Nixon also sought a way out, and he began to withdraw US troops later that year. For the United States, the punishing war in Vietnam cost significant international prestige, produced domestic political strife and severe economic strains at home, and impacted international and Anglo-American relations.[4] Combined with the devaluation of Sterling in 1967 and the planned withdrawal from East of Suez, Labour Prime Minister Harold Wilson had reneged on all three key United States desires of Britain in this period.[5]

Some scholars, such as John Baylis, would argue that in the mid-to-late 1960s we see a cooling of relations between Britain and the United States.[6] Fortunately for both the Labour Government and British defence policy, between 1964–1970, they had the combative personality of Denis Healey as Secretary of State for Defence to manage and mitigate against any difficulties.[7] Healey could, in his own words, transcend traditional criticisms of Labour MPs as being 'pacifists' or 'column-dodgers', on account of his

[2] Steininger, Rolf. 1997. "The Americans are in a hopeless position": Great Britain and the war in Vietnam. *Diplomacy & Statecraft*, 8:3, 238–240.

[3] Logevall, Fredrik. 2010. The Indochina wars and the Cold War, 1945–1975. In *The Cold War, Volume 2: Crises and Détente*, ed. Melvyn Leffler and Odd Arne Westad. Cambridge: Cambridge University Press, 299–300.

[4] Ibid.

[5] Young, John W. 2003. *The Labour governments 1964–70, Volume 2: International policy*. Manchester: Manchester University Press, 31–33, 69–81. Vickers, Rhiannon. 2016. Foreign and Defence Policy. In *Harold Wilson: The Unprincipled Prime Minister? Re-appraising Harold Wilson*, ed. Kevin Hickson and Andrew Crines. London: Biteback Publishing, 271.

[6] Baylis. *Anglo-American Relations Since 1939*, 146.

[7] Young. *The Labour Governments, Volume 2: International policy*, 9.

former military service.[8] As Defence Secretary, he did not shy away from confrontations or defence cuts. He maintained a constant drive for 'cost-effectiveness'.[9] It was under Healey, and in this turbulent Cold War environment eventually giving way to détente, that British defence officials turned to consider just how they would acquire a nerve agent capability. After assessing controversial nerve agent acquisition, this chapter will move on to analyse the political-military divide which brought CW policy to a standstill in the mid-1960s, and it will end by examining the impact and scale of a tide of publicity which struck the CBW field in 1968. Intense public scrutiny, reinvigorated following earlier public and parliamentary scrutiny from 1959 to 1960 and coinciding with more active public engagement in politics, would hold lasting and crucial ramifications for British CW policy.

Acquiring a Capability

The May 1963 Cabinet Defence Committee decision, explored at the end of Chap. 5 and which had approved the British development of a nerve agent capability and the revival of offensive weapons research, triggered a period of intense activity in the CW field. Shortly after the Cuban Missile Crisis, political approval had been granted for nerve agent weapons to be produced and stockpiled in Britain. In attempting to implement this new policy, the War Office, in consultation with the Royal Navy and the Air Ministry, expanded defensive research and created a two-phase programme for acquiring this new nerve agent capability.[10]

In CW defence, British experts increased research and they rapidly implemented a series of trials.[11] To test British defensive preparedness, the Army launched Exercise Tureen in late 1964, which consisted of a number of military exercises enacting potential war-time manoeuvres and

[8] Healey, Denis. 1989. *The Time of My Life*. London: Michael Joseph, 252. Healey's Conservative successor after 1970, Lord Carrington, placed him in the top rank of European politicians since the Second World War. See: Lord Carrington. 1988. *Reflect on Things Past: The Memoirs of Lord Carrington*. London: Collins, 225.

[9] Zuckerman. *Monkeys, Men and Missiles*, 376; Reynolds. *Britannia Overruled*, 215.

[10] TNA, DEFE 24/31, 'Restrictions on Discussion of Chemical Warfare', DEP, 3 December 1963, Attached draft report.

[11] TNA, DEFE 24/31, 'Soviet Tactical Chemical Warfare', Director of Military Intelligence, 26 April 1963.

activities, at Imber Range, near Salisbury.[12] As officials noted at the time, 'this appears to have been the first British Chemical Exercise with troops for about a generation'.[13] This greater focus on defensive measures was in direct response to the perceived Soviet nerve agent threat. British intelligence was still benefitting from the comparative rush of intelligence gains in the early 1960s, with sources such as Oleg Penkovsky providing insights into Soviet intent and capabilities.[14] The JIC believed the Soviet Union to possess warheads filled with VR-55, an analogue of VX, as well as multi-rocket launchers, artillery shells and land mines for the delivery of lethal CW agents.[15] The JIC also warned that Soviet satellite countries in the Warsaw Pact were beginning to receive Soviet-supplied and nerve agent-filled FROG and SCUD warheads.[16] Perhaps of greatest concern for British defence officials, though, were reports alleging that Soviet officials viewed chemical weapons as a 'normal feature of modern warfare', and that the country would not hesitate to use them.[17] This consolidating picture of the Soviet threat played a significant part in adding to the urgency with which British defensive preparations were expanded from 1964.

In meeting the 1963 Defence Committee approval for the acquisition of a lethal nerve agent capability, for Europe and to counter potential proliferation in the Third World, the War Office, the Royal Navy and the Air Ministry all agreed to a two-phase process.[18] Phase one would be the securing of an interim CW capability, likely from the United States, which

[12] TNA, DEFE 24/6, 'Exercise Tureen', DAEP, 30 September 1964, Annex.

[13] TNA, WO 188/2530, 'Report on Exercise Tureen', J. Constant, 23 November 1964.

[14] TNA, CAB 158/51, 'Soviet Bloc War Potential, 1964–68', Joint Intelligence Committee, 28 February 1964, Annex.

[15] Ibid; Spiers. *Chemical Weaponry*, 7.

[16] TNA, CAB 158/48, 'The Soviet Chemical and Biological Warfare threat', Joint Intelligence Committee, 26 March 1963.

[17] Ibid.

[18] TNA, DEFE 24/31, 'UK Chemical Warfare Retaliatory Capability', Final Draft, WEPC Paper, 18 November 1963, Attached report; TNA, DEFE 24/31, 'CW—Retaliatory Capability', DMGO, 24 August 1965; Ulf Schmidt. *Secret Science*, 324–325. In CW research and development, the United States Army also placed greater emphasis on the development of binary CW agents, particularly after 1968. A binary chemical weapon split a CW agent into two non-lethal components which were then combined before use, to form a lethal agent. While British officials were keen on adopting this technology in 1964, it never progressed at the rate of United States research, and it never acquired the same level of backing. See for example: TNA, WO 32/20163, 'CW Agents—Binary System', AEP 4, 28 October 1964.

would be operational from 1965–1969.[19] This initial requirement would be met by all three Services and cost just under £3 million.[20] For the Royal Navy and the Air Ministry, their chosen capability was an aerial spray system for the dispersal of lethal VX, which had been uncovered by British researchers in the 1950s.[21] This aerial system was to be designed for delivery by the Royal Navy's Buccaneer and the RAF's Hunter GA9, despite safety concerns that the plane could crash and release VX or that there could be an accidental release of VX due to human error.[22] For the Army, the interim capability would be the procurement of 22,000 sarin-filled 105 mm shells from the United States.[23] These capabilities were, however, only temporary solutions. They were to bridge the gap until the completion of phase two, the development of a domestic capability to meet long-term CW requirements from 1970.

A core part of this reinvigorated CW policy also came in the form of a series of war game exercises, sponsored by the War Office and the Ministry of Defence (MoD).[24] The war games, running from mid-to-late 1965, plotted the predicted use of chemical weapons on the Western front at the outbreak of war in Europe. They contained some surprising revelations.[25] The war games forecast that with the widespread use of chemical weapons, much of the West German countryside would be contaminated. The hazardous area would stretch back around 20 kilometres from the border. Even with this large area of contamination hindering a Soviet advance, officials did not believe that CW was substantially militarily beneficial for NATO use in Europe.[26] They noted that 'chemical warfare … does not appear to have any decisive effect in slowing down the rate of advance of a

[19] TNA, AIR 2/17792, 'An Aircraft Spray System for Retaliatory Chemical Warfare (CW)', Operational Requirements Committee, MoD, 4 February 1965; Balmer. Keeping Nothing Secret, 879–881.

[20] TNA, AIR 2/17792, 'An Aircraft Spray System for Retaliatory Chemical Warfare (CW)', Operational Requirements Committee, MoD, 4 February 1965.

[21] Ibid.

[22] TNA, AIR 2/17791, 'NASR.1199—Aircraft Spray System', A. Frank, 6 October 1964.

[23] TNA, DEFE 24/31, 'CW Retaliatory Capability', DMGO to the Deputy Chief of the General Staff, 24 August 1965; Balmer. Keeping Nothing Secret, 880.

[24] TNA, DEFE 5/142, 'Tactical Use of Chemical and Biological Warfare', Note by the Secretary of the Chiefs of Staff Committee, 19 August 1963, Annex.

[25] TNA, DEFE 48/34, 'Civilian Population at Risk in Chemical Warfare in North-West Europe 1968–73', Ministry of Defence, December 1965, Annex B.

[26] TNA, DEFE 48/1, 'Chemical Warfare and the Land Battle in Europe in the Period 1968–97', Ministry of Defence, July 1965.

well-equipped and trained force'.[27] This finding, reinforced by perceptions of increasingly advanced Soviet defensive preparations and equipment, was in itself acknowledgement of a substantial shift in attitudes towards the military utility of chemical weapons. Since the late 1940s, chemical weapons had been thought of as useful in hampering any Soviet advance and in countering the Soviet Union's numerical advantage in tanks and troops. Yet it was now ruefully noted that even VX did not cause casualties quickly enough, allowing Soviet troops to don protective equipment before fatalities became widespread. Existing capabilities would fail to adequately hinder the advancement of Soviet forces. This was all the more alarming given that the Russian CW agent VR-55 was 'believed to be' in the category envisaged for a genuinely effective CW agent.[28] This perception of the Soviet Union possessing an edge in CW capabilities further compounded the feeling that Britain was being left behind.

The war games revealed another major concern. As defence officials noted, 'any force that was ill-equipped or badly trained would be very vulnerable', yet British intelligence believed that Soviet forces were very well-equipped and very well-trained.[29] This left one group, which while not a force, certainly was within the conflict zone and ill-prepared for nerve agent exposure: civilians living in the Federal Republic of Germany. Shockingly it was predicted that after the outbreak of CW in Europe, around 240,000 to 300,000 West German civilians would be within the hazardous CW zone.[30] In the centre of a conflict between NATO and Warsaw Pact forces, they would be at serious risk of nerve agent exposure from chemical weapons employed by both sides, whether it be VX, VR-55, or sarin.[31] Stray attacks, wind-changes, and the various routes of fleeing civilians, on top of conventional attacks between the warring sides, would

[27] Ibid.

[28] Ibid.

[29] TNA, CAB 158/56, 'Soviet Bloc War Potential, 1965–1969', Joint Intelligence Committee, 18 February 1965; TNA, CAB 158/59, 'Likely Scale and Nature of an Attack on the United Kingdom in the Early Stages of General War up to the End of 1969', Joint Intelligence Committee, 12 August 1965, Annex; TNA, DEFE 48/1, 'Chemical Warfare and the Land Battle in Europe in the Period 1968–97', Ministry of Defence, July 1965; TNA, DEFE 48/34, 'Civilian Population at Risk in Chemical Warfare in North-West Europe 1968–73', Ministry of Defence, [exact date not provided] December 1965.

[30] Ibid.

[31] TNA, DEFE 48/1, 'Chemical Warfare and the Land Battle in Europe in the Period 1968–97', Ministry of Defence, July 1965; TNA, DEFE 48/34, 'Civilian Population at Risk in Chemical Warfare in North-West Europe 1968–73', Ministry of Defence, December 1965.

have led to chaotic scenes and countless casualties. For British defence officials, given that CW use did not offer any real military advantage in slowing down Soviet forces and that it would expose hundreds of thousands of civilians to lethal chemical weapons, the main aim was increasingly seen as preventing the outbreak of CW in Europe altogether, through deterrence.[32] Arguments supporting the acquisition of nerve agent weapons thus began to pivot in the 1960s, away from emphasising the utility and military value of CW and heavily towards deterring Soviet initiation.

Although the war games were crucially important for considerations of the impact of nerve agent use, the slightly alarming conclusion as to their military value led the CoS to further reconsider the role of chemical weapons in Europe. The need for chemical weapons in the Third World to counter possible proliferation remained unquestioned, even if beyond retaliation or deterrence the role and geographic focus of them remained unclear. The war games were limited to Europe and thus only directly affected considerations of a CW capability for continental defence. Over this European role, though, there emerged some disagreement within the CoS. Chief Scientific Advisor Solly Zuckerman, once again exerting considerable influence over British CW policy, argued that given the apparent lack of operational effectiveness of chemical weapons against Soviet forces and even with deterrence arguments, there was no justification for their acquisition for a European war.[33] This was despite his prior support for a capability. Even with the rest of the CoS accepting the apparent military shortcomings of nerve agent weapons in Europe, they were hesitant to commit to Zuckerman's interpretation. Placing much greater emphasis on deterrence and on the option of having a retaliatory capability, and believing the ability to use nerve agent weapons entailing other benefits, they resisted any questioning of a European capability.[34] Firstly, they believed that possession of chemical weapons would cause a degree of uncertainty in any Soviet decision to initiate CW.[35] Secondly, the ability to retaliate in kind was thought not only a deterrence factor but a morale booster for

[32] TNA, DEFE 11/660, 'Chemical and Biological Warfare', Chief of the Defence Staff to the Secretary of State, 28 October 1965.

[33] Ibid.

[34] TNA, DEFE 5/162, 'Chemical and Biological Warfare', Note by the Secretary of the Chiefs of Staff Committee, 28 September 1965, Annex A.

[35] Ibid.

British troops, particularly if they themselves were under CW attack.[36] The CoS thus used this morale argument in a slightly dubious way, with the benefits of boosting British troops at the potentially harrowing cost of NATO civilians and Soviet troops, not just a seemingly acceptable cost but a justification and reason for weapons acquisition and deployment.

In September 1965, even after re-affirming the need for a CW capability in the Third World and downplaying the impact of the war games, the CoS hesitated in following through and finalising nerve agent acquisition. By fostering a degree of doubt over the utility of chemical weapons in Europe a divide had emerged between Zuckerman and the rest of the CoS, which created some uncertainty over the direction of British CW policy as a whole.[37] Support for nerve agents in the Third World remained unchanged, but by undermining the role of chemical weapons in Europe, Zuckerman and the war games had inadvertently brought the entire nature of a British CW capability back into question.

In addition to this uncertainty surrounding the military role of chemical weapons in Europe, the CoS was increasingly wary of the significant political and financial costs involved in actually following through with the production of nerve agent weapons. Phase one depended on nerve agent acquisition from the United States, but difficulties, reminiscent of when Britain had attempted the same feat in 1953–1954, were again emerging over securing a nerve agent capability.[38] The deal had been thrown into doubt when, much to the surprise of British officials, the United States State Department blocked the sale of nerve agent weapons. This was even though the United States Army and the Department of Defense supported the selling of nerve agents to Britain. But the State Department maintained that no sales could be made as CW policy was also under review in the United States.[39] British officials dubbed this setback a 'bitter blow'.[40] Due to the decision being made mid-level in the State Department, a British Ministerial level intervention was thought necessary to secure this

[36] Ibid.

[37] In terms of actually considering a capability that could fulfil both roles, an aerial VX spray system was still thought usable in both European and Third World conflicts.

[38] TNA, DEFE 5/162, 'Chemical and Biological Warfare', Note by the Secretary of the Chiefs of Staff Committee, 28 September 1965, Annex A.

[39] TNA, DEFE 24/31, BDS Washington to M. C. Cornford, 11 November 1965.

[40] Ibid.

vital first stage and temporary capability from the United States.[41] This would have entailed asking the Labour Government, under Prime Minister Harold Wilson, to intervene and purchase nerve agents from the United States, thus implementing and approving Macmillan's 1963 decision to acquire nerve agent weapons. To overcome supply issues, and in tandem with concerns arising over the role of nerve agent weapons in Europe, Ministerial guidance was now thought vital. In October 1965, the CoS thus sought Ministerial clarification regarding how they should proceed in implementing Macmillan's support for nerve agent acquisition and for guidance on the future direction of CW policy under the Wilson Government in the Cold War.[42]

The Publicity Red-Line

This CoS request for political input and clarification was sent to the Secretary of State for Defence, Denis Healey. While Healey deemed continued investment in defensive measures 'unobjectionable', he believed that the production and purchase of nerve agent weapons, regardless of the 1963 Defence Committee decision, necessitated Prime Ministerial approval.[43] On 8 November 1965, Healey sought Harold Wilson's backing for nerve agent weapons. He reiterated earlier concerns by warning Wilson that the superpowers had pulled far ahead, and worse still, that the Soviet Union might supply chemical weapons to Third World countries 'to embarrass us outside of Europe'.[44] Britain being caught unprepared was thus viewed as potentially humiliating, especially if this was achieved by a country other than the Soviet Union. This line of justification was tapping into prestige, perceptions of military efficiency, and of reputational damage. Though Healey supported a limited capability for the Third World, he agreed with Zuckerman's analysis that in Europe, chemical weapons were unlikely to have a significant effect in slowing or

[41]TNA, DEFE 5/162, 'Chemical and Biological Warfare', Note by the Secretary of the Chiefs of Staff Committee, 28 September 1965; TNA, DEFE 24/31, 'Chemical Warfare', DAEP to the Deputy Chief of the General Staff, 29 November 1965.

[42]TNA, DEFE 11/660, 'Chemical and Biological Warfare', Chief of the Defence Staff to the Secretary of State, 28 October 1965; TNA, DEFE 24/31, 'Chemical Warfare', DAEP to the Deputy Chief of the General Staff, 29 November 1965.

[43]TNA, PREM 13/3464, 'Biological and Chemical Warfare', Denis Healey to Harold Wilson, 8 November 1965.

[44]Ibid; Balmer. Keeping Nothing Secret, 880–882.

hampering the Soviet juggernaut. His support for a CW capability was therefore conditional on it being restricted to limited wars outside of Europe, for deterring CW use by a Soviet-supported country in the Third World, and that it be secret until used, which again raised serious questions as to the possibility and viability of a secret deterrent. Although Healey addressed the contentious issue of publicity and deterrence, his insights perhaps confused the situation further, whilst revealing the extent of political aversion to any publicity over nerve agent weapons. Reflective of his appreciation of the wider public's aversion to the nerve agents, and to the potentially serious domestic and international backlash resulting from any publicity, Healey doubted whether it would be a viable political move to deter an enemy's first-use of chemical weapons. Deterring first-use necessitated publicity, which Healey thought unacceptable; as such, he informed Wilson that:

> Although our possession of this capability would not be publicised and might not, therefore, deter an initial use of chemical weapons by an enemy, it would enable us to provide a quick response in retaliation and thereby we hope deter further use.[45]

With the politically sensitive nature of chemical weapons, Healey did not want to deter the first-use of chemical weapons as British capabilities definitively 'would not' be publicised, a key component of effective deterrence. Consequently, Healey's stance represented a slightly bizarre interpretation of CW deterrence. Publicity was not initially needed, but the deterrent effect would come into play once British forces had already been exposed to lethal CW agents.[46] It was for this policy, of a limited CW capability to be kept secret and potentially used in retaliation against countries outside of Europe, which Healey sought Prime Ministerial approval for in November 1965. Though, for Healey and the CoS, their timing was unfortunate.

From 1964–1966, Wilson's first Labour Government only had a very narrow majority. This made it extremely vulnerable to both internal and external pressures.[47] It was dominated by political concerns, as the Prime

[45] TNA, PREM 13/3464, 'Biological and Chemical Warfare', Denis Healey to Harold Wilson, 8 November 1965.

[46] Ibid.

[47] Young. *The Labour governments, Volume 2: International policy*, 1–2.

Minister was preparing for another election to secure a larger majority, which took place in March 1966.[48] In the build-up to this second election, Wilson was keen to defer any decisions on politically sensitive topics. Unsurprisingly, he delayed any decision on controversial nerve agent weapons until things were 'a little quieter'.[49] Alongside pressing electoral concerns was the evolving and dangerous situation in Rhodesia, where a Unilateral Declaration of Independence had been declared in November 1965. These more pressing issues thus overshadowed any decision on CW policy, and it dropped down the list of priorities.[50] For although Healey initially backed the acquisition of nerve agent weapons and Macmillan's 1963 decision to acquire chemical weapons, he stood little chance of securing the approval of Wilson on such a controversial issue and in the face of such a substantial political backdrop.

However, this reluctance to engage with the nerve agent issue was not one solely limited to the pre-election period. After winning a significant Labour majority in the March 1966 election, Wilson continued to delay any decision. In this second Wilson Government, Healey, who had retained his position as Defence Secretary, was also increasingly sceptical as to the political costs of acquiring nerve agent weapons. In June 1966, Healey's Assistant Under-Secretary for Policy, Frank Cooper, echoed these concerns when he advised Healey that:

> The whole concept of an independent British CW capability seems reminiscent of the Pre-Suez era or even earlier ... Are we going to spray nerve gas on Africans, Indonesians or even the Chinese? Can there be a deterrent if we do not admit to having, let alone publicise, the weapon?[51]

For Cooper, merely the option of having a retaliatory capability was questionable, regardless of its actual use or role. He thought that an independent CW deterrent in mid-1960s British defence policy an outdated and unnecessary cost. If the political fallout from just the production and development of nerve agents was thought unacceptable, then how could

[48] Clarke, Peter. 2004. *Hope and Glory: Britain 1900–2000*. London: Penguin Books, 296.

[49] TNA, PREM 13/3464, J.A. Peduzie to W. K. Reid, 25 February 1966; TNA, AIR 8/2391, 'Chemical Warfare Weapons', W. D. Hodgkinson, 11 May 1966; Balmer. Keeping Nothing Secret, 881.

[50] TNA, DEFE 25/24, R. J. Andrew to AUS (POL), 15 June 1966.

[51] TNA, DEFE 13/846, 'Chemical Warfare Policy', Assistant Under Secretary (Policy) to the Secretary of State for Defence, 20 June 1966.

the Government justify actual use, especially in a limited war with a Third World country? This not only conjured up images of Britain aggressively attempting to flex its muscles abroad but also manifested a similar and comparable image to the hugely controversial Italian use of gas warfare in Abyssinia in the 1930s.[52] In a global war with nuclear weapons, CW could perhaps be justified, but this was certainly not the case for nerve agent weapons in a limited conflict. The envisaged military role of nerve agents to counter and deter the proliferation of Soviet chemical weapons in the Third World was therefore significantly at odds with concerns of negative publicity and political fallout. As Cooper's insights attest, deterrence and publicity constituted an increasingly contradictory and seemingly insurmountable bone of contention between politicians and defence officials.[53] While the political and public ramifications of the acquisition and brandishing of nerve agent weapons were too significant for politicians to contemplate, for military planners this same publicity was not a choice, but a necessity for successful deterrence.

Senior British defence officials remained wedded to the idea of publicising British nerve agent weapons, as this was viewed as an inseparable part of deterrence and thus integral to CW policy. This stance was in stark contrast to global events. Occurring at the same time as secret defence assessments on the need for a nerve agent capability and deterrent, the UN passed Resolution 2162 B.[54] The Resolution encouraged states to accede to the Geneva Protocol, and the Conference of the Eighteen-Nation Committee on Disarmament was tasked with seeking the cessation of the development and production of CBW.[55]

The substantial political backdrop also remained and international support for disarmament was showing signs of growing. But, defence officials still pushed for a political guarantee that they could publicise a British nerve agent capability before they acquired one. Without a guarantee regarding publicising a nerve agent capability, the acquired capability would likely have little deterrent value, which was now the main purpose of any capability in Europe. The Assistant Chief of the Defence Staff surmised in December 1966 that, regardless of any feared negative publicity,

[52] Spiers. The Geneva protocol, 328–336.

[53] TNA, DEFE 13/846, 'Chemical Warfare Policy', Assistant Under Secretary (Policy) to the Secretary of State for Defence, 20 June 1966.

[54] United Nations. 1966. Question of general and complete disarmament. 1484th plenary meeting, United Nations Resolution 2162 B (XXI), December 5.

[55] Ibid.

'the military need has been established' and the Government should acquire a nerve agent capability.[56] This was even after defence officials acknowledged that Ministers might find the 'political criticism' of producing nerve agents too substantial, and that when it comes to CW 'public opinion tends to be governed by emotion rather than reason'.[57] In July 1965, Michael Quinlan had already surmised the views of Chief of the Air Staff, and wider military views on CW deterrence, when he wrote that 'a deterrent weapon ... can deter only if the enemy knows about it'.[58] Quinlan represented the views of many defence officials when he argued that publicity was needed in some form and that for a weapon to act as a deterrent enemy countries not only needed to be made aware that Britain possessed CW stocks, but that Britain was willing to use them.[59] He also proposed a measure to both appease political dissatisfaction and meet military requirements when he suggested that the Government deliberately loosen security classifications in British CW policy.[60] This measure could have led to some limited information on British intent and policy seeping into the public domain, being picked up by Soviet intelligence, and thus fulfilling the deterrent requirement in some form. Although this path of assuming Soviet officials would discover British capabilities was not acted upon, it again revealed military officials attempting to find some wiggle room in the political red-line over publicising a nerve agent capability.

Even with these tentative attempts at compromise, the disagreement over publicity and deterrence was an impassable obstacle in the nerve agent debate. It exposed a stark divide in approach by Labour politicians and defence officials. In light of all these difficulties, controversies and contradictions associated with CW policy and nerve agent acquisition, rather than directly engage with the nerve agent decision, Healey decided

[56] TNA, DEFE 11/660, 'Chemical and Biological Warfare', Assistant Chief of the Defence Staff to the Vice-Chief of the Defence Staff, 22 December 1966.

[57] Ibid.; TNA, DEFE 13/846, 'Chemical Warfare Policy', Private Secretary to the Secretary of State for Defence, 20 June 1966. Defence officials ultimately thought that attempts at maintaining secrecy were unlikely to succeed, and thus publicity was thought an inevitable consequence of acquiring a capability. See: TNA, DEFE 25/24, R. J. Andrew to A.U.S.(POL), 15 June 1966. Also see: Balmer. Keeping Nothing Secret, 881–883.

[58] TNA, DEFE 25/24, Michael Quinlan to the Secretary of the Chiefs of Staff Committee, 7 July 1965.

[59] Ibid.

[60] Ibid.

to defer and marginalise it.[61] As a result of largely political considerations, nerve agent policy was essentially side-lined. This lack of senior political guidance and clarity also caused some confusion for civil servants within the Cabinet Office, who, by late 1966, were uncertain over how to deal with the matter.[62] Cabinet Office officials were in the dark as to whether they needed to 'continue to pigeon-hole' CW policy, or if the nerve agent question actually needed addressing.[63] For Healey, his frustration with defence officials insisting on publicity for CW deterrence would grow progressively worse, after the Defence Research Committee (DRC—formerly the DRPC) finished a substantial review of CBW policy for the CoS.

Completed in December 1966, the DRC review was intended to inform CoS CW policy after the setbacks and delays which had plagued it since the May 1963 Cabinet Defence Committee decision.[64] Defence officials believed that once the review was complete, and if it still supported the acquisition of nerve agent weapons, then they should make a fresh approach to Ministers. The review was also intended to answer 'mutterings' from within the defence establishment, which accused the Army of spending too much on CBW research.[65] However, as soon discovered by the DRC, even though it was intended to provide a cohesive report and inform Ministerial deliberations, it could accomplish little due to the uncertainty surrounding existing political thinking.[66] The DRC felt unable to recommend any substantial policy changes or recommendations due to what was in essence policy paralysis at the most senior level, with senior politicians refusing to engage with the topic and defence officials lacking a clear direction. As a result, its report mostly agreed with and reinforced

[61] For examples of the delays, deferrals, and isolation of CW policy, see: TNA, DEFE 13/846, 'Chemical Warfare Policy', Private Secretary to Secretary of State for Defence, 20 June 1966; TNA, PREM 13/3464, Michael Palliser to W. K. Reid, 17 December 1966.

[62] Ibid.

[63] Ibid.

[64] TNA, DEFE 25/24, 'Review of CW and BW Research Programme', Note by the Defence Research Staff, 1 December 1966; TNA, WO 32/21760, 'Review of CW and BW Research Programme', Defence Research Committee, 15 December 1966; Balmer. Keeping Nothing Secret, 881.

[65] TNA, WO 32/21760, 'Review of CW and BW Research Programme', Defence Research Committee, 15 December 1966.

[66] TNA, DEFE 13/557, 'Chemical Warfare', A. H. Cottrell to the Minister of Defence (Equipment), 13 September 1967.

existing military thinking.[67] One key area where the DRC review reverberated with prevailing tensions between politicians and defence officials was over publicity and the role of CW deterrence, with the DRC arguing that:

> We are always so secretive about having a lethal CW capability that it has no deterrent effect; all we have is the ability to retaliate. If we develop CW weapons, we should let the fact be known ...[68]

The DRC, like Quinlan before, also tentatively considered a possible half-way measure to bridge the political-military divide: to rely on foreign intelligence agencies uncovering British nerve agent capabilities.[69] This would have created the unusual situation where the domestic population would be deliberately kept in the dark but where defence officials actively wanted, and in fact for deterrence purposes needed, foreign intelligence agencies to unearth British CW capabilities.[70] The DRC, though, advised against taking this half-way approach. The chances of the domestic population finding out and of awkward parliamentary questions arising was deemed too high. There was also no guarantee that Soviet officials would keep this information secret and, as seen in previous cases, Soviet officials were not averse to using CBW as a publicity tool.[71]

When the DRC proposed publicising Britain's future nerve agent capabilities, it was representing the existing views of senior defence officials and the CoS readily endorsed its findings.[72] The CoS also envisaged a public relations campaign for British CW policy to pre-empt negative publicity. This was apparently in a similar vein to the United States Chemical Corps in the early 1960s, which had recruited civilian public relations consultants to improve the image of CBW.[73] Such a measure was thought

[67] TNA, WO 32/21760, 'Review of CW and BW Research Programme', Defence Research Committee, 15 December 1966.

[68] Ibid. Also see: TNA, DEFE 11/660, Minutes of Meeting, Defence Research Committee, 18 January 1967.

[69] TNA, DEFE 13/557, 'Chemical Warfare', A. H. Cottrell to the Minister of Defence (Equipment), 13 September 1967; Balmer. Keeping Nothing Secret, 882.

[70] TNA, DEFE 13/557, 'Chemical Warfare', A. H. Cottrell to the Minister of Defence (Equipment), 13 September 1967.

[71] A prime example of this can be seen in 1954, with questions previously arising over CBW and West German re-armament.

[72] TNA, DEFE 4/222, Minutes of Meeting, Chiefs of Staff Committee, 17 October 1967.

[73] Ibid.

necessary to temper the predicted political and public backlash resulting from publicising a nerve agent capability.

In October 1967, the CoS went one step further and expanded their request for nerve agent weapons. Not only did they request a CW capability for the Third World to counter Soviet proliferation, but they reiterated their earlier support for nerve agent weapons in an expanded role in Europe, despite the findings of the war games.[74] In reaching this additional conclusion, the CoS was aided with the removal of one significant obstacle: Solly Zuckerman had left his role as the MoD's Chief Scientific Advisor in 1966. The CoS was no longer divided.[75] This revived request for a significant nerve agent capability for Europe again had roots in perceptions of the Soviet CW threat, with British intelligence possessing 'considerable knowledge' of Soviet CW capabilities and Soviet intent to use chemical weapons in battle.[76] Deterrence was growing in importance. It was increasingly seen as being in NATO's interest that CW did not become a feature of war, especially with the fighting likely occurring in West Germany and with Soviet forces potentially better equipped.[77] A European CW deterrent, therefore, re-emerged as a primary purpose for a nerve agent capability.[78]

In late 1967, when Healey was again approached to approve the acquisition of a nerve agent capability, publicity for deterrence was an essential component.[79] The CoS request entailed Healey dropping his red-line over publicity, and accepting Britain broadcasting its nerve agent capabilities and willingness to use them globally.

When the nerve agent question reached Healey, this time, rather than just defer, he deferred and delegated. Initially, Healey deferred by informing the CoS that any Ministerial meeting on CW would detract

[74] Ibid.

[75] Ibid; Peyton. *Solly Zuckerman*, 147.

[76] TNA, DEFE 4/222, Minutes of Meeting, Chiefs of Staff Committee, 17 October 1967.

[77] Ibid; TNA, WO 32/21760, 'Chemical Weapons for General War in Europe', Army Combat Development Committee, 30 August 1967, Annex.

[78] The Army also justified a European CW requirement by framing it as a necessary component of 'a convincing fighting posture'. This convincing fighting posture entailed more than one NATO country possessing chemical weapons to retaliate against, and deter, Soviet first-use. See: TNA, WO 32/21760, 'Chemical Weapons for General War in Europe', Army Combat Development Committee, 30 August 1967, Annex.

[79] TNA, DEFE 4/222, Minutes of Meeting, Chiefs of Staff Committee, 17 October 1967.

from important ongoing discussions on British nuclear policy.[80] Yet after these Cabinet discussions had finished, Healey still refused to engage with the nerve agent question. He instead decided to delegate the issue to his Minister of Defence for Equipment, Roy Mason.[81] Mason, labelled an 'uncompromising' figure who frequently took a tough line, is predominantly known for his later dealings with the IRA as Northern Ireland Secretary.[82] In late 1967 and early 1968, he was responsible for setting the direction of British CW policy after Healey's decision to divest himself of the controversial issue. Mason, as Minister of Defence for Equipment, also wielded considerable influence over CW policy as Porton Down came under his purview.

In forming his decision on the future of British CW policy, Mason benefitted from the prior analysis of another Minister of Defence, Sir Joseph Percival William Mallalieu MP, then Minister of Defence for the Royal Navy.[83] Mallalieu, Oxford-educated and a journalist, fought in the Royal Navy during the Second World War.[84] When he had assessed CW policy in August 1967, he provided one of the most open, frank, and forthcoming considerations of CW policy in postwar Britain. He ruefully started his review by stating that 'it is a tall order to rush in where the Prime Minister declines to tread'.[85] Mallalieu further observed that Wilson had maintained a 'masterly silence' on the subject and that the Prime Minister had even managed to stifle Conservative MP Michael Hamilton's enquiries into CW research at Porton Down, despite Hamilton being the local MP. In terms of the controversial and divisive subject of CW deterrence, Mallalieu rather succinctly noted that:

[80] TNA, DEFE 13/846, 'Chemical and Biological Warfare', R. M. Hastie-Smith to AUS (POL), 19 October 1967; TNA, PREM 13/3464, Burke L. to Michael Palliser, 20 October 1967.

[81] TNA, WO 32/21761, 'Chemical and Biological Warfare, and the future of MRE and CDEE Porton', Roy Mason to the Secretary of State for Defence, 18 January 1968.

[82] *The Telegraph*, 'Lord Mason of Barnsley—obituary', 20 April 2015.

[83] TNA, DEFE 13/557, 'Chemical Warfare and Biological Warfare—The Future of MRE and CDEE, Porton', Sir Joseph Percival William Mallalieu, 22 August 1967.

[84] *The Times*, 'Admirals Hold Their Peace Until Tomorrow', 21 February 1966, 10.

[85] TNA, DEFE 13/557, 'Chemical Warfare and Biological Warfare—The Future of MRE and CDEE, Porton', Sir Joseph Percival William Mallalieu, 22 August 1967.

> "Deterrence" in nuclear terms means spending an enormous amount of money and making loud noises about it; in chemical and biological [warfare] terms, there is no need to spend much, and there are strong political objections to boasting about what we do spend.[86]

Mallalieu elaborated that while nuclear weapons were 'expensive and nasty', chemical weapons were 'cheap and nasty'.[87] Picking up on fears of proliferation in the Third World, just as Thorneycroft and Healey had done so before him, he crassly warned that unlike nuclear weapons any 'tin pot country' could acquire a CW capability.[88] Combining a sweeping judgement as to the cheapness of producing chemical weapons and the ability of countries to produce them, Mallalieu both downplayed the complexity of mass-producing chemical weapons and sought to undermine their acquisition. Nuclear weapons were judged a comparative luxury, and thus the right path for Britain, chemical weapons less so. Importantly for British CW policy, Mallalieu thus did not support the 1963 policy to acquire nerve agent weapons. He stated that British CW scientists should instead 'find out enough to be able to defend ourselves, but go no further'.[89] This, he believed, had already been the implicit policy for three years. He felt that unless the CoS really needed this implicit policy shift explicitly clarified to them, then there should be no further need for political guidance. All Mallalieu thought should happen was that the CoS should confirm what they should have long suspected, which was that while the policy of acquiring chemical weapons would not be acted upon, it would not be entirely or directly dismissed either. Britain would keep the nerve agent option open, in secret, but not act upon it and not officially re-define what its actual policy was.

Unfortunately for the CoS, Roy Mason took a similar line to Mallalieu's assessment.[90] In January 1968, while acknowledging that defensive research and expenditure should continue due to there being little political risk in the area, Mason warned that:

[86] Ibid.
[87] Ibid.
[88] Ibid.
[89] Ibid.
[90] TNA, WO 32/21761, 'Chemical and Biological Warfare, and the future of MRE and CDEE Porton', Roy Mason to the Secretary of State for Defence, 18 January 1968; Balmer. Keeping Nothing Secret, 884.

> It is quite unrealistic to suppose that we would get any support from the Prime Minister, for the foreseeable future, to spend more money to develop a CW retaliatory capability, let alone to publicise our doing so ...[91]

Mason took a tough line on the CoS request to acquire and publicise a nerve agent capability. He flatly refused to endorse the 1963 policy. He also admonished the CoS, informing them that the only reason policy was not explicitly reversed was due to it being in line with NATO requirements for member states to possess a CW capability.[92] Instead of attempting to secure greater funding for CW, he advised them to step up efforts in reducing expenditure in the field. Although avoiding the same fate as BW research, which was undergoing a process of 'civilianisation', Mason cautioned that there would be closer scrutiny of CW research expenditure.[93] In January 1968, the state of affairs was thus made abundantly clear to the CoS and defence officials, when Mason explicitly clarified British CW policy: publicity for deterrence was out of the question, acting on the 1963 decision to acquire chemical weapons was unrealistic, and CW research was fortunate not to be 'civilianised'.[94]

Even with this harsh rebuke, Mason stopped short of seeking Cabinet approval for a reversal of policy and a nerve agent capability was not indefinitely ruled out. Instead, mirroring the ongoing drift and marginalisation of nerve agent policy, he decided that CW policy would not be reviewed for another two years, implementing a period of abeyance.[95] After these two years had elapsed, as he surely knew, there would likely be a new Government in power to deal with the nerve agent problem. Mason therefore not only slapped down the CoS request, but he refused to address it

[91] TNA, WO 32/21761, 'Chemical and Biological Warfare, and the future of MRE and CDEE Porton', Roy Mason to the Secretary of State for Defence, 18 January 1968.

[92] Ibid. From 1967, the need for flexible response had led Britain to endorse the requirement that while NATO countries should principally rely on nuclear and conventional forces for deterrence, chemical weapons should be acquired for use in retaliation. See: TNA, DEFE 5/176, 'Chemical and Biological Warfare', Note by the Secretary of the CoS, 15 December 1967, Annex A; Spiers. *Chemical Warfare*, 198–200. Also see: Spiers. *Chemical Warfare*, 198–199.

[93] TNA, WO 32/21761, 'Chemical and Biological Warfare, and the future of MRE and CDEE Porton', Roy Mason to the Secretary of State for Defence, 18 January 1968.

[94] Ibid; Carter and Balmer. Chemical and Biological Warfare and Defence, 299; Balmer. Keeping Nothing Secret, 884.

[95] TNA, WO 32/21761, 'Chemical and Biological Warfare, and the future of MRE and CDEE Porton', Roy Mason to the Secretary of State for Defence, 18 January 1968.

further, leaving the final controversial decision regarding nerve agent weapons for his successor to address.[96]

Following Mason's intervention, instead of expanding CW capabilities and research, significant cutbacks looked likely. In view of this adverse political reaction, the CoS willingly accepted Mason's suggested two-year abeyance period.[97] The Army, in particular, was happy to accept the idea. Officials were now fearful that if the question of a nerve agent capability actually reached the Cabinet level, then it would be rejected and the direction of CW policy in fact resoundingly reversed, rather than supported, by Wilson.[98] The Army's Deputy Chief of Staff believed that such a policy reversal would have been immensely damaging to British CW policy, as it would have led to the 'irrevocable closure' of Britain's nerve agent pilot plant at Nancekuke and the cutting of vital funding for research.[99] In readily accepting the proposed abeyance period of two years, the CoS avoided the direct and unwanted attention of more senior political figures such as Wilson and Healey, while ensuring the continuance of CW research and keeping the possible future acquisition of a nerve agent capability open.[100] This shrewd CoS acceptance of two years of abeyance did not go unnoticed. Cabinet Office officials pondered the reasons for defence officials accepting abeyance when it ran so starkly against their strong wishes for a nerve agent capability. They were highly sceptical as to the motives of the CoS.[101] The CoS had strongly pushed for nerve agent weapons then suddenly accepted a firm, if slightly ambiguous, no. Cabinet Office officials correctly surmised that the CoS was, in fact, simply biding its time in preparation for a future approach to acquire chemical weapons, when the political climate was more sympathetic.[102]

[96] Ibid.; TNA, DEFE 4/224, Chiefs of Staff Meeting, 30 January 1968, Confidential Annex; TNA, DEFE 13/557, Chief of the Defence Staff to Roy Mason, 31 January 1968.

[97] TNA, DEFE 4/224, Chiefs of Staff Meeting, 30 January 1968, Confidential Annex; TNA, DEFE 13/557, Chief of the Defence Staff to Roy Mason, 31 January 1968.

[98] TNA, WO 32/21761, 'CW/BW and the Future of MRE and CDEE', DCS (Army), 25 January 1968.

[99] Ibid.

[100] TNA, DEFE 4/224, Chiefs of Staff Meeting, 30 January 1968, Confidential Annex; TNA, DEFE 13/557, Chief of the Defence Staff to Roy Mason, 31 January 1968.

[101] TNA, PREM 13/3464, Burke L. to Michael Palliser, 2 February 1968.

[102] Ibid.

Importantly for the future of CW policy, alongside Ministerial deferrals, there was also a wider re-orientation in British global defence commitments. In January 1968, coinciding with Mason's rebuke, the CoS recognised that the Labour Government's policy of significantly reducing British commitments East of Suez would undermine a central role and reason for the acquiring of a CW capability.[103] Thorneycroft in 1963, Healey in 1965 and the CoS throughout the 1960s had all emphasised the need for a CW capability outside of Europe, to deter and retaliate against chemical weapons proliferation in the Third World. A key part of this was to avoid being embarrassed or humiliated abroad in limited conflicts. However, in January 1968, the CoS noted that 'changes in our Defence policy East of Suez had considerably reduced the need for this capability outside of Europe'.[104] With the change in policy and the subsequent reduction in British commitments outside Europe, there was now no need to deter or retaliate with chemical weapons if British forces were not active in the region to be threatened by them.[105] Crucially, this left an offensive nerve agent capability only one viable potential role: deterrence and retaliation in Europe. The CoS was set on acquiring a CW capability, and after years of justifying the CW requirement as necessary for deterrence and retaliation in the Third World, it simply re-orientated its focus exclusively on the European deterrent dimension.[106] This re-orientation would have a lasting impact, for at the very heart of a nerve agent capability exclusively lay deterrence and publicity.[107]

[103] TNA, DEFE 4/224, Chiefs of Staff Meeting, 30 January 1968, Confidential Annex; Baylis, John. 1986. "Greenwoodery" and British defence policy. *International Affairs*, 62:3, 457.

[104] TNA, DEFE 4/224, Chiefs of Staff Meeting, 30 January 1968, Confidential Annex.

[105] Ibid.

[106] Alongside failing to meet the 1963 Cabinet decision to acquire lethal nerve agents, the incapacitating agent requirement was also in limbo, but for a very different reason. Britain was to be without an operational incapacitating agent as no 'wholly satisfactory agent' had be discovered for the role. While the United States Army had settled on BZ, this was deemed too hazardous for a British incapacitating agent and for the British 'minimal risk' requirement. See: TNA, WO 32/2176, 'CW/BW and the Future of MRE and CDEE', DCS Army, 25 January 1968.

[107] TNA, DEFE 5/176, 'Chemical and Biological Warfare', Note by the Secretary of the CoS, 15 December 1967, Annex A; TNA, WO 32/2176, 'CW/BW and the Future of MRE and CDEE', DCS Army, 25 January 1968.

A 'Campaign of Criticism'[108]

Political aversion towards approving a nerve agent capability and the publicity dimension was seemingly justified, when, from 1968, Labour Cabinet members faced a 'campaign of criticism' over British CW policy.[109] Grass-roots movements, student activism, the press and MPs all became involved and actively engaged with British CW policy. Similarly to the spate of publicity in 1959–1960, much of this attention was not triggered by British actions. Britain did not send troops to fight in Vietnam, but United States actions there had a pervasive impact on British domestic politics by fuelling a wider public appreciation and awareness of CBW.[110] United States forces initially relied on herbicides to counter North Vietnamese guerrilla warfare tactics, but this had escalated to incapacitating riot-control agents, such as CS gas, which were believed to be militarily highly effective.[111] In 1965, when seeking authorisation from President Lyndon B. Johnson for their use in Vietnam, United States officials had noted that despite predicted international criticism approving their use was 'common sense'.[112] The perceived advantages in using CS gas against North Vietnamese forces led to the procurement of it increasing by a factor of 24 from 1965 to 1969. In total, it is estimated that United States forces used just under 7 million kilograms of the agent in the Vietnam War.[113]

[108] TNA, DEFE 13/997, 'Porton—Publicity', Note of a meeting held by the Minister of Defence for Equipment, 12 June 1968.

[109] Ibid.

[110] Vickers. Foreign and Defence Policy, 266.

[111] Ellison, D. Hank. 2011. *Chemical Warfare During the Vietnam War: Riot Control Agents in Combat*. London: Routledge, 6; Martin. Norms, Military Utility, and the Use/Non-use of Weapons, 321; Spelling. Driven to Tears, 704.

[112] National Security Archive, Chemical and Biological Warfare, Box 13, 'Tear Gas in Vietnam', Memorandum for the President, 23 September 1965.

[113] The Harvard Sussex Program Archive, 'The use of CS in Vietnam', Matthew Meselson, July 1969; Meselson, Matthew. 1970. CB weapons: the facts. In *The supreme folly: chemical & biological weapons*, ed. Frank Barnaby. London: National Council of Labour Colleges Publishing Society for the Womens International League for Peace and Freedom, 3–6; Martin. Norms, Military Utility, and the Use/Non-use of Weapons, 332; Walker. *Britain and Disarmament*, 45. Coinciding with this rapid escalation of CS use, the United States Army also developed advanced delivery methods for the CW agent. A select few included the E8 launcher, chemical dispersal bombs and a bagged dispenser, which were all designed for the rapid dispersal of CS on the battlefield. For further details, see: Ellison. *Chemical Warfare During the Vietnam War*, 134.

Due to the nature of the war, which was broadcast into the homes and lives of citizens, the extensive use of riot-control agents by United States forces did not remain hidden for long. Domestic protests against CBW subsequently gained traction in the United States, and in 1967, 5000 scientists signed a petition calling for a review of United States CBW policy.[114] Many commentators were fearful that escalation would continue, and that lethal nerve agents might be used. Reflecting these concerns at the time, Elinor Langer wrote that what is 'unthinkable at one moment may be policy the next'.[115] Such fears and domestic pressures triggered a public statement by President Johnson, who claimed that United States forces would only ever use lethal chemical weapons with the explicit approval of the Commander-in-Chief.[116]

Ignited by this growing attention on the CBW field in the United States, very public questions began being asked about British CW policy and its involvement in CBW research. This coincided with a much wider and gradual shift taking place in the 1960s, away from science and scientists receiving a reverential positive tone towards more critical coverage in the press.[117] On 27 May 1966, a very early indicator of grass-roots opposition could be seen when the Committee of 100, an anti-war protest group, demonstrated outside Porton Down.[118] In 1963, members of the group had also forced their way into Porton Down, leading to raids on their

[114] *The New York Times*, 'United States Review Urged on Chemical War', 15 February 1967, 1; Guillemin, Jeanne, Meselson, Matthew, Robinson, Julian Perry and Sims, Nicholas. 2017. Witness Seminar: Origins of the Biological Warfare Convention. In *Biological Threats in the twenty-first Century: The Politics, People, Science and Historical Roots*, ed. Filippa Lentzos. London: Imperial College Press, 381. The petition, and public awareness of the dangers of CW, was given greater impetus by the allegations of Egyptian forces were using chemical weapons, and nerve agents, in Yemen in January 1967. See: TNA, DEFE 55/418, 'Yemen: Gas Warfare', Technical Notes on Meetings Held at the Central Intelligence Agency, Langley, Virginia, USA on March 11th and 12th, 1968. Despite British and United States authorities disagreeing over whether nerve agents were in fact used, the accusations drew worldwide attention and refocused attention on CBW.

[115] Langer, Elinor. 1967. Chemical and Biological Warfare(II): The Weapons and the Policies. *Science*, 155:3760, 299–303. See also: National Security Archive, Chemical and Biological Warfare, Box 12, 'United States Policy on Chemical and Biological Weapons: An Historical Precis, James C. Kellogg, 29 March 1967.

[116] Ibid.

[117] Gregory and Miller. *Science in Public*, 45.

[118] Hammond and Carter. *From Biological Warfare to Healthcare*, 225; Balmer. *Secrecy and Science*, 96; Schmidt. *Secret Science*, 384–385.

headquarters by Special Branch.[119] Immediately after the 1966 protest however, the Committee of 100 lost a degree of momentum and publicity on CW policy lulled until 1967, when in line with events in Vietnam, publicity in the United States, and alleged use of chemical weapons by Egyptian forces in Yemen, CW again attracted mainstream attention in Britain.[120] In January 1967, Prime Minister Harold Wilson informed the Commons that evidence 'suggests pretty strongly that poison gas may have been used' in Yemen.[121] Within two months of these public accusations that Egypt had used lethal chemical weapons, public attention in the CW field gained further traction with a BBC radio programme called 'Make a Desolation and Call it Peace', which covered CBW.[122] In April, BBC Horizon then released a T.V. documentary titled 'The Shape of War to Come'. It introduced to the audience different types of chemical and biological weapons, their effects, their availability and possible arguments for their use.[123] These revelations represented an increasing and crucial flow of information into the public domain. Before these key interventions and events, postwar public protest movements had been largely starved of information on CW, with the slight exception of 1959–1960. From 1967, though, events were to develop rapidly and drastically.

After hearing and watching some of this BBC coverage of CBW, one attentive listener was spurred into action.[124] Elizabeth Sigmund, referred

[119] Ibid.

[120] Salvia, Joseph. 1967. Gas in Yemen. *Scientist and Citizen*, 9:7, 149; Terrill. The chemical warfare legacy of the Yemen war, 109–114; Shoham, Dany. 1998. Chemical and biological weapons in Egypt. *The Nonproliferation Review*, 5:3, 48; Spiers. *A History of Chemical and Biological Weapons*, 86. For further newspaper and article sources on CW and Yemen, see: The Harvard Sussex Program Archive, D6: Conflicts since 1960, Yemen.

[121] Hansard, House of Commons, 'Aden', 31 January 1967, Vol.750, cc242–4; Meselson, Matthew. 1969. The Yemen. In *CBW: Chemical and Biological Warfare*, ed. Steven Rose. Boston, MA: Beacon Press, 99–102; Spiers. *A History of Chemical and Biological Weapons*, 86. The government's handling of these CW allegations would later be branded as a 'a great moral issue'. See: Hansard, House of Commons, 'Aden', 19 June 1967, Vol.748, cc1126–262.

[122] Synge, A. 1969. Chemical and Biological Warfare. *New Blackfriars*, 50:585, 256; Sigmund. *Rage Against the Dying*, 7–8. Further publicity in the United States also came in the form of special articles designed to inform the public, for example, see: Sidel, Victor and Goldwyn, Robert. 1967. Chemical Weapons: What They Are and What They Do. *Scientist and Citizen*, 9:7, 141–148.

[123] Synopsis found at: BBC Two England, Horizon: The Shape of War to Come, 25 April 1967. Also see: Balmer. *Secrecy and Science*, 91.

[124] Sigmund. *Rage Against the Dying*, 7–8.

to as the 'toxic avenger', initiated her campaign to hold British CBW policy to account.[125] For Elizabeth Sigmund, the emerging coverage and revelations tapped into a deeper personal aversion to CBW. As she later revealed, her views were in part motivated by her grandfather, a Colonel, suffering badly from gas exposure during the First World War, seeing photos of long-lines of gas victims, and reading Wilfred Owen's *Dulce et Decorum est*.[126] The legacy of the First World War, and changing views over time, also coincides with Richard Price's point regarding the generational tradition of the non-use of chemical weapons, which has strengthened over time and become a core component of the moral discourse relating to CW and international norms.[127] Stories and personal accounts of past horrors strengthened and solidified moral opposition to CW, passing from generation to generation. In addition to the legacy of the First World War and personal accounts of her grandfather, Elizabeth Sigmund's views were also shaped by her caring nature, a strong desire to help others, and frustration at a perceived injustice being done. All of these can be seen clearly in her later campaigning to acquire compensation for volunteers at Porton Down. These victims had been deliberately exposed to CW agents.[128] It was the collection of personal experiences, aversion, and sense of injustice, which the BBC played a part bringing to the fore through its reports. In her endeavour to hold government policy to account, Sigmund wrote, in her own words, a 'clumsy, but effective' letter to the *Observer* in April 1968.[129] In this published letter, she called for a united, grass-roots movement to scrutinise and raise awareness of CBW research. She received numerous responses, ranging from seasoned campaigners in the Campaign for Nuclear Disarmament (CND), to newcomers, shocked by revelations and by the nature of CBW.[130]

[125] Justine Picardie, 'The Toxic Avenger', *The Independent*, 30 September 1995. Also see: Schmidt. *Secret Science*, 413. For further details of her activities, see: Sigmund. *Rage Against the Dying*, 1–128.

[126] See Chapter 1, 'Sylvia in Devon', by Elizabeth Sigmund in: Sigmund, Elizabeth and Crowther, Gail. *2014. Sylvia Plath in Devon: A Year's Turning*. United Kingdom: Fonthill Media; *Dreadnought South West: Rebellious Sounds*, 'Interview with Elizabeth Sigmund', undated.

[127] Price. *The Chemical Weapons Taboo*, 175.

[128] Conversation between the author and William Sigmund, 'Elizabeth Sigmund and CBW', 16 June 2020. Sadly, Elizabeth passed away in 2017. These insights are based on interviews, archival records, and Elizabeth's own publications.

[129] Sigmund. *Rage Against the Dying*, 7–8.

[130] Ibid.

With an upsurge of support, and reflective of the wider groundswell of opposition against CBW at the time, Sigmund formed the Anti-Chemical and Biological Warfare Group.[131] The group, she recalled, 'was not a structured, carefully planned strategy for ending CBW in Britain', but a 'haphazard and pragmatic use of every piece of information that we could obtain'.[132] The Group relied on her mobilising and creating an impressive network and array of contacts, from fellow campaigners to academics, members of the press, and politicians.[133] She attained and mobilised emerging information, drew on scientific and medical expertise to support the movement, and highly effectively communicated and conveyed key messages to a range of interested parties. A core aim of the Anti-Chemical and Biological Warfare Group was to change the course of government policy through applying pressure, both behind the scenes through personal contacts and publicly.[134]

Following the establishment of the group, Sigmund went on to play an integral part in campaigning and raising awareness of British CBW research in the crucial period of May-June 1968. When, just months after Mason's decision to push the nerve agent question into abeyance, a tide of publicity and scrutiny struck. During this pivotal period, Sigmund represented a wider trend in British public perceptions of CBW, when the public mood shifted decisively against Government policy and when it morphed into a 'campaign of criticism'.[135] For Porton Down and British CW policy, 1968 became the 'year of the barricades'.[136] Other grass-roots movements which played a significant role included the CND, the Christian CND, and the hard-line Southampton Peace Action Committee.[137] As shown by Brian Balmer, the Women's International League for Peace and Freedom also played a key role in spreading awareness of CBW issues and in educating the public and its members.[138] The controversial nature of CBW and the

[131] Ibid.

[132] Ibid, 95.

[133] Conversation between the author and William Sigmund, 'Elizabeth Sigmund and CBW', 16 June 2020.

[134] Ibid.

[135] TNA, DEFE 13/997, 'Porton—Publicity', Note of a meeting held by the Minister of Defence for Equipment, 12 June 1968.

[136] Schmidt. *Secret Science*, 384.

[137] TNA, DEFE 13/997, MA to CGS, 29 May 1968; Balmer. *Secrecy and* Science, 98.

[138] Balmer, Brian. 2021. "Science was digging its own grave": the Women's International League for Peace and Freedom and the campaign against chemical and biological warfare. *The Nonproliferation Review*, [online first], 1–19.

mixing of protest movements also drew the attention of the British Humanist Association.[139] On 1 June 1968, protest groups converged on Porton Down for a four-day protest.[140] The presence of the CND was particularly troubling for the Labour Government, as the CND had embedded itself within the Labour Party.[141] Of great alarm for Wilson's Government was a story published by the CND's magazine *Sanity* on 8 June 1968, which contained aerial pictures of Britain's top-secret CW research facility at Nancekuke.[142] As *Sanity* proclaimed, they had 'the pictures no-one dare print'.[143] The story represented a massive security breach. Pictures of Nancekuke were published and shown in major news outlets across the country.[144]

Coinciding with this growing awareness of CBW at the grass-roots level was a more general rise in student protests; by mid-1968, they had become widespread.[145] As noted by Peter Clarke, anoraks, placards and loud-hailers marked the end of Britain's exemption from the student-led protests which had been rampant in Europe and the United States.[146] In Britain, student protestors, courtesy of the efforts of the BBC and other grass-roots movements, were now far more informed. And, in response to revelations in the CBW field, they escalated their involvement and engagement. An early example of student activism had been seen on 7 May 1968, when nearly 200 students at Essex University disrupted a talk by the visiting Dr T. Inch, a Porton Down official.[147] In the ensuing confrontation, one student allegedly sidled up to Dr Inch and sprinkled some Coleman's mustard on his sleeve, saying 'here, have some mustard gas yourself'.[148] In response, Essex University suspended three students. This then provoked

[139] Balmer, Brian. 2015. An Open Day for Secrets: Biological Warfare, Steganography, and Hiding Things in Plain Sight. In *Absence in Science, Security and Policy: From Research Agendas to Global Strategy*, ed. Brian Rappert and Brian Balmer. Hampshire: Palgrave Macmillan, 39–40.

[140] Cookson, John and Nottingham, Judith. 1969. *A Survey of Chemical and Biological Warfare*. London: Monthly Review Press Classics, 155; Schmidt. *Secret Science*, 413.

[141] Healey. *The Time of My Life*, 241.

[142] Balmer. *Secrecy and Science*, 102.

[143] *The Times*, 'Ministry of Defence to check CND pictures', 8 June 1968. As referred to in: Balmer. *Secrecy and Science*, 102.

[144] Ibid.

[145] Clarke. *Hope and* Glory, 290.

[146] Ibid.

[147] *The Guardian*, 'Path barred by students', 8 May 1968, 1.

[148] Sigmund. *Rage Against the Dying*, 16.

an even larger student counter-reaction, with students deciding that they would establish a 'free university' and hold a continuous meeting to discuss CBW.[149]

Growing student activism in the field was not the only level at which universities were engaged in the CBW debate, prominent scientists became involved. In May-June 1968, 21 scientists wrote to Harold Wilson requesting greater transparency over British CBW research.[150] This coincided with the efforts of other professionals and academics. One key contribution came in the form of the Stockholm International Peace Research Institute (SIPRI), a think-tank established by the Swedish Government. Other crucial interventions, representing more grass-roots orientated attempts to draw attention and increase awareness of the CBW field, can be seen with the engagement of the British Society for Social Responsibility in Science and the Bernal Peace Library sponsoring a conference on CBW in London. The conference attracted radio, T.V., and newspaper coverage, and led to questions by MPs in Parliament.[151] Other key interventions also came with academic works, including edited books by Nigel Calder and Steven Rose, the latter based on discussions and findings from the London conference.[152] Emerging sources of information and analysis were importantly developing awareness of controversial CBW capabilities in the public sphere. Information, previously scarce, was mobilised quickly by interested and engaged observers, members of the media, and participants of protest movements.

In May and June 1968, interlinking, fuelling, and feeding off these grass-roots movements was reinvigorated interest in British CBW policy by the press and the BBC. On 6 June 1968, the BBC released provocative coverage of CBW with a documentary titled 'A Plague on Your Children'.[153] Controversially, the programme showed the effects of the incapacitating agent BZ on animals, and it alluded to the immense secrecy surrounding British activities in the CBW field.[154] The programme, rather frustratingly

[149] *The Guardian*, 'Essex students decide on a free university', 14 May 1968, 1

[150] For some examples of the letters see: TNA, DEFE 13/997, letters to Harold Wilson dated 29 May 1968.

[151] Synge. Chemical and Biological Warfare, 256.

[152] Calder, Nigel (ed.). 1968. *Unless Peace Comes: A Scientific Forecast of New Weapons*. New York, N.Y.: Viking Press; Rose, Steven (ed.). 1968. *CBW: Conference on Chemical and Biological Warfare*, London: Harrap Publishing.

[153] Ibid, 9; Balmer. *Secrecy and Science*, 91.

[154] Sigmund. *Rage Against the Dying*, 9, 11.

for British officials, included a quote from a United States General referring to how nerve agents were useful for achieving a 'clean kill'.[155] Adrian Malone, the programme's producer, contended that the public reaction had been one of 'overwhelming horror and revulsion', but the Government did not find the exposure overly negative.[156] What some government officials did mind, though, was the continued fixation of the BBC on British CBW policy, and in particular the perceived bias of some BBC presenters. Increasingly alternative and critical views were finding a receptive audience, gaining traction and mainstream attention.

Shortly after, the BBC again refocused attention on the Government's CBW policy in 'Points of View', when host Robert Robinson discussed public responses to 'A Plague on Your Children'.[157] During the programme, Robinson revealed negative and emotional public responses to the information they had learnt of British CBW activities. Some British officials, in particular the Chief of Public Relations at the MoD, John Peters, were outraged. They condemned the coverage as overly negative and selective.[158] Although angered by the perceived bias of BBC reporters against British CBW research, John Morris, the then Minister of Defence for Equipment and thus responsible for Porton Down, wanted to avoid a public confrontation. He feared that if concerns were raised with the Chairman of the BBC, then the situation may escalate to the benefit of Robinson by giving him and the programme more attention.[159] Morris' private office urged officials to stay away from direct confrontation. They wanted to avoid 'a reaction in favour of the wretched Mr. Robinson—who is after all a not very important performer in a not very important programme'.[160] Ultimately, it was the continued probing questions of BBC presenters and the perceived outspoken nature of them which irked

[155] Following the programme was a live panel discussion, and despite officials wanting to get Zuckerman in to defend and rationalise CBW policy, his participation was thought 'extremely unlikely'. Instead the Government was represented by Dr. Eric Haddon, the retired head of CDEE in the programme. For further details, see: Balmer. *Secrecy and Science*, 100.

[156] For more details, see: Balmer. *Secrecy and Science*, 100–102.

[157] Ibid, 103.

[158] TNA, DEFE 13/997, 'Criticisms of Porton—BBC T/V Programme "Points of View"', John Peters, 14 June 1968.

[159] TNA, DEFE 13/997, '"Points of View" and Porton', John Groves, 24 June 1968.

[160] Ibid; Balmer. *Secrecy and Science*, 105.

government officials, who were accustomed to relative quiet in this top-secret area. Publicity was troublesome, but not yet disastrous.

British defence officials did imagine more constructive ways of dealing with this negative coverage, showing a degree of constructive spin in countering the negative narrative.[161] Government practice, as supported by Wilson, Healey and Morris, was to operate in a slightly grey area; to be as open as possible without breaching any security considerations.[162] A primary aim of these measures was to mitigate 'lively public interest' and criticisms of British CBW policy through minor concessions.[163] As noted by Brian Balmer, the protests and attention focused as much on the secret nature of CBW research programmes as the weapons themselves.[164] In a remarkably similar vein to 1960, government officials thus again selectively deployed limited disclosures in an attempt to gain some control over the flow of information. As Morris' private office noted, their target audience for this was not the 'irrational ones who follow the latest protest fashion' or the ones attempting to damage British defence policy, likely referring to the growing protest movements, it was 'the reasonable and responsible ones'.[165] It was thought that this latter group could be reached through scientific publications and further transparency outside of Parliament, alongside patient repetition inside of Parliament.[166]

In order to combat the negative coverage outside of Parliament, officials agreed to release unclassified Porton publications in British scientific journals.[167] This was an attempt to shift the debate to factual scientific considerations and draw attention to the defensive nature of British research. All releases were carefully vetted beforehand, and senior British

[161] TNA, DEFE 13/997, 'Porton—Publicity', Note of a meeting held by the Minister of Defence for Equipment, 12 June 1968.

[162] TNA, WO 32/21761, 'Chemical and Biological Warfare, and the future of MRE and CDEE Porton', Roy Mason to the Secretary of State for Defence, 18 January 1968; TNA, DEFE 13/997, 'PR Policy for Porton', T. C. James, 10 June 1968; TNA, DEFE 13/997, 'Porton—Publicity', Note of a meeting held by Minister (E), 12 June 1968.

[163] TNA, DEFE 13/997, 'PR Policy for Porton', T. C. James, 10 June 1968.

[164] Balmer. An Open Day for Secrets, 39–40.

[165] TNA, DEFE 13/997, 'Porton—Publicity', Note of a meeting held by Minister (E), 12 June 1968; Balmer. *Secrecy and Science*, 111, 114.

[166] TNA, DEFE 13/997, 'Criticisms of Porton—BBC T/V Programme "Points of View"', John Peters, 14 June 1968.

[167] TNA, DEFE 13/997, 'Porton—Publicity', Note of a meeting held by Minister (E), 12 June 1968.

officials like Solly Zuckerman were consulted.[168] In terms of reaching the wider public, select reporters were invited for tours of Porton Down, which produced favourable responses in *The Times*, *The Telegraph* and *The Guardian*.[169] These tours, however, largely focused on BW rather than CW research. After reporters had gone on the tour and been briefed, their stories were often 'unsensational and objective'; precisely the line the Government desired.[170] In essence, defence officials and Morris attempted to boil controversial and sensationalist stories down to a bland, scientific discussion devoid of drama.

Inside Parliament, establishing a united political position on CBW research proved difficult. Numerous backbench labour MPs had joined in with the damning public chorus and sided with the grass-roots movements, student protests and negative media coverage. For many Labour MPs, CBW research had long been a divisive issue, with some of them already having expressed grave concerns in 1959–1960. In 1968 however, their views coincided with the substantial public attention afforded to CBW, giving them a groundswell of support to confront and question the Government's CBW policy. Unfortunately for Prime Minister Harold Wilson, his Government's CBW research thus directly set him at odds with this surge of publicity and with many of his own supporters and MPs.

In response to public unease and in order to provide a political review of British research into science and technology, a Parliamentary Select Committee on Science and Technology had been formed, holding meetings and interviews in mid-1968. Running concurrently with the appearance of student CBW protests and press coverage, the Select Committee consisted of 14 MPs and part of its remit was to review and scrutinise aspects of British scientific defence research, including CW.[171] The findings and statements from the review were to be kept confidential. The 'witnesses' were allowed to clarify and proof their statements before final circulation. One individual to be interviewed on British scientific research

[168] For an example of later detailed consultation see: PREM 13/3465, A. R. Jaffray to P. J. Moon, 13 May 1970.

[169] TNA, DEFE 13/997, 'Porton—Publicity', Note of a meeting held by Minister (E), 12 June 1968.

[170] TNA, DEFE 13/997, 'PR Policy for Porton', T. C. James, 10 June 1968. One paper in which Government officials deliberately targeted was *The Sun*, as it had taken a tough and sensationalist stance on British CBW research.

[171] Galbraith, Russell. 2000. *Inside Outside: The Man They Can't Gag*. Edinburgh: Mainstream Publishing, 101.

was Solly Zuckerman. While Zuckerman was not afraid of confrontation and tended to speak his mind, he ensured the Select Committee had little to work with when he made over 130 changes to his original manuscript.[172] Zuckerman informed the Chair of the Select Committee, Arthur Palmer MP, that the steno-typist must have 'misheard' him when he was being interviewed and that he did not want his views to be distorted.[173]

While Zuckerman managed to stifle any controversy, the Government's fortunes did not hold. When the Select Committee turned to review CW policy in May 1968, one frustrated Labour MP intervened before any confidential manuscript could be proofed or edited. Tam Dalyell MP, an ardent critic and member of the Select Committee, wanted to bring greater transparency to British CBW research. He was irritated when he realised that officials could 'side-line' material from their reports.[174] To pre-empt this editing process, and to reveal more information on British CW policy, Dalyell leaked an un-proofed confidential manuscript on CBW to *The Observer*.[175] Containing confidential evidence from the top tier of British CBW experts, the leak included information supplied by the Director and Deputy-Director of CW research at Porton Down.[176] Unedited witness statements revealed the strengths and costs of CW research, the names of scientists, and the universities collaborating with Porton Down. Perhaps most damningly of all from the perspective of defence officials, it confirmed that Britain possessed no nerve agent stocks and no retaliatory CW capability.[177] A few days after his leak, and already knowing the answer, Dalyell asked in the Commons for a list of universities involved in British CBW research; it was reluctantly supplied.[178]

Though Dalyell focused on the role of universities, perhaps to capitalise on existing ill-will and student activism in the area, government officials

[172] TNA, CAB 168/169, Solly Zuckerman to Arthur Palmer, 14 November 1968.

[173] Ibid.

[174] For a more detailed account of the Dalyell incident see: Galbraith. *Inside Outside*, 101–120; Balmer. *Secrecy and Science*, 95–96.

[175] Laurence Marks and Joanna Slaughter, 'Biological warfare: Dons named', *The Observer*, 26 May 1968, 1–2.

[176] Although the former Director of CDEE, Dr. Haddon, had revealed that Britain possessed no stocks on the 6 June 1968 BBC show 'A Plague on Your Children', the Dalyell revelation was more politically concerning.

[177] Laurence Marks and Joanna Slaughter, 'Biological warfare: Dons named', *The Observer*, 26 May 1968, 1–2.

[178] Hansard, House of Commons, 'Chemical and Microbiological Warfare', 29 May 1968, Vol.765, cc226–7W.

remained divided on how best to deal with the leak and particularly on how to handle the public revelation that Britain possessed no chemical weapons capability or deterrent. The direction of policy again became an area of contention and disagreement between officials in the Foreign Office and the MoD.[179] Even with the leak, the MoD wished to only confirm that Britain possessed no chemical weapons as a last resort. A key reason why MoD officials stuck to maintaining attempts at secrecy, even though the story had already broken, was that they took comfort in the fact that 'potential enemy countries will tend to believe that this is a partial deception by the British Government'.[180] Along this line, with government officials resisting being drawn in and with the release of secret information originating from a Labour MP on a Parliamentary Committee, an attempt at deception could perhaps have been seen as a possibility to an external observer. Defence officials were also aware that one of their biggest secrets had remained hidden: the 1963 Cabinet Defence Committee decision, which the CoS still wanted to enact, to acquire a lethal nerve agent capability.[181] As defence officials surely realised, a positive public response to the confirmation and news that Britain did not possess chemical weapons would also have made their future acquisition that much harder.

Foreign Office officials, on the other hand, believed that there would be every advantage to Britain publicising its lack of chemical weapons, especially given the ongoing 'public agitation'.[182] One key supporter of this stance was Fred Mulley, the Minister for Disarmament, who was keen on publicising and capitalising on the revelation.[183] Mulley, at this time, was also busy forming and submitting a working paper on microbiological warfare disarmament at Geneva, the location of the UN's Eighteen-Nation Committee on Disarmament.[184] In the CW field, the Foreign Office and Mulley both believed that Dalyell's leak was, in essence, a PR coup which

[179] Balmer. Keeping Nothing Secret, 887–890. The Foreign Office would become the Foreign and Commonwealth Office (FCO) in October 1968.

[180] TNA, DEFE 13/998, 'Select Committee on Science & Technology—Porton Evidence', L. Dunnett, 11 June 1968.

[181] Ibid; TNA, DEFE 13/997, 'Porton—Publicity', Note of a meeting held by the Minister of Defence for Equipment, 12 June 1968.

[182] TNA, DEFE 13/998, 'Stockpiles of Chemical Weapons, R. C. Hope-Jones to K. T. Nash, 9 July 1968.

[183] TNA, DEFE 13/999, 'Stockpiles of Chemical Weapons', Fred Mulley to Denis Healey, 14 August 1968.

[184] Spelling. Driven to Tears, 706

needed building upon.[185] Here Britain could distinguish itself from the superpowers, claim the moral high ground, and at the same time take the sting and momentum out of ongoing public criticism. Yet despite these potential political advantages, a strange middle path was again taken, partially revealing Britain's lack of CW capabilities. Morris confirmed in Parliament that 'the only stocks of nerve gases currently held are small quantities necessary for the development and testing of defensive measures'.[186] But, this statement seemingly had little impact. The Government neither completely covered up CW capabilities nor trumpeted or benefitted from the revelation and the positive aspects of disposing of nerve agent weapons.[187] Attempts to muddle through under intense scrutiny, characterised by limited concessions and hoping interest would die down after being starved of further controversial information, continued.

Government resistance and ability to stay the course was, though, slowly eroding. One of the most frustrating political facets of the Dalyell saga was not the actual content of the leak, but that for Wilson and the Labour Party the story did not disappear, it lingered in national newspapers for almost two months. The sense of drama over the whole situation was given greater weight by parliamentary procedure. When MPs, including Wilson, voted to subject Dalyell to a review by the Committee of Privileges.[188] The Committee subsequently found Dalyell guilty of breaching parliamentary privilege through deliberately leaking confidential information, and Wilson again voted in favour of punishing the Labour backbencher.[189] The Prime Minister also implemented a three-line whip on the vote, showing his strong feelings on the matter.[190] For Wilson, this tough line had two clear political benefits. Firstly, as he surely planned, a three-line whip would deter many other Labour MPs from siding with their colleague against the Government. Secondly, Wilson not only needed to prevent others from joining Dalyell, but he also had to send a clear

[185] TNA, DEFE 13/999, 'Stockpiles of Chemical Weapons', Fred Mulley to Denis Healey, 14 August 1968.

[186] Hansard, House of Commons, 'Nerve Gases', 12 June 1968, Vol.766, c221.

[187] Hansard, House of Commons, 'Chemical and Biological Weapons (Information)', 24 June 1968, Vol.767, c26W; Balmer. Keeping Nothing Secret, 887–890.

[188] Galbraith. *Inside Outside*, 109, 115.

[189] Hansard, House of Commons, 'Committee of Privileges (Mr. Tam Dalyell)', 24 July 1968, Vol.769, cc587–666; Wilkinson. *Secrecy and the Media*, 356.

[190] Galbraith. *Inside Outside*, 115.

message that there were repercussions to taking Dalyell's hard-line against Government policy.

Notably, even with this tough line, there were still numerous backbench Labour MPs who voted against disciplinary action on 24 July 1968, with the vote passing 244–52.[191] The 52 noes, which on the surface seems low, can actually be seen as unusually high given that Edward Heath, the Leader of the Conservative Opposition, voted with Wilson and against Dalyell. In the preceding debate, numerous Labour backbenchers voiced their support for Dalyell, and in one instance Labour MP Emrys Hughes used the opportunity to directly attack Wilson's CBW policies.[192] Although Wilson won the vote and secured the parliamentary condemnation of Dalyell for the 'gross contempt of the House', his party had not escaped unscathed, and CBW continued to divide the Labour Party.[193]

In September 1968, after a brief respite in publicity and coverage of CBW, the CND's *Sanity* magazine again came to the fore. It reported on 'Britain's first nerve gas victim'.[194] Following an earlier story by the *Observer*, the magazine reported the allegations that Flt. Lt. William Cockayne had been exposed to lethal nerve agents when he worked at Porton Down in 1954.[195] Cockayne had made the connection between his long-term physical and psychological health issues and his work at Porton Down after watching BBC coverage of British CBW research.[196] In addition to the *Observer* and *Sanity* covering the story, *Private Eye* reported on it, and Elizabeth Sigmund attempted to secure compensation for Cockayne.[197] Cockayne also had the support of his local Labour MP, James Dickens, who had voted in support of Dalyell in Parliament two months before.[198] Ultimately, though, the situation was dealt with comparatively

[191] Hansard, House of Commons, 'Committee of Privileges (Mr. Tam Dalyell)', 24 July 1968, Vol.769, cc587–666.

[192] Ibid.

[193] Ibid.

[194] Petherbridge, John. 1968. Britain's first nerve gas victim. *Sanity: the voice of CND*, September, 1–2.

[195] TNA, DEFE 13/999, 'Flight Lieutenant Cockayne', Private Office (Minister of Defence for Equipment) to D. Security (Army), 26 September 1968; Petherbridge. Britain's first nerve gas victim, 1–2.

[196] Schmidt. *Secret Science*, 389.

[197] TNA, DEFE 13/999, 'Flight Lieutenant Cockayne, Private Office (Minister of Defence for Equipment) to D. Security (Army), 26 September 1968.

[198] Foot, Monica. 1968. Nerve Gas Victim. *Freedom: Anarchist Weekly*, 29:26, 24.

quietly, and the MoD stonewalled any claims for compensation.[199] Despite it representing a breach of security, with the articles over Cockayne potentially breaking the law, Morris again showed his publicity savvy, if morally questionable, approach. He concluded that 'any action taken on security grounds would merely give the articles in question, and Cockayne, much more public interest than they at present enjoy'.[200]

Caution and the desire to avoid any negative publicity whatsoever impacted British CW research. In October 1968, serious doubts were raised over the possibility of British officials conducting trials on even CW defensive measures. Heightened public interest had direct consequences. With all the public attention on CW policy, Healey requested that Morris reconsider the timing of Operation Fearless, the testing of CW defences which included the use of an aerial spray.[201] Although all the trials were to take place out at sea, out of sight of land and passing shipping, the sheer volume of public criticism and attention afforded to CW policy entailed a serious re-think. In reviewing the situation, Morris decided that the defensive trials were to be postponed 'in order to gauge the strength of press reaction and public opinion'.[202]

Coverage of British CW policy did ease slightly after the rush of mid-1968, but it did not disappear altogether. There emerged a steady trickle of parliamentary questions on the links between Britain and the United States in the CW field, chemical and biological weapons disarmament and wider questions on environmental pollution, animal testing, and safety in CW policy.

In June 1969, while public and parliamentary discussions continued at a slower pace, the constant pressure and attention afforded to CW policy by campaigners, the BBC, newspapers and MPs did eventually draw more concessions from the Government.[203] On 18 June 1969, in response to questions reflecting public concerns over the transportation and storage of

[199] Schmidt. *Secret Science*, 390.

[200] TNA, DEFE 13/999, 'Flight Lieutenant Cockayne, Private Office (Minister of Defence for Equipment) to D. Security (Army), 26 September 1968.

[201] TNA, DEFE 13/999, 'FEARLESS—CW Trials', Roy Mason to U.S. of S. (RN), 14 October 1968.

[202] Ibid.

[203] Another noteworthy area worth further exploring is the emotional and public impact of animal testing for CBW research. For further details on the internal debates and strife, along with the different vested interests in publicity, see: Balmer. Keeping Nothing Secret, 884–891.

nerve agents, Morris confirmed in Parliament something which Dalyell had leaked a year before and that defence officials wished kept quiet. Morris categorically informed Parliament that 'there are no poisonous gases and substances in the United Kingdom for use in biological and chemical warfare'.[204] Surprisingly, this concession again did not appear to draw the shock-waves that might have been expected. Instead, despite this categorical statement and major concession, public criticism continued.

In July 1969, Dr Steven Rose, who regularly drew public attention to Government CBW policy and was driven by scientific, moral, and political opposition to CBW developments, highlighted the pollution effects of CW and he alleged that in Britain VX was transported by road.[205] Rose's article in *The Times* also referenced ongoing incidents in the United States, with the Skull Valley incident—a trial resulting in the accidental poisoning of thousands of sheep, increasing CS gas use in Vietnam, and the decision to dispose of 27,000 tons of surplus CW agents.[206] After Rose's article, questions again arose in Parliament, with Peter Bessell MP, fearful of the risks to his local constituents, querying whether nerve agents would be transported through his constituency.[207] These concerns over safety and of the nerve agent threat lingered in the consciousness of the public and the press, culminating in Morris again reiterating in August 1969 that Britain held no stockpiles of chemical weapons.[208] Adding to Morris' admission, the Government's Army Minister, Ivor Richard MP, responded to continued interest in December. He also reiterated that 'elaborate safety mea-

[204] Hansard, House of Commons, 'Biological and Chemical Warfare', 18 June 1969, Vol.785, c107W.

[205] Steven Rose, 'The heavy price of advancing military technology', *The Times*, 4 July 1969, 9. For more on Rose's activities, see: Rose, Steven and Lentzos, Filippa. 2016. Interview: Unconventional Weapons and Activist Scientists. In *Biological Threats in the twenty-first Century: The Politics, People, Science and Historical Roots*, ed. Filippa Lentzos. London: Imperial College Press).

[206] James Reston, 'The Pentagon: Nerve Gas in Okinawa', *The New York Times*, 20 July 1969, E12 (Editorial); Tucker. *War of Nerves*, 203–212; Allen, Steven. 2007. An Analysis of Factors Leading to a U.S. Renunciation of Biological Weapons. PhD diss., George Mason University, 499–501.

[207] Hansard, House of Commons, 'Chemical and Biological Warfare', 21 July 1969, Vol.787, cc1230–11,230; Hansard, House of Commons, 'Adjournment', 25 July 1969, Vol.787, cc2234–80. Bessell was also referencing the sarin incident at Okinawa, see: Hunt, Richard. 2015. *Melvin Laird and the foundation of the post-Vietnam Military, 1969–1973*, Vol.7. Secretaries of Defense Historical Studies. Washington, DC: Office of the Secretary of Defense, 336–337.

[208] *The Times*, 'Assurance given on nerve gas transport', 13 August 1969, 2.

sures were in place' and that Britain did not possess stockpiles of chemical weapons.[209] After Dalyell's leak and multiple Ministerial interventions stating that Britain did not possess CW stocks, any pretence or attempts at deception were severely undermined. Defence officials had previously taken some comfort in there being ambiguity over Britain's CW capabilities in 1968, but by late 1969 this position was threadbare.[210]

Remarkably, even with all the negative publicity and scrutiny, which had led some defence officials to bemoan the requirement for 'these politically difficult devices', the CoS still desired a British nerve agent capability.[211] In January 1968 the CoS had tactfully agreed on the need for a two-year deferral. In early 1969 they initially wished to be ready for Ministerial review in January 1970, which would be when Mason's period of two-year abeyance would end.[212] After again reviewing policy, and despite the United States proving unforthcoming in providing 'firm assurances' that they would use their CW stocks on behalf of British forces, the CoS wisely decided against pursuing Ministerial approval.[213] This was not a result of any military argument, for the need for chemical weapons for deterrence and retaliation remained, especially with the United States proving non-committal to British requests. Rather, it was due to publicity and political pressures leading to the CoS self-imposing an extension to the two-year abeyance period. As Sir Ian Hogg, the Vice-Chief of the General Staff, astutely observed:

> The Government will be starting to occupy itself with electoral thoughts. Any Government would be most reluctant to grasp a disagreeable nettle in any case at such a time ... The time to tackle this problem is with a Government having a good majority and a long lease of life.[214]

[209] *The Telegraph*, 'Nerve gas base safe, says Minister', 23 December 1969, 1.

[210] A key intervention and event was also seen when President Richard Nixon renounced the offensive use of chemical and biological weapons in November 1969, which has been included in Chapter 7, with it playing an important part in the rise of disarmament.

[211] TNA, WO 32/21761, 'Chemical Warfare', W.B. Lord, 9 September 1968; TNA, WO 32/21761, 'United Kingdom Retaliatory CW Capability', S. M. Abraham, 6 February 1969; TNA, WO 32/21761, 'United Kingdom Retaliatory CW Capability', Vice-Chief of the General Staff, 4 March 1969.

[212] Ibid.

[213] TNA, WO 32/21761, 'United Kingdom Retaliatory CW Capability', S. M. Abraham, 6 February 1969.

[214] TNA, WO 32/21761, 'United Kingdom Retaliatory CW Capability', Vice-Chief of the General Staff, 4 March 1969.

British defence officials thus again read the political situation, let it dictate the drift of CW policy and temporarily override their desire to possess nerve agent weapons. The unpopularity of chemical weapons now meant that defence officials needed to be even more careful when seeking political approval. As Hogg clarified, any political decision to acquire chemical weapons would 'likely cause a row' and cost votes, whereas a decision to rule out the acquisition of chemical weapons altogether would likely win votes.[215] Due to the negative publicity surrounding chemical weapons and the politically costly toll of approving the acquisition of nerve agents, the CoS maintained that the best course of action was to continue to defer.[216] While the campaign of criticism in the late 1960s was a substantial dent to the CoS request for nerve agent weapons, it was by no means the end of the road for defence officials; nerve agent acquisition would again resurface in the early 1970s.[217] The decision to wait until after the election and to delay was not a change in policy, but a deliberate decision to wait for a more sympathetic political climate in order to acquire chemical weapons, displaying full recognition of the difficulties involved in reaching the desired policy outcome. As ensuing events were to prove, this CoS decision to quietly continue the period of abeyance was, for defence officials and supporters of an offensive CW policy at least, the right call.

Throughout the 1960s, after Healey's deferrals and Mason's intervention, it was increasingly evident that politically, publicity for the purpose of deterrence was a red-line and an insurmountable obstacle. For the Labour Government, acquiring nerve agent weapons was predominantly a political, not a military decision, and a highly unwelcome one at that. Even with the best efforts of the CoS, and despite the positive outcome of the 1963 Cabinet Defence Committee decision, CW policy was effectively 'pigeonholed' for almost five years. Wilson, after coming into power in 1964, had not been able to contemplate following through with the decision to acquire nerve agent weapons. Yet, instead of outwardly reversing policy, he settled for drift and abeyance as CW policy gradually slipped down the

[215] Ibid.

[216] Ibid.

[217] For examples of later considerations, see: TNA, WO 32/21761, 'CW—Possible UK Retaliatory Capability', P. J. Bayne to CDS, CNS, CGS and CAS, 18 February 1970; TNA, DEFE 13/1000, 'Chemical Warfare Policy', Note by the Secretary of the Chiefs of Staff Committee, 27 July 1972, Annex; TNA, DEFE 4/272, CoS 24th Meeting, 19 July 1972, Confidential Annex; TNA, DEFE 5/193, 'Chemical Warfare Policy', Chiefs of Staff Committee, 27 July 1972, Annex A.

list of Ministerial priorities. British CW policy was ignored by Wilson, disowned by Healey, and delegated to Mason, who ultimately left the problem for his successor. Despite not pro-actively changing nerve agent policy, they still had a substantial impact on it.

From 1967, compounding political aversion to publicity, British CW policy was thrust into the limelight on an unprecedented scale. This unintended publicity, in part brought about by United States actions in Vietnam, took the form of grass-roots movements, student activism, concerned experts, media coverage and parliamentary scrutiny. This profusion of sources forced the Government to repeatedly concede that Britain possessed no nerve agent capability, and it revealed proactive public and parliamentary aversion towards chemical weapons. The experience of publicity in a formerly immensely secret area occurred during a period of policy abeyance. It would have a lasting influence and impact. Adverse publicity confirmed that political concerns and fears over publicising a nerve agent capability, for the purposes of deterrence, were well-founded. It also served as a stark warning for defence officials, bringing into doubt the viability of publicity for the purpose of deterrence, especially in view of a likely hostile public reception. Even after 1969 and despite numerous concessions, publicity would not fade. Another storm was about to strike. When, just six months before a general election, the Wilson Government decided publicly to clarify its interpretation of the 1925 Geneva Protocol.

CHAPTER 7

From the CS Debacle to the Rise of CW Disarmament, 1970–1976

It is now 50 years since the First World War, but men still shudder at the idea of gas warfare.
—Frank Hooley MP (Hansard, House of Commons, 'C.S. Gas', 19 March 1970, Vol.798, cc807–21,807)

European and superpower détente, continuing in the early 1970s, built upon the momentum established by superpower rapprochement in the late 1960s. Accompanying these developments, and very much a part of them, were disarmament negotiations. Key markers of changing attitudes and of reduced tensions were seen with strategic arms limitations talks, leading to the SALT I agreement and Anti-Ballistic Missile Treaty in May 1972, and the summit meetings between Soviet Premier Leonid Brezhnev and President of the United States Richard Nixon.[1] In Europe, epitomising signs of greater cooperation, was West German Chancellor Willy Brandt and his policy of *Neue Ostpolitik*.[2] This period of détente between the superpowers and their respective European blocs culminated in 1975,

[1] Spohr, Kristina and Reynolds, David. 2016. *Transcending the Cold War: Summits, Statecraft, and the Dissolution of Bipolarity in Europe, 1970–1990*. Oxford: Oxford University Press, 4–5, 8–9.

[2] Kemp-Welch, A. 2008. *Poland under Communism: A Cold War History*. Cambridge: Cambridge University Press, 174–176.

W. King, *Nerve Agents in Postwar Britain*, Britain and the World,
https://doi.org/10.1007/978-3-030-70474-2_7

when the Conference on Security and Cooperation in Europe met and produced the Helsinki accords.[3] The accords were to re-shape the nature of the Cold War with the inclusion and recognition of human rights as an aspect of international security.[4] Détente as a whole was not, however, a smooth and steady affair only heading in one direction. Conflicts and confrontations were still present. Fighting continued in Vietnam, the Indo-Pakistani War of 1971, the Arab-Israeli War of 1973, and the Angolan Civil War in 1975–1976.[5] In all cases, superpower rivalry often added an uncertain dimension to détente, as while securing some key milestones in disarmament the question regarding its long-term viability and sustainability remained.

For Britain, navigating the changing pace of détente, both in Europe and globally, was problematic. This was especially so as, alongside a shifting international climate, British domestic politics was in a period of flux. In June 1970, Conservative victory in the general election led to Edward Heath replacing Harold Wilson as Prime Minister, and Lord Carrington taking over from Denis Healey as Secretary of State for Defence.[6] After four years of Conservative rule, Labour would return in February 1974, Wilson again at the helm but with a minority Labour Government, which preceded another General Election in October and a Wilson Government with a slender majority.[7] Reflective of these significant and comparatively fast-paced changes in political leadership, British defence policy was also in transition from 1970.[8] For British defence officials, the coming of détente and the emergence of disarmament talks, particularly in the area of biological weapons, brought with it both new opportunities and new challenges. A key marker of change, and of British officials pressing and playing

[3] Davy, Richard. 2009. Helsinki myths: setting the record straight on the Final Act of the CSCE, 1975. *Cold War History*, 9:1, 1–2.

[4] Hanhimäki, Jussi. 2003. "They can write it in Swahili": Kissinger, the Soviets, and the Helsinki accords, 1973–75. *Journal of Transatlantic Studies*, 1:1, 37–38.

[5] Ibid, 152; Schulzinger, Robert. 2010. Détente in the Nixon-Ford years, 1969–1976. In *The Cold War, Volume 2: Crises and Détente*, ed. Melvyn Leffler and Odd Arne Westad. Cambridge: Cambridge University Press, 385.

[6] Robb, Thomas. 2014. *A strained partnership? United States-UK relations in the era of détente, 1969–77*. Manchester: Manchester University Press, 2.

[7] Young. *Cold War Europe*, 123.

[8] Baylis. *British Defence Policy*, 33–34, 153.

a leading role in global disarmament, was the Biological Weapons Convention (BWC) of April 1972. The BWC built on a British proposal put forward in 1969, and established precedent and practices which would eventually spread to, and dominate, British considerations of CW policy.[9]

Facilitated by détente, and influenced by achievements in BW disarmament, chemical weapons disarmament negotiations culminated in Britain producing and submitting to the UN a draft Chemical Weapons Convention in 1976. Before the ascendency of disarmament, though, this chapter will explore a politically damaging encounter British politicians had with the CBW field in early 1970, when Wilson thought it necessary to publicly qualify Britain's interpretation of the widely accepted 1925 Geneva Protocol. Disagreements over interpretations of the Geneva Protocol, which had also taken place between Foreign Office and defence officials in secret during the 1950s, were to be revealed publicly and with damning results. This chapter will then move on to analyse the rise of disarmament and how this irrevocably fuelled and shaped British CW policy, as well as international moves towards achieving successful CBW disarmament. A crucial zenith in this transition to disarmament can be seen in 1976, with the culmination of years of disagreements and debates in Britain over the acquisition and role of nerve agent weapons. This is the year when acquisition and deterrence finally gave way to disarmament.

The CS Debacle

In December 1969, adding to the previous waves of negative publicity which had swept across the CW field since 1967, a substantial and politically costly CBW issue arose for Harold Wilson's Government.[10] On this occasion, unlike with previous cases seen in 1959–1960 and 1967–1968, negative publicity and political fallout were entirely self-inflicted. An attempt by the Government at openness in the CW field drew widespread

[9] Freedman, Lawrence. 1977. *U.S. Intelligence and the Soviet Strategic Threat*. London: The Macmillan Press, 153–168; Burr, William and Rosenberg, David. 2010. Nuclear competition in an era of stalemate, 1963–75. In *The Cambridge History of the Cold War, Volume 3: Endings*, ed. Melvyn Leffler and Odd Arne Westad. Cambridge: Cambridge University Press, 108.

[10] Walker. *Britain and Disarmament*, 31.

international criticism. A controversial debate over riot-control agents and the Geneva Protocol was deliberately thrust into the public spotlight.[11]

The disagreement, just six months before the June 1970 general election, centred on whether riot-control agents such as CS should be considered under the 1925 Geneva Protocol as illegal for use in war.[12] It coincided with United States use of CS in Vietnam, and followed shortly after controversial CS use in Northern Ireland in August 1969 and the resultant and ongoing Himsworth Committee, which had been established to investigate safety concerns surrounding the use of CS.[13] It also came during the period of abeyance which had been imposed on any decision to acquire nerve agent weapons, which the CoS, while supporting acquiring a nerve agent capability, had accepted due to fears of an adverse political decision. In late 1969, the Geneva Protocol again become a source of contention in British CBW policy, but on this occasion, surprisingly, defence officials would have more success.

Wilson's Cabinet was split over whether chemical riot-control agents could be considered under the Geneva Protocol as either 'poisonous' or as an 'other gas'.[14] This ambiguity fuelled a stark divide, reflecting ongoing disagreements between FCO and defence officials.[15] The central argument provided by Denis Healey and officials in the MoD, who were keen to keep the option of military use open, was that CS was not a gas, but smoke, and thus not a part of the 1925 Geneva Protocol. They also contended that 'science has provided the Army and the Police with a humane and harmless alternative to older methods'.[16] This argument

[11] While the CS debate is hugely important, this author does not wish to repeat the arguments made elsewhere, which are far more detailed than could be managed in this sub-section of the manuscript. For more information on this saga in British CW policy, please see the highly informative: Spelling. Driven to Tears, 701–725.

[12] TNA, CAB 129/146, 'The Geneva Protocol and the Use of Riot Control Agents in War', Note by the Prime Minister, 16 December 1969, Annex C. For a brief overview of the linkages between CS and the Geneva Protocol of 1925, see: Meselson, Matthew and Robinson, Julian Perry. 1969. Escalation of chemical warfare. *New Scientist*, August, 14–15.

[13] 15 For further details on the Himsworth Committee, see: Balmer, Brian, Spelling, Alex and McLeish, Caitríona. 2018. Tear Gas Epistemology: The Himsworth Committee and Weapons as Drugs. In *Chemical Bodies: The Techno-Politics of Control*, ed. Alex Mankoo and Brian Rappert. London: Rowman & Littlefield International, 103–124.

[14] Ibid; Spelling. Driven to Tears, 701–703.

[15] Ibid.

[16] TNA, CAB 129/146, 'The Geneva Protocol and the Use of Riot Control Agents in War', Note by the Prime Minister, 16 December 1969, Annex B; Spelling. Driven to Tears, 704.

reinforced the view that CS was an advanced and sophisticated scientific breakthrough, and that it was more humane than previous capabilities that were available when the Geneva Protocol was formed.[17] Healey, when presenting his case to Cabinet colleagues, also believed that the Government's decision over CS had far-reaching consequences for internal security.[18] He warned that if CS were included under the Geneva Protocol, then Britain would be using a chemical agent in peace-time that was barred from use in war. In contrast to nerve agent weapons, CS had a slightly more acceptable role, as it was used in dealing with localised riots and 'dangerous criminals'.[19] This encompassing of the internal security dimension, building upon earlier debates within government over CS incapacitating agents and points regarding the 'humane' nature of CS, set it apart from other chemical weapons. Healey had some seemingly persuasive arguments.

On the opposing side to Healey and the MoD was, once again, the FCO. Unsurprisingly, it strongly opposed the exemption of CS from the Geneva Protocol.[20] While the FCO accepted that there would indeed be consequences from stating that CS was exempt from the Geneva Protocol, it believed that these would be overwhelmingly negative and international, rather than the domestic security concerns referred to by Healey.[21] Officials in the FCO were troubled by the prospect that any public qualification of the widely accepted Geneva Protocol would 'gravely prejudice our standing in the international disarmament negotiations'.[22] During a period of emerging détente, these negotiations at both the UN and bilateral levels were crucial in addressing weapons proliferation and arms races in controversial weaponry. It was also feared Healey's recommendations would expose Britain to accusations that it was only changing policy to please the United States, which had not even requested that Britain do so.[23]

[17] TNA, CAB 128/44, Minutes of Meeting, Cabinet Conclusions, 18 December 1969.

[18] TNA, CAB 129/146, 'The Geneva Protocol and the Use of Riot Control Agents in War', Note by the Prime Minister, 16 December 1969, Annex B.

[19] Ibid.

[20] TNA, PREM 13/3465, 'The Geneva Protocol and the Use of Riot Control Agents in War', Solly Zuckerman to Harold Wilson, 24 September 1969.

[21] TNA, CAB 129/146, 'The Geneva Protocol and the Use of Riot Control Agents in War', Note by the Prime Minister, 16 December 1969, Annex A.

[22] Ibid.

[23] Ibid.

Supporting this strong FCO rebuttal was the Attorney-General, Sir Elwyn Jones, who argued that incapacitating agents, including CS, were significantly harmful and deleterious to man, and that they were included within the remit of the 1925 Geneva Protocol.[24] Solly Zuckerman, never shy to make his opinion known, also wrote personally to Wilson on the subject. He appealed for a more nuanced take on the CS debate which avoided publicity altogether.[25] Zuckerman believed that the issue had nothing to do with the use of CS for protest- and riot-control, and that any public qualification or exemption would only cast doubt on Britain's good faith in ongoing biological weapons disarmament negotiations. Any publicity would trigger 'formidable political repercussions'.[26] Unfortunately for Wilson, he did not heed Zuckerman's warnings or the advice of the FCO and Attorney-General. Instead, he put the decision before Cabinet on 18 December 1969.[27]

In the Cabinet meeting, members had before them an introductory text by Wilson. He put forward two options, both involving parliamentary statements. They were reflective of attempts to increase transparency in defence science and justify any policy qualification.[28] In his submission, the Prime Minister informed Cabinet that in light of domestic and international pressures, Britain had to qualify whether or not its military forces would be able to use riot-control agents in war.[29] In terms of the two options he provided, the first was to reaffirm the Geneva Protocol without qualification, which was supported by the FCO and the Attorney-General. The second, more drastic step supported by the MoD and Healey, was to reaffirm the Geneva Protocol but to explain that in the eyes of the British Government CS was exempt from it.[30] Oddly for postwar debates over CW policy and the Geneva Protocol, this time it was defence officials who

[24] TNA, CAB 129/146, 'The Geneva Protocol and the Use of Riot Control Agents in War', Note by the Prime Minister, 16 December 1969, Annex C; Spelling. Driven to Tears, 708.

[25] TNA, PREM 13/3465, 'The Geneva Protocol and the Use of Riot Control Agents in War', Solly Zuckerman to Harold Wilson, 24 September 1969.

[26] Ibid.

[27] TNA, CAB 128/44, Minutes of Meeting, Cabinet Conclusions, 18 December 1969; Spelling. Driven to Tears, 708–710.

[28] TNA, CAB 129/146, 'The Geneva Protocol and the Use of Riot Control Agents in War', Note by the Prime Minister, 16 December 1969.

[29] Ibid.

[30] Ibid.

came out on top.[31] The majority of the Cabinet supported releasing a public statement that Britain would take a more nuanced interpretation of the widely accepted Geneva Protocol, with the qualification that riot-control agents, such as CS, were exempt from it.[32] Arguments over military utility, specifically in relation to Northern Ireland, scientific sophistication, and the belief that incapacitating agents were suitably different to other forms of chemical and biological weapons persuaded the Cabinet to support Healey.[33] To a degree adding salt to the already open wound, it was Michael Stewart, the Secretary of State for Foreign and Commonwealth Affairs, who would make the announcement in Parliament.[34]

On 2 February 1970, when responding to a parliamentary question on disarmament, Stewart informed the House that 'we regard CS and other such gases … as being outside the scope of the Geneva Protocol'.[35] As anticipated and forewarned by Zuckerman and the FCO, this announcement triggered a tirade of criticism. Domestically, the announcement caused dismay. As Stewart noted:

> The decision has led to a far greater volume of correspondence from members of the public, again virtually unanimous in disapproval, than any other single development in the disarmament negotiations in recent years.[36]

Stewart elaborated that many who opposed the Government's qualification were, in fact, Labour supporters. As seen with earlier public protest movements against CBW, Labour Party MPs and members played an active role. Criticism of the Government's position was not only coming from 'disreputable organisations', but organisations such as the UN Association, the Quakers, and the British Council of Churches.[37] The strong domestic reaction was further seen through a barrage of parliamentary questions,

[31] TNA, CAB 128/44, Minutes of Meeting, Cabinet Conclusions, 18 December 1969; Spelling. Driven to Tears, 709–712.

[32] TNA, CAB 128/44, Minutes of Meeting, Cabinet Conclusions, 18 December 1969.

[33] Spelling. Driven to Tears, 710–712.

[34] TNA, CAB 129/146, 'The Geneva Protocol and the Use of Riot Control Agents in War', Note by the Prime Minister, 16 December 1969; Spelling. Driven to Tears, 701, 713.

[35] Hansard, House of Commons, 'Disarmament', 2 February 1970, Vol.795, cc17-8W17W.

[36] TNA, PREM 13/3465, 'CS and the Geneva Protocol', Michael Stewart to Harold Wilson, 15 April 1970.

[37] Ibid.

personal letters, criticism in the press and conference resolutions.[38] Labour MP Frank Hooley surmised the views of many when he stated that 'it is now 50 years since the First World War, but men still shudder at the idea of gas warfare'.[39] Any hopes of publicly and successfully justifying a nuanced take on the Geneva Protocol and CS gas had quickly faded.

On the international front, the policy clarification did not fare much better. At the UN Disarmament Conference in Geneva, Britain came under heavy criticism from India, Sweden, and several other countries. It was accused of carrying out a 'unilateral re-interpretation' of the internationally recognised Geneva Protocol.[40] Nicholas Sims, then a consultant on disarmament, concluded that 'the credibility of British policy in the field of chemical and biological warfare had been very severely damaged'.[41] Yet despite this domestic pressure, international reaction, and appeals by Stewart, Wilson refused to alter course. As astutely observed by Alex Spelling, the entire incident was handled with great awkwardness.[42] Perhaps of greatest concern for defence officials, though, was that this substantial blowback and negative publicity was not even over lethal nerve agents, but riot-control agents.[43] It again highlighted how controversial the CW field was.

In mid-February 1970, with the backlash against the Government's decision to clarify its interpretation of the Geneva Protocol ongoing, British defence officials again turned to review nerve agent acquisition. However, even with the two-year period of abeyance ending, which had

[38] PREM 13/3465, A. R. Jaffray to P. J. Moon, 13 May 1970.

[39] Hansard, House of Commons, 'C.S. Gas', 19 March 1970, Vol.798, cc807–21,807.

[40] John Myers, 'India & Sweden attack Britain on CS gas', *The Sunday Telegraph*, 13 March 1970, 5.

[41] TNA, PREM 13/3465, Record of a Meeting between Lord Chalfont and a Deputation of the United Nations Association, 13 April 1970.

[42] Spelling. Driven to Tears, 702.

[43] TNA, PREM 13/3465, Alun Chalfont to Harold Wilson, 5 January 1970; TNA, PREM 13/3465, 'CS and the Geneva Protocol', Michael Stewart to Harold Wilson, 15 April 1970. Even with a change in government, and despite the MoD being bombarded with questions on the Governments CS policy, Lord Carrington, the new Secretary of State for Defence, agreed with Healey's prior stance. The Government's stance that CS was exempt from the 1925 Geneva Protocol thus remained in place, remarkably it appears until the drafting and signing of the Chemical Weapons Convention, in 1993. For letters to the MoD see: TNA, DEFE 24/404. For Carrington see: TNA, DEFE 13/1055, 'CS and the 1925 Geneva Protocol', A. R. Jaffray to J. A. Graham, 6 October 1970. For the cementing of the CS stance see: Walker. *Britain and Disarmament*, 37; Spelling. Driven to Tears, 718–721.

started in January 1968 and was imposed by Roy Mason, defence officials remained hesitant to revive Ministerial discussions over a nerve agent capability. Reinforcing this judgement was the observation that throughout the two-year abeyance period publicity in the CW field had been rampant and unrepentant.[44] Since 1967, coverage of CBW had largely proved bruising and negative. Growing and active engagement in the anti-CBW movement had attracted widespread support and attention from influential grass-roots campaigners, such as Elizabeth Sigmund, to MPs, students, academics, scientists, the press, and international organisations. In 1970, the negative attention surrounding and enveloping CW policy had not ebbed, but increased with the Geneva Protocol clarification. This led the CoS to conclude that:

> In political terms, there have been numerous developments heightening UK and international interest in and concern about CW, and … [there can be] little doubt that Ministers would expect public and Parliamentary opinion to view the present as a remarkably strange and inopportune time for new UK re-armament measures in this field.[45]

Accepting that for two years British CW policy had come under significant domestic and international pressure, the CoS still maintained that Britain needed a nerve agent capability.[46] This was despite the unfavourable political climate.[47] The military requirement for nerve agent weapons and a CW deterrent thus remained unaltered, and the CoS again confirmed that 'the military case for having a retaliatory capability still holds good'.[48] This steadfast commitment to acquiring a nerve agent capability had now been the aim of the CoS for almost eight years, since they changed course in 1962 and Cabinet Defence Committee approval was granted in 1963. The tide of negative publicity could not in itself undo or change the perspectives of defence officials as to the military utility and deterrence

[44] TNA, WO 32/21761, 'CW—Possible UK Retaliatory Capability', P. J. Bayne to CDS, CNS, CGS and CAS, 18 February 1970.

[45] Ibid.

[46] Ibid. Also as referred to in: TNA, DEFE 13/1000, 'Chemical Warfare Policy', Note by the Secretary of the Chiefs of Staff Committee, 27 July 1972, Annex.

[47] TNA, WO 32/21761, 'CW—Possible UK Retaliatory Capability', P. J. Bayne to CDS, CNS, CGS and CAS, 18 February 1970.

[48] Ibid.

value of nerve agent weapons.[49] As in 1969, the CoS simply continued to allow CW policy to drift and remain under the Ministerial radar by unilaterally self-imposing an extension to the abeyance period.[50] By avoiding Ministerial attention when the public and political situation was adverse, the CoS kept alive the increasingly faint but future possibility of acquiring nerve agent weapons.

The Rise of Disarmament

At the same time as the CoS were maintaining the military necessity of a nerve agent capability, and alongside the CS debacle, was the rise of disarmament talks addressing CBW. These important talks were facilitated by superpower détente. Here, important decisions taken by Wilson's Labour Government were to have a lasting impact on the CW field. British discussions and considerations of chemical weapons disarmament has a long history, spanning decades before Britain signed the Chemical Weapons Convention (CWC) in 1993.[51] Early Cold War attempts at chemical and biological weapons disarmament had stuttered. It was only with the coming of détente that chemical weapons disarmament gained significant traction. These earlier, less successful attempts included UN discussions on the Geneva Protocol and chemical and biological weapons disarmament in 1952; a year in which defence officials had pushed for a revision in first-use policy. Participants at the UN discussion were, however, slightly distracted when contestants in that year's Miss Universe competition were granted access to view the debate.[52] Although the Soviet delegation made proposals, for chemical and biological weapons disarmament, an international

[49] Ibid.

[50] Ibid.

[51] The history of British chemical weapons disarmament of course has a much longer history than this, stretching long before the Cold War, for an earlier example, see: Spiers. Gas disarmament in the 1920s, 281–300. For more details on the international drive for chemical weapons disarmament and the origins of the Chemical Weapons Convention, see: Robinson, Julian Perry. 1993. Origins of the Chemical Weapons Convention. In *Shadows and Substance: The Chemical Weapons Convention*, ed. Benoit Morel and Kyle Olson. Boulder, CO: Westview Press, 37–54.

[52] *The Manchester Guardian*, 'Commotion at the U.N.', 21 June 1952, 7.

consensus and genuine superpower desire for chemical and biological weapons disarmament was lacking. This was until the late 1960s.[53]

An indicator of Wilson's commitment to chemical weapons disarmament was seen in 1968, when the Government supported and promoted a report by a 'UN Group of Consultant Experts on CBW'.[54] The findings of the report were to be put before the Eighteen-Nation Disarmament Conference at Geneva, and each nation was to have one representative expert. In January 1969, with the support of Wilson, Minister for Disarmament Fred Mulley and Cabinet Secretary Burke Trend, Solly Zuckerman was selected as Britain's expert.[55] As Mulley noted, Zuckerman had already been involved in the UN expert panel on nuclear weapons, where he had done a 'magnificent job'.[56] Not explicitly acknowledged, Zuckerman was also particularly suited to the role given his active influence and engagement with CBW policy in the 1960s. Even with Zuckerman's involvement, though, the UN CBW report was soon found to be more complex than expected. The Soviet representative delayed and deferred decisions.[57] As Zuckerman noted, the previous UN nuclear weapons panel had been 'child's play' in comparison to the one on chemical and biological weapons.[58] In overcoming Soviet obstructiveness, Zuckerman took an active role in discussions and the formulation of the report.[59] After much

[53] *The Manchester Guardian*, 'Banning Germ Warfare', 16 August 1952, 5.

[54] TNA, PREM 13/3468, 'Chemical and Bacteriological (Biological) Weapons UN Secretary General's Consultant Experts', Solly Zuckerman to Harold Wilson, 6 January 1969; TNA, PREM 13/3468, 'UN Study on Chemical and Biological Warfare', Fred Mulley to Harold Wilson, 7 January 1969. Pre-dating this in the mid-1960s, was also the important support of the Geneva Protocol at the UN in 1966, see: Humphrey, J. H. and Meselson, Matthew. 1969. Preventing CBW. In *CBW: Chemical and Biological Warfare*, ed. Steven Rose. Boston, MA: Beacon Press, 157–171. The UN had also turned to revisit the Geneva Protocol in 1966, but in the late 1960s there emerged significant traction for disarmament.

[55] TNA, PREM 13/3468, 'Chemical and Bacteriological (Biological) Weapons UN Secretary General's Consultant Experts', Solly Zuckerman to Harold Wilson, 6 January 1969; Meselson, Matthew. 1969. Why Not Poison? *Science*, April 25, 164: 413.

[56] TNA, PREM 13/3468, 'UN Study on Chemical and Biological Warfare', Fred Mulley to Harold Wilson, 7 January 1969.

[57] Zuckerman. *Monkeys, Men and Missiles*, 344–347.

[58] Ibid., 344–346.

[59] University of East Anglia, Solly Zuckerman Collection, CBW/1, Minutes of Meeting, Group of Consultant Experts on Chemical and Bacteriological (Biological) Weapons, 29 April 1969. Also see the meetings, in the same file, on 4, 5, 12 June 1969.

negotiating, the broad substance of a report was accepted by all sides. British officials dubbed it 'comprehensive and not unduly bias[ed]'.[60]

The final UN experts report on CBW was released in July 1969. It urged member states to stop the development, production and stockpiling of chemical and biological weapons.[61] With a foreword by U Thant, the Secretary-General of the UN, the report also outlined how chemical and biological weapons disarmament would slow down the arms race and reduce Cold War tensions.[62] Zuckerman was keen to include in the document the quote 'Armis bella non venenis geri', roughly meaning 'war is waged with weapons, not poisons'.[63]

Reflective of the growing aversion towards chemical and biological weapons, as well as the strengthening of détente, disarmament proposals continued to gain significant international support. In Britain, this growing focus on chemical and biological weapons disarmament was even reflected in the Queen's Speech of October 1969, which detailed that:

> My Government will strive for further progress on nuclear and nonnuclear arms control and disarmament. They will be particularly concerned with chemical and biological weapons, and will follow up with vigour the proposals they have put forward for a complete ban on biological methods of warfare.[64]

Coinciding with this increased focus on disarmament, was a crucial intervention by the US President, Richard Nixon. In November 1969, Nixon announced the ending of offensive biological weapons development in the United States and explicitly re-instated the policy of the

[60] TNA, DEFE 25/609, 'UN Experts Report on Chemical and Bacteriological (Biological) Weapons', P. Nailor, 3 July 1969; Zuckerman. *Monkeys, Men and Missiles*, 344–347.

[61] For the full report, see: United Nations. 1969. *Chemical and Bacteriological (Biological) Weapons and the Effects of Their Possible Use*. Report of the Secretary General. New York, NY: United Nations. For a summary, see: Langer, Elinor. 1969. U.N.: Experts' Report on CBW Supports Disarmament Effort. *Science*, 165:3889, 163–4; Spiers. *Chemical Warfare*, 175–176.

[62] United Nations. 1969. *Chemical and Bacteriological (Biological) Weapons and the Effects of Their Possible Use*. Report of the Secretary General. New York, NY: United Nations, xi-xii.

[63] Ibid, 1. Although Zuckerman was unsure if the term had actual historic use, he was nevertheless keen to include in the text. For communications, see: University of East Anglia, Solly Zuckerman Collection, CBW/1, as well as boxes CBW/2 and CBW/3.

[64] Hansard, House of Lords, 'The Queen's Speech', 28 October 1969, Vol.790, cc4–7.

no-first-use of lethal chemical weapons. This public declaration brought the United States clearly in line with the 1925 Geneva Protocol.[65]

In late 1969, Britain also pushed forward with strong proposals in the CBW field. And, it took a leading and controversial role in chemical and biological weapons disarmament when it proposed separating the two fields and tackling them in isolation.[66] In this vein, British officials tabled a draft 'Convention for the Prohibition of Biological Methods of Warfare', which the UN General Assembly acknowledged in December 1969.[67] The decision to separate a Biological Weapons Convention from a Chemical Weapons Convention was down to officials believing that in order to successfully and effectively secure the signing of any chemical and biological weapons disarmament treaty, the two fields had to be tackled individually.[68] In addition to the complexities of effectively attempting to negotiate two treaties at the same time and the matter of establishing some momentum, with hopes that disarmament of one would lead to the other, there was

[65] The Harvard Sussex Program Archive, 'The United States and the Geneva Protocol of 1925', Matthew Meselson, September 1969, 1–13; Meselson, Matthew. 1970. Behind the Nixon Policy for Chemical and Biological Warfare. *Bulletin of the Atomic Scientists*, January, 23–34; Robinson. Chemical Arms Control, 524. Nixon also banned the production of unitary chemical weapons, and he approved the closing down of the United States VX facility at Newport, Indiana. To circumvent this restriction in policy, United States planners substantially increased investment into binary chemical weapons. With a binary chemical weapon, two analogue mixtures would be combined before use to form the lethal nerve agent. As surmised by Tucker, after Nixon's intervention, United States expenditure on binary chemical weapons increased from but a few percent of the total CW budget, to roughly two-thirds by 1973. For further details see: Spiers. *Chemical Warfare*, 143; Spiers. *Weapons of Mass Destruction*, 15; Tucker. *War of Nerves*, 223–226; Allen. U.S. Renunciation of Biological Weapons, 497–501; Goldman, David. 2009. The Generals and the Germs: The Army Leadership's Response to Nixon's Review of Chemical and Biological Warfare Policies in 1969. *The Journal of Military History*, 73:2, 531–569.

[66] Howlett, Darryl. 1990. UK arms control and disarmament policy on chemical and biological weapons. In *UK arms control in the 1990s*, ed. Mark Hoffman. Manchester: Manchester University Press, 156–159; Spiers. *A History of Chemical and Biological Weapons*, 62; Wilkinson, Mark. 2018. *Before Intelligence Failed: British Secret Intelligence on Chemical and Biological Weapons in the Soviet Union, South Africa and Libya*. London: C. Hurst & Co., 32.

[67] For more details on the origins of the Biological Weapons Convention, see: Walker. *Britain and Disarmament*, 49–72.

[68] Spiers. *A History of Chemical and Biological Weapons*, 62; Utgoff. *The Challenge of Chemical Weapons*, 99.

also a military angle to this division.[69] In a military sense, chemical weapons were still widely developed and deployed by the superpowers, and the CoS still wished to possess a nerve agent capability, whereas at this stage the CoS had little desire to acquire biological weapons.[70] Tackling the BW treaty first thus also represented the preferable option in terms of any perceived military loss resulting from a disarmament treaty.[71]

The successful British attempt to split chemical and biological weapons disarmament, and guide the international approach in this direction, was a decision which caused a degree of political backlash. And, it was a decision which would haunt future British considerations of chemical weapons disarmament. On the domestic front, despite proposing biological weapons disarmament negotiations in the UN, Mulley came under criticism from backbench Labour MPs who believed that the Government was pursuing chemical and biological weapons disarmament too slowly.[72] Shortly after the CS debacle had gone public in February 1970, Wilson was also informed that the decision to separate CW and BW had 'won us few converts in public'.[73] While dealing with the barrage of negative publicity resulting from the clarification of the Geneva Protocol, Wilson was also advised to immediately begin consideration of chemical weapons disarmament, even if this was 'only on a defensive basis' in order to fend off criticism for separating the two fields.[74]

With Wilson's approval, the FCO drafted a tentative list of options for chemical weapons disarmament which Britain could put forward at the UN Eighteen-Nation Disarmament Conference in Geneva. Although in part driven by a sense of obligation, Wilson making a tactical political decision to fend off criticism, and the Cold War climate, there were also genuine hopes for chemical weapons disarmament gaining momentum. The options included a total ban on chemical weapons, a non-proliferation treaty, and the establishment of a chemical weapons free-zone in Europe. As cautioned

[69] For more details on the origins of the BW and CW separation, see: Walker. *Britain and Disarmament*, 49–72.

[70] TNA, WO 32/21761, 'CW—Possible UK Retaliatory Capability', P. J. Bayne to CDS, CNS, CGS and CAS, 18 February 1970.

[71] Conversely, it could also be said that an effective chemical weapons treaty with verification would have been more militarily beneficial for Britain, through removing or mitigating a Soviet advantage and somewhat alleviating British 'vulnerability' in this area.

[72] *The Guardian*, 'Demand for prohibition of chemical warfare', 30 September 1969, 8.

[73] PREM 13/3465, 'Possible Arms Control Measures in the Field of Chemical Warfare', Burke L. to Harold Wilson, 17 February 1970.

[74] Ibid.

by FCO officials, revealing the sensitive public and political nature of chemical weapons, the chemical weapons free-zone would likely 'provoke considerable parliamentary trouble' for the Government.[75] If the zone were established, United States chemical weapons stored in West Germany would potentially need to be relocated to Britain. Such a move, coinciding with the spate of negative publicity over incapacitating agents and the separation of chemical and biological weapons disarmament, would have revitalised public protests, grass-roots movements, and Labour MPs in opposition to Government CW policy. A chemical weapons free-zone, no matter how appealing, would have left the Labour Government in the unenviable position of being sandwiched between strong domestic opposition and the pressures of maintaining close Anglo-American relations.

Similarly to the CS debate, when it came to these chemical weapons disarmament discussions Healey and the MoD again took a hard-line towards, and a dim view of, early FCO proposals.[76] Healey attempted to stifle and limit these early disarmament considerations. He warned that the utmost care had to be taken with the United States, as the United States did not possess any stockpiles of biological weapons but it did possess chemical weapons.[77] A British proposal for chemical weapons disarmament could therefore run counter to the wishes of the United States, and prove particularly damaging given that Britain was reliant on United States nerve agent capabilities for CW deterrence in Europe. Healey was sceptical of the chemical weapons-free zone, as this involved the delicate and troublesome issue of verification.[78] British intelligence also believed that Warsaw Pact countries possessed stockpiles of chemical weapons. Any agreement had to ban not just Soviet chemical weapons from Europe, but those of Eastern European countries as well. Healey thus believed that dealing with just the Soviet Union would prove a 'useless' exercise.[79] In the CW field, there were numerous active countries, in Eastern Europe, the Middle East and Asia, and allies such as France and the United States.[80]

[75] Ibid.

[76] TNA, DEFE 11/672, 'Chemical Warfare—The Geneva Protocol and the Disarmament Negotiations', Note of a meeting between the Defence Secretary and Mr. Mulley, 22 September 1969.

[77] Ibid.

[78] Ibid.

[79] Ibid.

[80] While the MoD did approve the 'sounding out' of NATO countries on chemical weapons disarmament, it was unwilling to commit fully to disarmament. The ideas of a CW free-

In his resistance to the progression of chemical weapons disarmament discussions, Healey's sceptical views coincided with those of United States officials from the Arms Control and Disarmament Agency. United States officials expressed serious doubts over Britain's proposed chemical weapons disarmament discussions, counselling that it might give the impression that Britain was having second thoughts over splitting chemical and biological weapons disarmament. They cautioned this would reduce momentum and British credibility in disarmament talks. They also explicitly warned that such a drive for chemical weapons disarmament would undermine ongoing biological weapons disarmament negotiations, by slowing them down.[81] United States officials feared that countries in the Non-Aligned Movement would be bogged down and distracted by any new proposals.[82] With this external pressure from the United States and internal pressure from Healey and defence officials, exploratory chemical weapons disarmament discussions were temporarily shelved. The British decision to separately pursue biological weapons disarmament had seemingly obligated the FCO to initially consider chemical weapons disarmament as well, especially in light of the political criticism. But, the separation ultimately led to a delay in chemical weapons disarmament negotiations, with concerns that they might undermine successful biological weapons disarmament.

In late 1971, and in tandem with the desire to avoid any impeding of attempts at biological weapons disarmament, unwavering military support for a nerve agent capability continued the hiatus and malaise in CW policy. Throughout these biological weapons disarmament negotiations and discussions, the CoS remained committed to the acquisition of nerve agent weapons. Their costings had shown that Britain could now acquire a modern VX capability for around £9 million.[83] Four years after Mason had imposed a period of two-year abeyance in 1968, the CoS still continued to discuss the issue of acquisition in secret. But, they shunned political approval for fear of rejection. Seeking Ministerial approval would likely

zone and a ban on the transfer of chemical weapons attracted little initial support from NATO officials. As MoD officials noted, UK chemical weapons disarmament proposals 'attracted little attention when first raised and their demise is unlikely to be mourned'. See: TNA, CAB 148/71, 'Chemical Warfare and the Disarmament Negotiations' Note by the Secretaries of the Sub-Committee on Disarmament, 17 November 1969, Attached memorandum.

[81] TNA, FCO 66/220, 'CW', D. L. Benest, 9 March 1970.

[82] Ibid.

[83] TNA, WO 32/21761, 'Chemical Defence Policy', G. N. Gadsby, 12 November 1971.

have thrown CW policy into further decline as the political situation was still not conducive to nerve agent acquisition. Self-imposed abeyance and isolation thus continued. This was to be the de facto state of affairs in nerve agent policy until the Biological Weapons Convention opened for signatories in April 1972. This momentous occasion led to a former obstacle to chemical weapons disarmament turning into a key founding and supporting pillar.[84]

Successes in biological weapons disarmament spurred on British considerations of chemical weapons disarmament. The ground-breaking Biological Weapons Convention (BWC) was naturally focused on BW, with it binding signatories to 'never in any circumstances to develop, produce, stockpile or otherwise acquire or retain ... biological agents', but it also contained an article on chemical weapons.[85] In the BWC, of which Britain was a key driver, Article IX stated that:

> Each State Party to this Convention affirms the recognised objective of effective prohibition of chemical weapons and, to this end, undertakes to continue negotiations in good faith with a view to reaching early agreement on effective measures for the prohibition of their development, production and stockpiling and for their destruction ...[86]

The BWC bound Britain to continue disarmament negotiations in the CW field; momentum was seemingly building. As even though there was a substantial delay in the signing of the BWC in 1972 and the Chemical Weapons Convention in 1993, the two are heavily interlinked. Officials in the FCO had been pressing for chemical weapons disarmament since the separation of biological and chemical weapons disarmament talks, and with the success of the former, they moved with alacrity to the latter. The commitment of Article IX was taken particularly seriously by the FCO and by Members of Parliament, such as Lord Chalfont the former Minister of

[84] Croddy, Eric, Perez-Armendariz, Clarisa and Hart, John. 2002. *Chemical and Biological Warfare: A Comprehensive Survey for the Concerned Citizen*. New York, NY: Copernicus Books, 237–248. For details on how this was portrayed, and British press coverage of BW during this period, see: Balmer, Brian, Spelling, Alex and McLeish, Caitríona. 2016. Preventing "A Virological Hiroshima": Cold War Press Coverage of Biological Weapons Disarmament. *Journal of War & Culture Studies*, 9:1, 74–90.

[85] United Nations General Assembly. 1971. 'Convention on the Prohibition of the Development, Production and Stockpiling of Bacteriological (Biological) and Toxin Weapons and on their Destruction', 26th Session, Resolution 2826, December 16. Annex.

[86] Ibid.

State for Foreign and Commonwealth Affairs.[87] Chalfont called on Parliament to address the 'urgent need to get rid of this dreadful weapon', with chemical weapons disarmament being in line with both 'world opinion' and the BWC.[88] He further counselled parliamentary colleagues that 'not even the soldiers want this weapon'.[89]

Despite the BWC and very much contrary to what Chalfont believed, the CoS still wanted a nerve agent capability. British defence officials 'saw no reason' to change their request for acquiring lethal nerve agent weapons.[90] Gradually though, from July 1972, signs started to emerge that this strict and rigid stance was beginning to thaw. Defence officials slowly recognised that in the age of détente and with adverse publicity towards chemical weapons, holding out for a nerve agent capability for deterrence, especially if it involved publicity, was not feasible.[91] This change in approach was also to a degree facilitated by a compromising FCO, which made it 'absolutely clear' that it 'had no intention of ignoring our own defence requirements and [that] nothing would be proposed at Geneva without prior consultation with the MoD'.[92] Adding to this FCO drive were two major and, from the perspective of the CoS, alarming superpower actions in the chemical weapons disarmament field.

In July 1972, the CoS assessed the impact of the superpowers heading in the direction of further disarmament, as well as domestic trends and hostile public opinion. It analysed how these evolving pressures could shape or influence the possibility of acquiring a nerve agent capability. As the CoS noted, the first major superpower involvement came in the form of a Soviet proposal to the UN Conference on Disarmament at Geneva, which recommended a complete ban on the development, production and

[87] Jones, Matthew. 2017. *The Official History of the UK Strategic Nuclear Deterrence, Volume II: The Labour Government and the Polaris Programme, 1964–1970*. London: Routledge, 16.

[88] Hansard, House of Commons, 'Chemical Weapons', 7 June 1972, Vol.331, cc312–62.

[89] Ibid.

[90] TNA, DEFE 4/272, Minutes of Meeting, Chiefs of Staff Committee, 19 July 1972, Confidential Annex; DEFE 13/1000, 'Chemical Warfare Policy', B. G. Stanbridge, 27 July 1972, Annex A; DEFE 13/1000, 'Chemical Warfare Policy', B. G. Stanbridge, 27 July 1972, Annex A, Appendix 3.

[91] Ibid.

[92] TNA, DEFE 4/272, DEFE 4/272, Minutes of Meeting, Chiefs of Staff Committee, 19 July 1972, Confidential Annex; TNA, DEFE 13/1000, 'Chemical Warfare Policy', CDS to Secretary of State, 27 July 1972.

stockpiling of chemical weapons.[93] The second was a paper by the United States, which proposed a 'freeze' on existing stocks and a ban on the future production of chemical weapons.[94] In both cases, the CoS warned that unless the proposals contained stringent methods for verification, Britain would be left at a substantial disadvantage vis-à-vis the Eastern bloc and the Soviet Union if it conformed to them.[95] Without assurances that the Soviet Union had disposed of its CW stockpiles through stringent verification methods, Britain would have effectively unilaterally given up the option of ever acquiring a CW capability. This was all the more important given that the JIC now believed that the Soviet Union possessed the 'largest offensive CW capability in the world'.[96] If Britain were banned from production, then it would never be able to acquire a domestic nerve agent capability and bridge the substantial gap in capabilities which had emerged during the Cold War. Any 'freeze' on existing stocks would have prohibited Britain from developing chemical weapons, leaving it at a fixed and substantial disadvantage until Soviet and Warsaw Pact CW stocks expired.[97] The CoS thus recommended that Britain should not agree to any superpower initiative in chemical weapons disarmament which left the country at a military disadvantage; in their current forms, both initiatives did. Defence officials and the CoS also advocated the hindering and obstruction of any superpower chemical weapons disarmament negotiations which did not remove the Soviet CW threat through stringent verification or the need for a British CW deterrent.

Coinciding with these superpower actions was a growing and grudging acceptance that public and political attitudes towards CW were likely to remain hostile, adverse, and insurmountable. In terms of the negative publicity surrounding CW policy, the CoS accepted that:

> One cannot … ignore the highly emotional reaction which any reference to offensive CW provokes in some sections of public opinion in this country and elsewhere in the West. The CS gas issue, for example, illustrated how

[93] TNA, DEFE 4/272, DEFE 4/272, Minutes of Meeting, Chiefs of Staff Committee, 19 July 1972, Confidential Annex.

[94] Ibid.

[95] Ibid; TNA, DEFE 13/1000, 'Chemical Warfare Policy', CDS to Secretary of State, 27 July 1972.

[96] Ibid.

[97] Ibid.

sensitive was the feeling in some quarters towards the use of even incapacitating agents in war.[98]

While continuing to recognise, as they had throughout the 1960s, that the acquisition of nerve agents was politically an extremely troubling and emotional subject, in July 1972 defence officials were forced to fully acknowledge and confront the immense political difficulties surrounding the acquisition of a nerve agent capability. The CoS recognised that even when it came to those chemical weapons which were portrayed as 'humane' and 'non-lethal', such as with CS gas, then there was still public uproar.[99] Any public announcement on the production or possession of highly lethal nerve agents would have likely triggered more significant political difficulties and a far stronger public backlash. The military logic of lethal nerve agent acquisition and deterrence was thought of as rational and sensible by defence officials, whereas public criticism was branded highly emotional. The language of emotional reactions also seemed to imply that these groups could not be reasoned with, further compounding the seemingly hopeless situation. They could not be converted. Lessons learnt and observations made during periods of intense publicity in 1959–1960, 1967–1968, and the more recent CS debacle, were increasingly holding sway over nerve agent policy.

The impact of negative publicity was further reinforced by how defence officials continued to justify their request for nerve agent weapons, which was still for CW deterrence and retaliation in Europe.[100] In the eyes of the CoS, CW deterrence still relied upon other countries knowing that Britain possessed chemical weapons and that it was willing to use them.[101] This approach all but removed the possibility of keeping the production and possession of controversial nerve agent weapons a secret, which remained a red-line for politicians.[102] The political hurdle of publicity thus remained unresolved. In essence little had changed since Healey dismissed publicity

[98] TNA, DEFE 5/193, 'Chemical Warfare Policy', Chiefs of Staff Committee, 27 July 1972, Annex A.

[99] Ibid.

[100] TNA, DEFE 13/1000, 'Chemical Warfare Policy', CDS to Secretary of State, 27 July 1972.

[101] TNA, DEFE 5/193, 'Chemical Warfare Policy', Chiefs of Staff Committee, 27 July 1972, Annex A.

[102] It must also be observed that even if the decision had been taken to produce chemical weapons domestically, secretly, and for them not to fulfil a deterrence role at this time, the

for deterrence in the mid-1960s. And, in many ways, the political situation had become even more difficult for defence officials. The CoS, though, could not afford to separate deterrence from the nerve agent requirement, as this was their primary role and function. Defence officials were stuck. The military requirement for nerve agent weapons remained, but the hostile public environment and political aversion rendered their acquisition and publicity extremely unlikely and implausible. With publicity still viewed as a necessary part of deterrence and an integral part of policy, the CoS could not realistically seek Ministerial approval. Rather than risk negative reprisals, they again continued in silence and in limbo. By simply bringing the existence and nature of CW policy to the attention of Ministers, and the nerve agent request, it was still feared that the opposite result would occur—severe cutbacks to CW policy and a firm commitment to not possess these weapons. Adverse publicity and political pressure continued to play a significant part in enforcing abeyance, through ensuring that the CoS remained hesitant to bring nerve agent requirements to the attention of Ministers.

With the mounting and multifaceted shift in favour of chemical weapons disarmament, both domestically and internationally, the CoS did not explicitly drop its request for lethal nerve agents. But, in a small step, it did acknowledge that in light of these seemingly insurmountable obstacles they simply desired that 'the option of providing an offensive CW capability for our forces ... not be foreclosed'.[103] This subtle shift represented a change from over a decade of policy, as from late 1962 the CoS had always sought the acquisition of lethal chemical weapons. But in 1972, they just wanted the future option of acquiring them left open and not taken off the table. The policy since 1962 was more active, a pursual and request for nerve agent weapons that was simply on hold and waiting for the right political moment to be acted upon, the latter 1972 shift represented a more resigned acceptance. It was more passive, showing recognition that

public discovery of such a policy would likely have occurred anyway given the intense public and parliamentary interest in CBW weapons during this period.

[103] TNA, DEFE 13/1000, 'Chemical Warfare Policy', CDS to Secretary of State, 27 July 1972. The CoS continued to take this line, as in light of the perceived Soviet CW threat there was still support for nerve agent acquisition, particularly from the RAF. See: TNA, AIR 2/17296, 'Chemical Warfare', K. Williamson to D. Robertson, 19 January 1973. For the Soviet threat appreciation in 1973, see: TNA, DEFE 13/1000, 'Soviet Offensive Chemical Warfare Capability', Intelligence Note by DGI, 4 May 1973.

future political approval of acquisition was unlikely and that the CoS just wanted the remote option left open and not entirely foreclosed.

The CoS had given some limited ground on their desire for a nerve agent capability, but with the coming of superpower détente and chemical weapons disarmament discussions, this change was not enough. Serious questions emerged over even keeping open the future possibility of acquiring nerve agent weapons, especially if this clashed with disarmament discussions. As Lord Carrington, Healey's Conservative successor as Defence Secretary was warned, CoS resistance to Soviet and United States chemical weapons disarmament proposals was 'dogmatic' and 'negative'.[104] Carrington was advised that it would be completely unrealistic for Britain to independently oppose superpower disarmament negotiations. This was regardless of the military requirements or desire to keep the option of acquisition available.[105]

Before Carrington made a definitive decision on his own stance towards chemical weapons disarmament, and whether this involved strict verification measures as the CoS demanded or more flexibility as he was privately advised, he wanted to see the outcome of bilateral chemical weapons disarmament talks with the United States.[106] On 27–28 July 1972, British officials from the MoD and the FCO met United States officials from the Pentagon and the Arms Control and Disarmament Agency.[107] One of the main reasons for this bilateral meeting was that United States officials were concerned by the rival 'unacceptable' Soviet chemical weapons disarmament proposal and they wished to counter it with a united Western effort.[108] Disarmament itself had also turned into a form of Cold War competition. The United States, for domestic political reasons, was 'committed to going it alone' if necessary, even without British support.[109] With this keen and proactive United States stance, Carrington decided to take a more flexible approach to disarmament negotiations. While he accepted CoS concerns over the Soviet threat, he cautioned that:

[104] TNA, DEFE 13/1000, R.M. to the Secretary of State, 27 July 1972.

[105] Ibid.

[106] Ibid.

[107] TNA, DEFE 13/1000, 'CW: FCO/MoD Discussions in Washington July 27/28', J. Roberts, 31 July 1972.

[108] Ibid.

[109] TNA, DEFE 13/1000, 'CW', S. S. Crew to the APS of the Minister of State, 25 August 1972.

> It would be unrealistic to completely rule out British acceptance of a CW agreement, which although not entirely suitable from our point of view, nevertheless commanded a wide measure of international support ... If it is not possible to modify the United States position on a convention freezing stocks, we should adopt a more flexible fall-back position.[110]

From 1972–1974, with Carrington's tentative support, or at least without his outright opposition, the drive for chemical weapons disarmament gained further traction. Without Healey obstructing disarmament negotiations and with the formulation of the BWC no longer hampering efforts but fuelling them, British consideration of chemical weapons disarmament continued at pace. Disarmament officials in the FCO, responding to this change in environment shifted much more of their attention towards CW policy. They began asking serious questions over why CW deterrence with nerve agent weapons was needed. Building on the momentum and optimism established from the successful BWC, FCO officials believed that it was 'time that the voice of the UK was heard' in chemical weapons disarmament talks at Geneva.[111] They also noted that every effort should be made to ensure that Britain did not appear as 'American stooges'.[112] This increasing FCO attention and involvement in chemical weapons disarmament continued to grow throughout 1973. It was FCO officials in the Arms Control and Disarmament Department who took a tough line with the CoS request for a British retaliatory capability and of CW deterrence through the threat of retaliation.[113] In December 1973, they argued that if British forces were defensively well-equipped against CW then there would be little need for a British retaliatory capability.[114] In this endeavour, they used the military trials and war planning exercises conducted by defence officials in the 1960s in support of their claims. In part FCO officials used these assessments to argue that a well-equipped force could heavily mitigate against the military benefits of CW use. If British forces were

[110] TNA, DEFE 13/1000, 'Chemical Warfare Policy', Carrington to A/CDS, 10 August 1972.

[111] TNA, FCO 66/391, 'CW: UK Statement', H. C. Hainworth to D. M. Summerhayes, 29 March 1972.

[112] Ibid.

[113] TNA, FCO 66/497, D. E. Blatherwick to P. E. Gerahty, 4 December 1973.

[114] Ibid.

defensively well prepared then there would be no benefit to CW use by an enemy, thus deterring its initiation.[115]

In theory, this defensive approach to CW deterrence could work, especially if used in tandem with disarmament measures. However, in practice and in isolation, such an approach would be laden with immense difficulties.[116] Part of the problem of this form of deterrence is that a key feature of nerve agent weapons is not only their military and psychological effects, which could indeed be mitigated somewhat with advanced defensive measures and preparations, but perceptions of their near-myth-like status and effects.[117] Any successful deterrent, from a purely defensive posture, would therefore have the added difficulty of potentially altering, or at the very least undermining, an enemies' perception of the military utility of their own chemical weapons and of combatting this myth-like image. This perception of nerve agent weapons and their utility had long been ingrained in the minds of defence officials during the Cold War. Without such a measure, defence planners would have to hope that their rival counterparts had reached the same conclusions with regards to the effectiveness of defensive measures and that they had not been drawn in by the hype. Appreciating that defensive measures would militarily counter Soviet use of chemical weapons was one thing, convincing an enemy that your defensive measures nullified their capabilities and persuading them not to believe in the perceived effectiveness of their own nerve agent weapons, was another.

As British officials found, dispelling the superweapon nerve agent myth was no easy matter. They had in part attempted such a feat with minimal success in 1968 through patient repetition of the scientific arguments. Domestically, while the attempt to counter the at times inflated perception of the effectiveness of CW agents did have some success in calming public criticisms, achieving a similar outcome internationally and with a sceptical

[115] Mikulak, Robert. 1978. Preventing Chemical Warfare. In *Chemical Weapons and Chemical Arms Control*, ed. Matthew Meselson. New York, NY: Carnegie Endowment for International Peace, 70.

[116] For further discussion on deterrence through effective defensive measures see: The Harvard-Sussex Program Archive, 'The Role of Chemical Defense in Chemical Warfare, Chemical Deterrence, and Chemical Disarmament', adapted text of a Keynote Address by Matthew Meselson, 13 March 1991.

[117] For further details on this line of thinking on the 'superweapon myth', see: Meselson, Matthew. 1991. The Myth of Chemical Superweapons. *The Bulletin of the Atomic Scientists*, 47:3,12–15.

Soviet Union was an entirely different prospect. Unless Soviet officials had thus also dispelled the nerve agent myth or believed that advanced defensive measures would largely mitigate the potential military benefits of their nerve agent weapons, then deterrence through just advanced defensive measures would have likely failed. Although the success or viability of such an alternative approach to CW deterrence can be questioned, due to the military fixation with retaliatory deterrence this potential avenue was not given much meaningful consideration by defence officials.

FCO officials struck at the heart of the very role and plausibility of CW deterrence. They observed that if Soviet forces invaded with overwhelming conventional forces and used chemical weapons, NATO's response would be to use nuclear weapons, making chemical weapons all but redundant.[118] Even if Soviet forces used chemical weapons in a limited war, FCO officials thought that as the Soviet Union would have chosen to initiate CW then they would likely have already taken the necessary protective steps against NATO reprisals with chemical weapons.[119] Any retaliatory CW use by British forces would therefore not have necessarily recovered the lost ground, offered any real tangible military benefit, or even deterred subsequent Soviet CW attacks. Critical FCO officials sagely warned that CW use by either side, and retaliation, would create the most casualties and difficulties for civilians in NATO countries, not Soviet or NATO military forces.[120] In their eyes, chemical weapons lacked a clear deterrent role, when alternatives were potentially viable, and even a military one.[121]

The divergence between defence officials and FCO officials was primarily based on their faith in disarmament negotiations and their interpretations of CW deterrence. Both acknowledged the scale of the threat and the need to prevent the outbreak of CW. While the CW threat existed, to prevent the outbreak of CW, the CoS favoured deterrence through the threat of retaliation. A Chemical Weapons Convention could mitigate this requirement by potentially removing the threat altogether, but it would need stringent verification methods.[122] Without the removal of the threat,

[118] TNA, FCO 66/497, D. E. Blatherwick to P. E. Gerahty, 4 December 1973.

[119] Ibid.

[120] Ibid.

[121] In line with Meselson's argumentation over the superweapon myth, even the impact of CW agents on civilians could be questioned, with gas masks, staying indoors and public awareness also mitigating against CW use. Primarily those most heavily affected would be those without access to gas masks and other forms of CW defensive measures. See: Meselson. The Myth of Chemical Superweapons, 12–15.

[122] Both CW deterrence and disarmament are in effect attempts to remove or significantly reduce the possibility of CW use. See: Spiers. *Chemical Warfare*, 207.

the CoS viewed retaliatory CW deterrence as the only viable means of preventing the use of chemical weapons. This was thought best accomplished through the possession of a credible retaliatory capability in the form of nerve agent weapons. The retaliation-in-kind form of deterrence had long been an ingrained policy and belief in defence circles. It was also a core pillar of CW policy during the Second World War and even for Prime Ministers such as Clement Attlee and Anthony Eden in the early Cold War. For FCO officials, though, preventing the outbreak of CW could be accomplished through disarmament negotiations and through removing the military benefits of Soviet first-use with advanced defensive training and equipment. This form of CW deterrence was intended to discourage CW use through removing the military benefits of an enemies' nerve agent weapons. This would leave only the normative and political costs of use.

Complicating the disarmament picture and exacerbating this divide between FCO and defence officials, was the involvement of numerous other countries in international disarmament negotiations. On top of the tentative United States and Soviet proposals, in 1971 the Non-Aligned Movement had also presented a joint memorandum on chemical weapons disarmament to the UN. While the memorandum did not receive much attention, and as British officials critically observed it contained 'no original thinking', it did reflect broader feelings of international support for chemical weapons disarmament.[123] This international support was given greater weight when, building on its ventures in nuclear weapons disarmament, Japan submitted a detailed draft Convention on chemical weapons disarmament at the Geneva Conference in April 1974.[124] The Japanese draft, as outlined by Edward Spiers, advocated a gradualist approach to disarmament. It included the establishment of an international verification agency to observe the destruction of CW agents and, importantly, on-site inspections.[125] Surprisingly these robust verification methods were very much in line with the wishes of the British CoS. An independent verification agency and on-site inspections mitigated against fears that Britain would face chemical weapons without the possibility of acquisition or retaliation. With these thorough checks, the need for a CW deterrent

[123] TNA, FCO 66/312, 'The Non-Aligned Joint Memorandum on CW of 28 September 1971', Arms Control and Disarmament Department, [exact date not included] 1971.

[124] United States Arms Control and Disarmament Agency. 1976. *Documents on Disarmament: 1974*. Washington, DC: United States Government Printing Office, May, 106–113; Utgoff. *The Challenge of Chemical Weapons*, 106–107.

[125] Spiers. *Chemical Warfare*, 179.

would be reduced as there would be greater assurances over a potential enemy not possessing stockpiles of chemical weapons.

The Japanese draft, commended for its basic principles, reinvigorated international chemical weapons disarmament efforts. Despite the reluctance of the United States and the Soviet Union, growing international pressure pushed them to release a joint communiqué on the need for a bilateral initiative on chemical weapons disarmament in July 1974.[126] After Nixon resigned in August 1974, President Gerald Ford continued the bilateral drive for chemical weapons disarmament.[127] Even after the entry of Ford, though, United States officials could still not overcome Leonid Brezhnev's reservations over chemical weapons disarmament and particularly his concerns over the issue of verification. The Soviet Premier was hesitant to fully commit to on-site inspections and thorough verification measures.[128]

In the wider international sphere, there continued to be more success. Coinciding with stalling bilateral discussions and the Japanese draft were a plethora of interventions and statements by national representatives at the UN Disarmament talks in Geneva in favour of chemical weapons disarmament.[129] Among those to raise the issue were representatives from France, Poland, and Canada.[130] Growing attention afforded to chemical weapons disarmament, spanning across all divides in the Cold War from the Eastern bloc to NATO countries and the Non-Aligned Movement, culminated in the UN General Assembly approving Resolution 3256 in December 1974.[131] The resolution noted that the process of détente was conducive to further disarmament measures in the CW field, and it urged states to successfully conclude negotiations on the complete prohibition of the development, production, and stockpiling of chemical weapons.[132]

Complementing and reflecting this shift in favour of chemical weapons disarmament, President Ford also continued to build on bilateral chemical

[126] Goldblat, J. 1975. Chemical disarmament negotiations in 1974. In *Chemical Disarmament: New Weapons for Old*, A SIPRI Monograph. Stockholm: Almqvist & Wiksell, 1; Meselson, Matthew. 1975. What Policy for Nerve Gas? *Arms Control Today*, 5:4, 1.

[127] Ibid.

[128] Gerald R. Ford Library Papers, National Security Adviser, President's Subject File, Box 1, Chemical and Biological Weapons, 'Soviet Draft Chemical Weapons (CW) Convention', 22 August 1974. Found at: The Harvard Sussex Program Archive.

[129] United States Arms Control and Disarmament Agency, *Documents on Disarmament: 1974*, vi-xv.

[130] Ibid.

[131] Ibid, 790–792.

[132] Ibid.

weapons disarmament talks and Nixon's earlier work in the disarmament field. In January 1975, in a major milestone, Ford ratified the 1925 Geneva Protocol.[133]

The End of an Era

Throughout 1975–1976, the growing international commitment to chemical weapons disarmament heavily impacted British discussions over CW policy, especially after Porton Down had again come under public criticism and 'unfavourable attacks'. British research at Porton Down remained a 'politically sensitive' area.[134] The domestic political scene had also changed. The Labour Party, still under Harold Wilson, returned to office after winning the general election of October 1974.[135] It was after the return of Wilson and Labour that the British push for chemical weapons disarmament was to gain substantial impetus.[136] In opposition, Labour MPs such as John Morris had continued to pursue British progress in chemical weapons disarmament talks with parliamentary questions.[137]

With a leadership even more sympathetic towards disarmament, and growing international pressures and continued negative domestic publicity, FCO officials were maintaining their strong push for chemical weapons disarmament. Under this increasing weight, the CoS yielded and drastically changed direction in 1976, going far beyond the minor concessions they had made in 1972. On 2 March 1976, the CoS finally

[133] Gerald R. Ford Library Papers, NSA, Box 1, Chemical and Biological Weapons, 'Signing Ceremony for the Geneva Protocol and the Biological Weapons Convention', Henry Kissinger, 21 January 1975. Found at: The Harvard Sussex Program Archive; Spiers. *Chemical Warfare*, 144.

[134] University of East Anglia, Solly Zuckerman Collection, General Communications: C, Chemical Defence Establishment, R. G. Watson to Solly Zuckerman, 13 March 1975. This ongoing criticism was coincided with greater public engagement and awareness of British defence policy in general. See: Baylis, John. 1983. Introduction. In *Alternative Approaches to British Defence Policy*, ed. John Baylis. London: Macmillan Press, 1–2.

[135] The first election of February 1974 had resulted in a hung Parliament, and so another was called in October 1974.

[136] Young. *Cold War Europe*, 123.

[137] Other Labour figures involved were Lord Chalfont and Tam Dalyell. For examples see: Hansard, House of Commons, 'Treaty on Chemical and Biological Warfare', 19 April 1971, Vol.815, cc804–5805; Hansard, House of Lords, 'Chemical Weapons', 7 June 1972, Vol.331, cc312–62,312; Hansard, House of Commons, 'Defence', 19 March 1973, Vol.853, cc44–170.

acquiesced. Defence officials relinquished their request for nerve agent weapons, stating that there was:

> No requirement for the United Kingdom to have a capability to produce offensive chemical weapons for deterrent and retaliatory purposes: neither [is] there a need to retain the potential for resurrecting such a capability.[138]

This resounding dismissal of nerve agent acquisition represented the end of a policy almost 14 years old. Since 1962 the CoS had supported the acquisition of nerve agent weapons for deterrence and retaliation, when Solly Zuckerman and defence officials drastically reversed the 1956–1957 decision to abandon stocks and halt plans for the mass-production of chemical weapons. But, with the advent and rise of CBW disarmament talks, negative publicity and political aversion, the CoS was finally forced to change course. In certain areas, the CoS even went further than simply removing their request for nerve agent weapons, with 1976 also representing a major contraction in British CW policy as a whole. In addition to scrapping the nerve agent requirement, now the CoS supported substantial cuts to British CW research, the closure of the Nancekuke facility, the reduction of all staff involved in CBW research by around 700, and the curtailing of research and development work at Porton Down.[139] And, most importantly for the FCO, the green light for more detailed consideration of chemical weapons disarmament was given.[140] These substantial concessions and reductions led to an almost irreversible decline in Cold War British CW policy, and an important shift in stance.

To mitigate against these substantial cutbacks to CW research, as with much of the postwar period and as seen in 1953–1954 and 1958, Britain

[138] TNA, DEFE 4/282, Minutes of Meeting, Chiefs of Staff Committee, 2 March 1976, Confidential Annex; TNA, DEFE 13/1056, 'The Future of MRE and CDE', Chief of the Defence Staff and Acting Chief Executive (Procurement) to the Secretary of State for Defence, 16 March 1976. For later policy and confirmation of the decision, see: TNA, DEFE 13/1818, 'Chemical Warfare Policy', Note by the Secretary of the Chiefs of Staff Committee, 3 April 1980, Annex.

[139] TNA, DEFE 4/282, Minutes of Meeting, Chiefs of Staff Committee, 2 March 1976, Confidential Annex; TNA, DEFE 13/1056, 'The Future of MRE and CDE', Chief of the Defence Staff and Chief Executive (Procurement) to the Secretary of State for Defence, 3 March 1976.

[140] The actual greenlight and approval for the draft Chemical Weapons Convention would occur in august 1976. See: TNA, DEFE 4/282, Minutes of Meeting, Chiefs of Staff Committee, 5 August 1976.

again turned to the United States. Britain's lack of CW preparedness and its ever-dwindling CW effort led to the CoS explicitly accepting full reliance on the United States for CW deterrence, retaliation, offensive research, and development.[141] On this occasion, unlike in the 1950s, the decision to forego nerve agent weapons and to become completely dependent on the United States was communicated to senior United States officials. After initially expressing doubts, United States officials proved receptive to and understanding of the British decision.[142] This new level of reliance stretched beyond weapons and a retaliatory capability for deterrence, with the CoS even considering the purchase of defensive equipment from the United States.[143] For much of the Cold War this reliance on the United States for CW deterrence and a retaliatory capability had been more of an implicit policy for the CoS, yet in March 1976 this reliance and dependence was expanded, embraced, and explicitly acknowledged.

Reliance and dependence, however, came at a price. Officials in the United States stipulated that Britain should not make its drastic cuts to CW policy public, as the United States CW programme was seeking extra Congressional funding.[144] If Britain publicly proclaimed its reductions in CW research, it would have undermined these attempts by CW advocates in the United States. Serious questions would have been raised in Congress as to why the United States needed greater funding for chemical weapons when close allies were drastically curtailing their activities in the field. A condition of British dependence on the United States was therefore further secrecy in the short-term, meaning that eager FCO disarmament officials were unable fully to benefit from the significant concessions made by the CoS in early 1976. Publicity, of substantial and unilateral British reductions in CW policy and the renunciation of offensive weapons research and development, would have undoubtedly aided in garnering

[141] TNA, DEFE 4/282, Minutes of Meeting, Chiefs of Staff Committee, 2 March 1976, Confidential Annex; TNA, DEFE 13/1056, 'The Future of MRE and CDE', Chief of the Defence Staff and Chief Executive (Procurement) to the Secretary of State for Defence, 3 March 1976; TNA, DEFE 13/1056, 'The Future of MRE and CDE', Chief of the Defence Staff and Acting Chief Executive (Procurement) to the Secretary of State for Defence, 16 March 1976.

[142] Ibid.

[143] TNA, DEFE 4/282, Minutes of Meeting, Chiefs of Staff Committee, 2 March 1976, Confidential Annex.

[144] Ibid; TNA, DEFE 13/1056, 'The Future of MRE and CDE', Chief of the Defence Staff and Acting Chief Executive (Procurement) to the Secretary of State for Defence, 16 March 1976.

international support for a British chemical weapons disarmament proposal. This potentially rich vein of moral and political arguments thus initially went untapped for the sake of Anglo-American relations and was the price of reliance.

In line with the minimal publicity condition placed upon Britain by the United States, the British decision to greatly reduce CW expenditure initially attracted relatively scant attention.[145] Tucked away in the March 1976 parliamentary defence estimates, it was revealed that 'significant economies' would be made in British CBW policy along with the closure of the Nancekuke facility.[146] British officials downplayed this major moment in postwar policy. There was little fanfare or attention afforded to the decision to renounce offensive weapons research, development, and possession. Predictably there was some backlash from the hundreds of scientists now unemployed from the policy shift, as they expressed frustration and warned that such a reduction was 'unwise' given the scale of the Soviet threat.[147] This was quietened, however, with Government assurances that while the threat of the Soviet Union was fully appreciated the country needed to make economic cutbacks. Government officials stated that Britain would not be adversely affected by these reductions due to its close cooperation with the United States.[148] Any disquiet was swiftly stifled. British CBW researchers struggled to find politicians sympathetic to their plight. Conversely, some MPs, such as Michael Hamilton, used the cutbacks to draw even further attention to the need for chemical weapons disarmament, with Hamilton cautioning colleagues that the 'equally Ugly Sister' of BW, that of CW, still needed a disarmament treaty.[149]

While the United States encouraged secrecy, which limited disclosures on this significant policy shift and hampered the ability of FCO officials to fully capitalise on the situation, British defence officials had one final say over the direction of chemical weapons disarmament in mid-1976. The CoS and defence officials had a condition of their own for FCO disarmament negotiators: if Britain were to forego the option of acquiring a future

[145] James Wightman, 'Redundancies pledge on defence cuts', *The Daily Telegraph*, 21 February 1976, 7.

[146] House of Commons, 'Statement on the Defence Estimates 1976', Presented to Parliament by the Secretary of State for Defence, March 1976.

[147] Dennis Johnson, 'Scientists oppose closure of plant', *The Guardian*, 31 March 1976, 6.

[148] Hansard, House of Commons, 'Biological and Chemical Warfare', 1 April 1976, Vol.908, Cc1756–64.

[149] Ibid.

CW capability, then any disarmament treaty had to contain strict verification measures.[150] British defence officials were understandably keen to minimise any enemy's CW capabilities and the chance that British forces would be exposed to lethal chemical weapons.[151] This condition was seemingly made more urgent, and deemed essential by the CoS, after it had ruled out any British aspirations of attaining a nerve agent capability earlier in the year.

From the diplomatic standpoint, stringent verification measures would make a treaty slightly harder to negotiate considering Soviet objections to on-site inspections, as seen with its reservations over Japan's earlier proposals. Nevertheless, similarly to the BWC, Britain now also had a chance to show its genuine commitment to thorough and verified disarmament by pursuing a tough line. Officials could build on the momentum and optimism that had been fostered and created by the BWC. From the perspective of defence officials, these rigorous checks in any draft Convention were intended to ensure that even if Britain did not possess chemical weapons, then it would be unlikely to face a Soviet Union which did. With the removal of the Soviet CW threat, there would have been no need for a CW deterrent. On 3 August 1976, defence officials thus concluded that:

> As a state with no offensive or retaliatory CW capability of our own and with no intention of acquiring such a capability, a genuine CW disarmament measure would be to our advantage.[152]

By taking one piece out of the game, the CoS tactfully recognised that they could, in fact, level the playing field through a disarmament treaty. Having no intention of acquiring nerve agent weapons and with the Soviet Union long thought to have a significant advantage over Britain in CW capabilities, the CoS swung fully behind chemical weapons disarmament and rigorous verification measures.[153] Defence officials also appreciated that a British attempt would have the added benefits of countering criticism from Non-Aligned Countries. It would show Britain to be proactive in chemical weapons disarmament and it would form a part of the broader

[150] TNA, DEFE 13/1056, 'Draft Chemical Warfare Convention', Chief of the Defence Staff to the Secretary of State for Defence, 5 August 1976.

[151] Ibid.

[152] TNA, DEFE 13/1056, 'Draft Chemical Weapons Convention', A. P. Hockaday to PS/ Secretary of State, 3 August 1976.

[153] Ibid.

Western effort to make the world 'more secure'.[154] The CoS acquiescing to further chemical weapons disarmament talks, but insisting on stringent verification measures in any Convention actually coincided with the mood in much of the wider international community which had in sight the same goals: the complete, thorough, and verifiable dismantling and destruction of CW stockpiles and the banning of development and production.

One major drawback from tough verification measures in an international disarmament treaty, from the military perspective, was that it would 'result in the loss of the United States CW stockpile, which at present provides the only credible CW deterrent available for use in the NATO area'.[155] This created the slightly strange situation of Britain actively seeking reliance on the United States and its CW deterrent, while supporting disarmament talks which were attempting to scrap that very same deterrent.[156] Though this fear of losing the CW deterrent of the United States was not enough to override the rise of chemical weapons disarmament in British policy, with the CoS simply re-iterating that strict verification measures were of the 'greatest importance' in any disarmament treaty.[157] In 1976, defence officials ultimately thought it near-impossible to reverse or resist the growing surge of support for chemical weapons disarmament. They could only hope to guide its course and achieve the best possible outcome through stringent verification measures. The finale of this ongoing negotiation between the FCO and defence officials over strict verification would be seen on 6 August 1976, when Britain tabled a draft Chemical Weapons Convention at the UN Disarmament Conference in Geneva.[158] The draft Chemical Weapons Convention, after years of effort and debate,

[154] Ibid.

[155] TNA, DEFE 13/1056, 'Draft Chemical Warfare Convention', Chief of the Defence Staff to the Secretary of State for Defence, 5 August 1976.

[156] In this sense is was strange state of affairs in terms of Anglo-American relations, as until Britain had secured a disarmament treaty, reliance on the CW deterrent of the United States was the next best thing in terms of attempting to prevent Soviet use of chemical weapons. Spiers. *Chemical Warfare*, 207.

[157] TNA, DEFE 13/1056, 'Draft Chemical Warfare Convention', Chief of the Defence Staff to the Secretary of State for Defence, 5 August 1976.

[158] United States Arms Control and Disarmament Agency. 1978. *Documents on Disarmament: 1976*. Washington, DC: United States Government Printing Office, May, 520–526; Sims, Nicholas. 1987. *International Organization for Chemical Disarmament (SIPRI)*. Stockholm: Almqvist & Wiksell, 21; Howlett. UK arms control and disarmament policy, 159.

had the backing of the CoS and it contained strict verification measures.[159] It was seen by many, campaigners and foreign governments alike, as a welcome and important move towards disarmament.[160]

In the draft Convention, Britain recommended the prohibition of the 'development, production and stockpiling of chemical weapons'.[161] The draft included stringent measures for verification, with it calling for 'strict and effective international control', on-site verification, and the listing of all CW stockpiles.[162] This aspect of the draft Convention entailed state assent to onsite inspections and the verification of the destruction of stockpiles.[163] These strict verification methods were stronger than in the Japanese draft and they entailed potentially intrusive inspections from external officials. This British effort deviated greatly from the earlier Soviet draft, that had opposed on-site inspections by external officials, and from the original United States line of freezing stocks and halting production. Although described by Mark Allen, the British Ambassador to the UN Disarmament Conference, as a 'synthesis' of previous drafts put forward by the Soviet Union, Non-Aligned Countries, Japan and Canada, the draft Convention represented and signalled a major shift in British attitudes towards CW policy and nerve agent weapons.[164] Even though in parts the draft Convention read like a 'rough draft' that was rushed out in inexplicable haste, perhaps by FCO officials keen to make the most of the international climate and the CoS drastically changing course, it represented a critical turning point.[165] After years of internal debates and discussions, during 1976 defence officials fully transitioned to supporting chemical weapons disarmament and abandoned aspirations of acquiring nerve agent weapons.

Despite the draft Chemical Weapons Convention representing a considerable shift in British CW policy, with disarmament overriding

[159] TNA, DEFE 4/282, Minutes of Meeting, Chiefs of Staff Committee, 5 August 1976.

[160] Murphy et al. *No Fire, No Thunder*, 80.

[161] United States Arms Control and Disarmament Agency, *Documents on Disarmament: 1976*, 520–526; Sims. *International Organization for Chemical Disarmament*, 21.

[162] United States Arms Control and Disarmament Agency, *Documents on Disarmament: 1976*, 520–526.

[163] Sims. *International Organization for Chemical Disarmament*, 21–23; Spiers. *Chemical Warfare*, 179–181.

[164] Rod Chapman, 'Chemical war ban urged by Britain', *The Guardian*, 13 August 1976, 3; *The Daily Telegraph*, 'Britain's Draft Treaty on Chemical War', 13 August 1976, 4.

[165] Sims. *International Organization for Chemical Disarmament*, 23–24.

deterrence, internationally Britain's draft Convention was overtaken by other developments in the disarmament field. The British draft Convention was well-received and debated, with it remaining 'on the table', but it was overshadowed by reinvigorated bilateral talks on chemical weapons disarmament by the two superpowers.[166] These bilateral disarmament discussions were revitalised after the United States indicated that it was willing to pursue a 'two-track' approach. Alongside UN negotiations the United States would negotiate bilaterally with the Soviet Union.[167] Reflective of wider trends in détente and evolving international and domestic attitudes towards chemical weapons disarmament, United States officials had changed tack from their earlier line, and in 1976 they were far keener on verification and inspections from international representatives. This placed the United States much more in line with the thinking of British, Japanese, and other non-Eastern European countries in relation to strong verification measures. In view of these ongoing and seemingly promising bilateral negotiations, Britain held off pressing ahead with its own draft Chemical Weapons Convention at Geneva, wishing to avoid impeding or detracting from bilateral superpower talks.[168]

After abiding by United States requests to delay the publicity aspect of the March 1976 CoS decision and with ongoing superpower bilateral talks, Britain's dramatic shift in CW policy was publicly confirmed in October 1976, months after it had submitted the draft Chemical Weapons Convention.[169] Parliament was finally and explicitly informed that Britain would only focus on defensive research, Nancekuke would be closed and that this contraction would be mitigated by the capabilities of the United

[166] Foreign Relations of the United States, 1969–1976, Volume E–14, Part 2, Documents On Arms Control and Nonproliferation, 1973–1976, '194: Telegram 3110 From The Mission in Geneva to The Department of State, Geneva', 23 April 1976, 1514z; Sims. *International Organization for Chemical Disarmament*, 24; Spiers. *Chemical Warfare*, 179; Adams. *Chemical Warfare, Chemical Disarmament*, 173.

[167] Foreign Relations of the United States, 1969–1976, Volume E–14, Part 2, Documents On Arms Control and Nonproliferation, 1973–1976, '194: Telegram 3110 from the Mission in Geneva to the Department of State, Geneva', 23 April 1976, 1514z; Meselson, Matthew and Robinson, Julian Perry. 1980. Chemical Warfare and Chemical Disarmament. *Scientific American*, 242:4, 38–39, 45–47.

[168] Spiers. *Chemical Warfare*, 180; Adams. *Chemical Warfare, Chemical Disarmament*, 173.

[169] House of Commons, 'Cuts in Defence Expenditure: Ministry of Defence', Expenditure Committee (Defence and External Affairs Sub-Committee), 12 October 1976; David Fairhall, 'Porton's future in doubt after cuts', *The Guardian*, 20 November 1976, 20.

States.[170] This was even after some last-gasp protestations from supporters of a CW capability and those fearful of the Soviet threat. A potential 'chemical warfare onslaught by the Warsaw Pact' was reported and some Conservative MPs criticised the decision to significantly reduce CW expenditure and research.[171] This criticism, however, originated from a select few MPs. The vast majority of MPs raised no complaints. The last gasp paled in comparison to the prior campaigns against Government CW activities during 1959–1960 and 1967–1970, with outspoken supporters of such a controversial area few and far between in 1976. The minor backlash against the massive reductions in British CW policy failed to prevent or reverse the direction of British policy, which had orientated behind chemical weapons disarmament.

The legacy of this 1976 compromise over disarmament, with the abandoning of the request for a nerve agent capability and support for strict verification measures, would stretch far beyond the 1970s and throughout the remaining Cold War. The compromise between the FCO and defence officials represented a lasting and irrevocable shift in British attitudes towards nerve agent weapons. Even though bilateral superpower talks did eventually break down over the controversial issue of verification, the British draft Chemical Weapons Convention highlights a dramatic turning point in Britain's military outlook.[172] This substantial shift, in the face of the rising tide of disarmament, would remain a constant influence. British officials, committed to chemical weapons disarmament and strict verification, continued to work towards a Chemical Weapons Convention long after 1976.[173]

The foundation established and represented by the 1976 draft Chemical Weapons Convention would be built upon and updated with repeated FCO

[170] House of Commons, 'Cuts in Defence Expenditure: Ministry of Defence', Expenditure Committee (Defence and External Affairs Sub-Committee), 12 October 1976; Hansard, House of Commons, 'Chemical Warfare', 29 November 1976, Vol.921, c35W; Hansard, House of Commons, 'Chemical Warfare', 23 December 1976, Vol.923, cc895–6.

[171] *The Daily Telegraph*, 'Chemical Warfare', 15 November 1976, 4; Hansard, House of Commons, 'Chemical Warfare', 29 November 1976, Vol.921, c35W; Hansard, House of Commons, 'Chemical Warfare', 23 December 1976, Vol.923, cc895–6.

[172] Soviet officials were adamant that the technical verification processes, like that of soil sampling, could only be carried out by local national authorities. Such a measure would have left a substantial hole in a treaty designed to prevent stockpiling, with the effectiveness and reliability of any findings undoubtedly questionable.

[173] TNA, DEFE 24/1739, United States/UK Bilateral Meeting on CW, 30 June 1977; Sims. *International Organization for Chemical Disarmament*, 25.

attempts for chemical weapons disarmament in the late 1970s and 1980s.[174] In 1984 Britain tabled a further three chemical weapons disarmament papers, and in 1986 Britain again attempted to push for disarmament negotiations.[175] This ongoing British disarmament push culminated with the CWC, containing strong verification measures, opening for signatories in 1993.[176] The CWC has been ratified or acceded to by 193 states, and remains resilient to this day.[177] Across these chemical weapons disarmament attempts, up to and including the signing of the all-important CWC in 1993, British officials remained committed to strict verification measures. The origins of this stance can be traced to 1976, when the FCO and the CoS compromised over disarmament talks, when Britain's explicit commitment to strict verification measures solidified, and when defence officials finally dropped their request for nerve agent weapons.

[174] Ibid. For an account of bilateral superpower CW negotiations in this period, see: Spiers. *Chemical Warfare*, 179–190

[175] Adams. *Chemical Warfare, Chemical Disarmament*, 173–182; Spiers. *A History of Chemical and Biological Weapons*, 66–68. For further details on the early 1980s stumbling bloc of binaries and verification in chemical weapons disarmament talks, see: Spiers, Edward M. 1984. Bargaining with Binaries. *NATO Review*, 32:5, 20–25.

[176] Abe, Tatsuya. 2017. Challenge inspections under the Chemical Weapons Convention: between ideal and reality. *The Nonproliferation Review*, 24:1–2, 170–172. For the actual CWC Treaty see: Organisation for The Prohibition of Chemical Weapons, 'Convention on the Prohibition of the Development, Production, Stockpiling and Use of Chemical Weapons and On Their Destruction'.

[177] Costanzi & Koblentz, Controlling Novichoks after Salisbury, 599; Costanzi, Stefano and Koblentz, Gregory D. 2020. Updating the CWC: How We Got Here and What Is Next, *Arms Control Today*, April, 19.

CHAPTER 8

Conclusion: From Discovery to Disarmament

From acquiring German nerve agents to fully supporting and pursuing chemical weapons disarmament, nerve agent policy in postwar Britain was tumultuous and controversial throughout. In the immediate aftermath of the Second World War, the nerve agent discovery sent shockwaves through British defence policy. In the utmost secrecy, the Services jostled for control of new nerve agent weapons, and the search began for a new nerve agent capability to deter any potential enemy. With the Army emerging as the chief proponent of a nerve agent capability, a more clearly defined role for nerve agent weapons developed: that of a CW deterrent and a tactical weapon that was complementary to atomic weapons. In 1950, reflecting this growing military support, and spurred on by the views of Prime Minister Clement Attlee and the outbreak of the Korean War, the Labour Government approved the domestic production of sarin. Yet while approval for nerve agent acquisition was granted, policy implementation coincided with the 1952 Global Strategy Paper, which was to prove a crucial turning point for CW policy in the Cold War. The 1952 Global Strategy Paper revealed arguments and divides between military utility and normative factors, with defence officials attempting, and failing, to change British first-use policy. This setback, in a period of economic constraints, represented a bitter blow to defence officials who were advocates of a British nerve agent capability.

The resulting backlash over first-use and the 1952 Global Strategy did not however permanently solve the matter, for in 1955 defence officials

W. King, *Nerve Agents in Postwar Britain*, Britain and the World,
https://doi.org/10.1007/978-3-030-70474-2_8

again attempted to push for a review of first-use policy. On this occasion, as with the last, the Foreign Office again shut down the ambitions of defence officials. The ramifications of this second rebuttal would prove hugely significant, when, despite the military need for a CW deterrent and even after the best efforts of Prime Minister Anthony Eden, without a clear role and with no guarantee of use, nerve agent policy was significantly curtailed and reduced. In 1957, Britain disposed of existing CW stockpiles, plans for domestic mass-production were cancelled, and offensive weapons development was halted. The ensuing drift and decline lasted until the early 1960s, when, alongside rising fears of CW proliferation in the Third World, a substantial and holistic reappraisal of nerve agent weapons was conducted. Importantly, this period also witnessed some key interventions in the public sphere. Investigative journalists such as Henry Chapman Pincher revealed the British V-agent discovery and some information became available on British CW policy. These limited disclosures encouraged even greater care and secrecy amongst defence officials. It was in this climate that a secret review of CW policy took place. This was a crucial, if at times fractious, review. It coincided with the arrival of Solly Zuckerman, who played a significant part in pushing for a considerable reverse course in the CW field. In 1962, behind closed doors, the CoS again swung fully behind a British CW deterrent and a nerve agent capability, a policy stance which was to last until 1976. In 1963, though, and acting on the advice of the CoS, Prime Minister Harold Macmillan, like Attlee in 1950, approved British nerve agent acquisition.

Substantial changes in British politics were to impact the acquisition of nerve agent weapons, with the arrival in office of Labour Prime Minister Harold Wilson in 1964. Wilson was far more sensitive to public opinion and to the actions of his backbench MPs. As such, CW policy went through a period of drift and marginalisation in the mid-1960s, with defence officials seeking to acquire a nerve agent capability but with Labour politicians deferring and shutting down any request. The crux of this politico-military divide and policy malaise centred on interpretations of successful and effective deterrence. Military officials were demanding publicity, but politicians were seeking to avoid publicity at all costs. The political costs of a nerve agent deterrent were thought too great. This debate and ongoing divide over interpretations of deterrence resulted in a period of abeyance on the nerve agent question from 1968, which was intended to last two years, but which ended up lasting eight. Coinciding with this early abeyance in 1968 was a rise in publicity, growing public interest in CBW, political opposition to British CBW research, and the convening of international disarmament talks. Domestic criticism was particularly vocal and consistent. It included but was not limited to students, members of the press, scientists, academics, campaigners, and MPs. These groups and key individuals, such as Elizabeth Sigmund and Tam Dalyell, exerted considerable pressure on Government policy. From the political perspective, domestic and international developments compounded the impossible nature of publicising a British CW deterrent, yet the CoS and defence officials remained wedded to publicity for deterrence. Only in 1976 did the CoS finally give substantial ground, for that is the year in which defence officials abandoned their nerve agent request, when they agreed to significant reductions in CW policy, and when they acquiesced to the submission of a British draft Chemical Weapons Convention. The marker of 1976, therefore, represents a key milestone in British CW policy. It is when British consideration of nerve agent acquisition and deterrence clearly gave way to the pre-eminence of disarmament talks.

While covering a long time period, this book has predominantly focused upon the higher levels of nerve agent policy formulation. It has taken a largely top-down approach focusing on the history, actions and motives of senior government officials and defence officials when it came to the nerve agent question and defence policy in postwar Britain. This has facilitated an analysis of the driving factors of nerve agent policy and British considerations of nerve agent weapons for both deterrence and use. It is also important to recognise that this book in itself is not a complete history of the British nerve agent experience, but a complementary contribution to a much larger field. A field which I hope, and am sure, will grow even further in the future. Further avenues could readily be explored which would provide additional and valuable insights for both the CW field and British history. In a purely chronological sense, increasingly available archival sources will soon allow for a thorough understanding of the evolution of chemical weapons disarmament talks in the 1980s, and of British involvement in the formation of the Chemical Weapons Convention. This may well reveal another clash between defence and Foreign Office officials, or disagreements between the Anglo-American partners, especially if there were pressures from President Ronald Reagan's re-armament plans (including in the CW field). These sources would also allow for a detailed analysis of British CW policy at the end of the Cold War, and the reaction to the potential proliferation of CW weapons after it.[1] Public opinion

[1]These events, alongside growing proliferation in the Third World and with the comparative revival of the United States CW programme under President Reagan, would provide

regarding CW policy, which although addressed in this book and by other excellent accounts, could also be broadened out yet further, especially through the prism of emotions in history and more detailed accounts of the campaigners themselves. Additional areas that could build upon this project include activities in NATO and the Warsaw Pact regarding CW, transnational and comparative studies, and more directly the role of international organisations such as the UN and Third World actors through the Non-Aligned Movement. While these other avenues would provide further insights, this project has provided many of its own.

In line with the first aim of the book, the remarkable seriousness with which nerve agent weapons were taken by British government officials and by politicians has been revealed. These were not weapons always left on the periphery, nor an unconventional extra. The role and value of nerve agent weapons was repeatedly discussed at the highest echelons of the defence establishment, as well as in senior and influential committees such as the CoS, the DRPC and the all-important Cabinet Defence Committee. This level of attention was reflective of the active military interest in nerve agent weapons, with defence officials keen on CW deterrence and arguing for the acquisition of nerve agent weapons which could be used against Soviet forces. For defence officials, the perceived necessity of chemical weapons, whether through their tactical employment or deterrence value, overrode legal, political, and moral considerations of this abhorrent form of warfare. These costs were thought a worthwhile price to pay in order to gain access to nerve agent weapons. It is important to note that CoS support for the acquisition of nerve agent weapons after 1945 lasted for a total of almost 25 years. Nerve agent weapons were discussed, debated, and their acquisition approved on multiple occasions at the very highest levels of British defence policy.

Throughout the Cold War, senior politicians seriously considered and waded into these controversial waters. Political engagement regularly spanned party lines and drew the attention of Prime Ministers. In the immediate postwar period, it was the Labour Prime Minister, Clement Attlee, who supported CW deterrence through an effective retaliatory capability. It was under his watch that Britain took the bold steps of

fascinating insights. At the time of writing, currently CW policy can be traced in detail up to around 1985–86. This has also been greatly aided by the switch to a 20-year release rule, which should soon yield substantial dividends in terms of mapping CW policy at the end of the Cold War and through the negotiation of the Chemical Weapons Convention.

acquiring and storing German nerve agent weapons in 1945–1946 and of approving the domestic mass-production of sarin in 1950. The complete reverse of this was later seen with another Labour Prime Minister, Harold Wilson. In 1968, Wilson oversaw CW policy going into abeyance, believing the political costs of CW deterrence too great. And, in early 1976, he was again at the helm when defence policy dramatically shifted in favour of full chemical weapons disarmament.

Similarly, in the Conservative Party, from 1952 Conservative Prime Minister Winston Churchill actually did little to aid nerve agent acquisition. Although in 1954 he approved nerve agent production, his condition that items of defence be proven essential hampered CW policy. This pushed defence officials to repeatedly seek a change in first-use policy, which Churchill then did not support. Churchill's involvement in British nerve agent policy should thus not be pre-judged, as while he was interested in the field and the possibilities it represented, he was by no means the active or vocal supporter of chemical weapons that can sometimes be found in previous parts of his career (such as in earlier conflicts or during the Second World War). Of all the Conservative Prime Ministers, it was surprisingly Anthony Eden who was the most outspoken and confrontational in his support of a nerve agent capability. In doing so, he even went against his own Cabinet and Defence Secretary. By contrast, even though Harold Macmillan approved nerve agent acquisition in 1963, he was not an ardent believer or advocate of CW deterrence in the same manner as Eden.[2]

Often the nerve agent issue was dubbed important enough for Prime Ministers to attempt to take personal control of policy. At times, they even sought to consult directly with Presidents of the United States. Churchill and Eden were particularly keen on using the Anglo-American relationship to secure either nerve agents from the United States, or assurances that the United States would retaliate and deter on behalf of Britain. Even in 1963, when Macmillan approved the expansion of CW policy and the acquisition of nerve agents, he did so in part to maintain close Anglo-American ties and to secure British access to advanced United States CW research. This seeking of support from the United States, and of Prime Ministers willingly taking up the nerve agent question at the highest levels of Anglo-American relations, is further reflective of the seriousness with

[2] Conversely Macmillan also implemented the substantial policy cutbacks of 1957 and he presided over the slump and drift of British CW policy from 1957 until 1962.

which the nerve agent question was taken by Prime Ministers and defence officials alike.

Though Prime Ministers and defence officials followed a variety of paths and approaches when it came to the nerve agent question, what they all had in common was an attempt to adhere to some degree of secrecy. The second key finding of this book shows that secrecy, publicity, and parliamentary opinion were at times serious constraints on the ambitions of defence officials. This amalgamation of factors often led to surprising revelations and to confusing and contradictory policies. Constant pressure from interested and active groups and individuals also eventually forced even to CoS to recognise that publicity for deterrence was an immensely troublesome and difficult endeavour, and they even began to factor public opinion and likely adverse emotional reactions into their decision-making. This publicity aspect also reveals a broader theme, that of critical public and parliamentary reaction to almost any form of Government policy or statement on CW, as well as tactical and calculating Government responses to limit any adverse publicity. These responses included limited disclosures in an attempt to control and stifle the flow of information. Adverse publicity also increased over time and mirrored increasing public and parliamentary awareness of the nerve agents and what they represented. This trend reflected a broader shift in the coverage of science and politics in Cold War Britain. During the immediate postwar period, political statements were mostly moderate or even supportive of the success of wartime CW deterrence, yet as information perforated into the public domain from the mid-1950s, there was a clear shift in public attitudes and awareness. A key part of nerve agents in postwar Britain is thus the handling of, and adaptation to, this increasing public awareness and comprehension of the nerve agents.

While keen on secrecy, steering or limiting the public debate often came at a price. The price seemed to increase alongside public awareness and the level of scrutiny. In 1960, for example, mounting public interest pushed the Government into revealing surprising weaknesses in Britain's defensive posture, such as the country possessing no CW capabilities, in order to try and stifle and limit the level of public attention. Likewise, in the late 1960s, public and political pressures again pushed officials into revealing that Britain possessed no stockpiles of chemical weapons. In both cases, limited disclosures, in an attempt to quieten adverse publicity, seemingly trumped strict secrecy. The flip-side of this was that this strict adherence to some form of secrecy, as well as military pressure, also prevented government officials from fully capitalising on Britain's decision not to produce or acquire nerve agent weapons at an earlier date. Even though they had made these concessions, British officials did not fully mine a potentially hugely beneficial moral and political vein or proclaim what was in effect unilateral disarmament. A key reason for this was the pressure of defence officials keen to maintain some semblance of security in the Cold War, which coincided with a legacy of secrecy that had been ingrained over time. A public announcement would also likely have committed Britain to a set course long before defence officials were willing to do so. This would take until 1976. Publicity thus played a large part in repeatedly pushing nerve agent policy down a strange middle path. Limited disclosures revealed British weakness, but adherence to some form of secrecy prevented access to the substantial domestic and international political benefits which could have been accrued from unilateral disarmament.[3] Campaigners and those critical of government policy, including backbench Labour MPs, could have been won over with admissions of non-possession and greater transparency, and in theory Britain could have taken an even stronger and more vocal role in international disarmament talks.

A core reason for maintaining some form of secrecy, in addition to secrecy ingrained over time, security concerns, and the aspirations of defence officials, was that Britain was often tied in to follow strict secrecy measures for the sake of maintaining close Anglo-American relations. This factor of alliance networks and publicity was clearly emphasised in 1976. When, even after military approval for disarmament talks, détente, and with serious cutbacks to CW policy, Britain was initially unable to fully promote its abandonment of offensive weapons. British publicity, for the purposes of capitalising on its lack of chemical weapons, would have drawn significant dissatisfaction from colleagues in the United States, who themselves had a substantial CW programme and who were often seeking greater funding for it. As Britain was reliant on the retaliatory and deterrent CW capability of the United States and promulgating its opposition to such weapons would have proved untenable. Publicity thus represented a constraint and a complex web. International alliances as well as internal security pressures prevented Britain from fully utilising the political and moral benefits of its lack of chemical weapons, but with public and parliamentary pressures demanding disclosures.

[3] Recent research is also showing the strategic use of the chemical weapons taboo, see: Bentley, Michelle. 2014. Strategic taboos: chemical weapons and US foreign policy. *International Affairs*, 90:5, 1033–1048.

During critical stages in nerve agent policy, British officials themselves were also not averse to withholding crucial information from the United States. In 1956–1957 and in 1963, the United States was not initially officially informed of substantial shifts in British policy. In 1957, this was fuelled by concerned British officials who feared the ramifications of senior United States officials becoming aware of British reliance, and in 1963 it was a result of fears of negative publicity and the leaking of secret information. These extraordinary measures, of withholding information on substantial policy changes from a close ally, were largely taken on the grounds of secrecy. This withholding of information and adherence to secrecy was, though, contradictory. Much of British CW research and policy was predicated and dependent upon a close and open working relationship with the United States. The United States and Britain also took vastly different approaches to CW publicity, which did not aid cohesion. With United States officials viewing publicity more as a potential tool to secure greater funding and with British officials keen to avoid almost any form of it, at least when it concerned deterrence and an actual capability. There was a clear degree of friction between the two approaches throughout much of this Cold War period.

Differing approaches to publicity also represented a particularly troublesome and divisive issue for British defence officials and politicians, which came to dominate considerations of CW deterrence. This divide came to the fore in the mid-1960s. It was largely a result of defence officials deeming a CW deterrent and a nerve agent requirement as vital, and insisting that for deterrence to work Britain needed to publicise its CW capabilities and intent. From a political perspective, any public disclosure over CW research, preparedness, or intent, even for deterrence, was anathema. Here, it was not so much 1976 which represented the crucial turning point for many politicians as to a nerve agent capability and deterrent, but the legacies of both 1959–1960 and 1967–1970. In the latter period, government officials faced mounting pressure from grass-roots movements, the media, backbench MPs and increasing disarmament talks, and as a result political aversion to nerve agent acquisition solidified. The political acceptability of nerve agent publicity was thus shattered in this period. From 1970, this is not so much a political account, but one of defence officials attempting to deal with the aftermath, aversion, and drift resulting from politicians effectively abandoning nerve agent acquisition in the late 1960s.

Even though secrecy and publicity did lead to clashes and contradictory policies, especially in terms of deterrence, it does point towards the third significant finding of this book: despite there being no unanimous interpretation as to what CW deterrence entailed or how it could be accomplished, it was an active, conflicting, and important part of defence policy in postwar Britain.

For defence officials, successful CW deterrence entailed the threat of, and ability to respond with, a like-for-like retaliatory capability. This was very much in line with other areas, such as nuclear deterrence, which often followed a similar mould. A core part of the origins of this approach to CW deterrence stems from CW experiences in the Second World War (an interesting further study could be to also trace these deterrence views to earlier times, such as the First World War). For defence officials and politicians alike, it was the example of wartime CW deterrence which was seen as a success and a template for CW deterrence in the Cold War. Preparedness and the threat of retaliation were thought central to deterring German first-use. Wartime deterrence of the German first-use of chemical weapons was also seen as an almost reassuring experience. It could be seen to show that even at the height of a world war deterrence could prevent the outbreak of a repugnant form of warfare. This past success of CW deterrence thus provided defence officials with a persuasive, re-enforcing, and acceptable interpretation of what effective deterrence with WMDs entailed in the Cold War.

In many ways, though, CW policy was also severely confined and straight-jacketed by this legacy of the Second World War and by ingrained conceptions of CW deterrence. Military officials were uncompromising over the idea of deterrence as only achievable through possession of a capability for retaliation-in-kind. This pervasive and lasting belief in the success of wartime CW deterrence played a significant part in defence officials remaining so committed to their interpretation of CW deterrence. For Eden, the perceived prior success of CW deterrence during the Second World War left such a mark, and was felt so persuasive an approach, that he willingly and unilaterally attempted to halt drastic changes to British CW policy. For defence officials, it meant that they were always committed to deterrence through the acquisition of a retaliatory capability and to publicity, just like in the Second World War. This wartime legacy and commitment to retaliatory deterrence thus led to a substantial degree of rigidity in policy. In 1976 defence officials believed that they had a vaguely credible alternative to retaliation-in-kind deterrence, as based on their experiences and outlook the only real alternative was verifiable and thorough disarmament, rather than other forms of deterrence proposed by FCO officials and by Denis Healey.

This focus on retaliation-in-kind deterrence thus led to the overlooking of alternative forms of deterrence, which could have removed CoS demands for nerve agent publicity or a capability at an earlier date. These other interpretations of CW deterrence also vary greatly from nuclear deterrence. At times the meaning and understanding of CW deterrence were far more nuanced and varied than in other aspects of defence policy. For example, few defence experts would advise either allowing an enemy to use a capability first, then deterring its further use as Healey had suggested, or deterring the use of a weapon of mass destruction through purely defensive measures/equipment, which was seriously considered and advocated by FCO officials. Successful CW deterrence through purely defensive measures and equipment is a particularly interesting prospect, which was perhaps not given enough credence at the time due to the dominance of retaliatory deterrence.

The entrenched interpretation of the role and value of CW deterrence and of nerve agent weapons did, however, provide the CW field with a distinct advantage over the closely related field of BW. While the two fields were to a degree overlapping, they followed very different paths in the postwar period. Unlike biological weapons, chemical weapons were consistently seen as genuinely complementary to nuclear weapons. Tactical chemical weapons were viewed as a useful ancillary option. With biological weapons, though, officials believed there to be some overlap with nuclear weapons. Biological weapons were also seen as a strategic weapon to inflict causalities on a massive scale. This was not unique to the BW field, as even when CW policy was perceived to overlap with the strategic dominance of nuclear weapons, it was also curtailed. Advocates of the strategic potentials of CW, particularly in the Air Ministry, were unable to secure serious and sustained considerations of a strategic CW capability. A core difference in the varying and alternative experiences of CW and BW policy is that advocates of nerve agent weapons could fall back on their perceived tactical and complementary role and on the need for a nerve agent deterrent, while BW advocates struggled to separate and isolate the role of biological weapons from nuclear weapons.

Intelligence on the Soviet threat also played a role, albeit a slightly mixed one, in the perceived necessity of a CW deterrent capability. Often, policy decisions were taken without an accurate intelligence picture, due to a lack of verifiable information. Early intelligence reports largely relied on wartime information, a process of mirroring British or US efforts, or of extrapolating or assessing what Soviet defensive preparations meant for its offensive preparations and intent. This difficulty was not so much a failure of intelligence; it more reveals the severe difficulty of gathering verifiable intelligence on CW capabilities and the stringent levels of secrecy present in the Soviet Union. Notably, there were some intelligence gains, such as in the late 1950s, which played a role supporting the 1962–1963 decision to acquire nerve agent weapons, especially regarding fears of proliferation in the Warsaw Pact and Third World. But, intelligence on Soviet capabilities more often influenced policy through stressing that a serious threat existed which needed to be deterred, but that its exact nature or form remained unknown.

The fourth finding of this book relates to often turbulent but extremely close Anglo-American cooperation. The dominance and scale of the United States CW programme in part facilitated a decline in British CW policy, by providing the option of complete dependence and continued access to advanced research. In stark contrast to the almost complete dependence seen in 1976, in the immediate postwar period British CW policy was in a comparatively strong position. The country held an independent CW capability, a large stockpile of captured nerve agents, and an advanced research effort. At this early stage of the Cold War, British CW policy benefitted from tripartite cooperation, but it was not wholly reliant on this network. As the Cold War progressed, however, and alongside economic pressures and political aversion, Britain gradually shifted from independence to interdependence. By 1976, it was almost entirely dependent on the United States for most areas of nerve agent policy. Certainly, for nerve agent retaliatory capabilities and deterrence, Britain was dependent on the United States from 1957. Even though Ministerial approval was given for the acquisition of nerve agent weapons in 1963, it was never acted upon. The nerve agents in postwar Britain is thus also, underneath the internal debates and discussions, an account of British nerve agent policy in decline and of growing British reliance on the United States.

Tripartite and Anglo-American collaboration in the CW field, though, occurred with noteworthy intensity and closeness, with CW cooperation perhaps only rivalled by BW cooperation in its depth and scope in British defence policy. This close cooperation, while having strong roots in the wartime experiences of the countries involved, was deepened and shaped during the Cold War. Regular conferences, the division of research, and coordinating committees all further developed collaboration and led to a striking level of cooperation. This closeness allowed Britain to reduce expenditure in the CW field. It also helped avoid the unnecessary duplication of resources. Close Anglo-American ties also held open the option of Britain acquiring nerve agent weapons from the United States. The Anglo-American relationship was not always smooth or completely open, but it did still facilitate the pooling of resources which was of vital importance to Britain, especially in light of the perceived Soviet CW threat and given the research burden that came with the nerve agent age. And, even though there was turbulence in cooperation and the scale of the United States effort provided the option of British reliance, Anglo-American and tripartite relations were central to keeping Britain at the forefront of CW research in the Cold War.

The fifth finding is that despite the significant importance attached to a nerve agent capability, military necessity, perceptions of the threat, and substantial technological developments, were seldom enough to override fears of adverse publicity, economic pressures, moral aversion, or the pre-eminence of internationally recognised arms control treaties. In the Cold War, arguments over military utility continually failed to overcome the constraints to their acquisition and possession. In 1946, in the postwar period, retention of German stocks had been approved. In 1950, with the Korean War raging, nerve agent development was given the green light. In 1954, after the nerve agent deal with the United States had fallen through, nerve agent production was again initially approved. And, in 1963, with fears of Third World proliferation, nerve agent acquisition was yet again granted political support. Yet despite these policies, remarkably, Britain never domestically mass-produced nerve agent weapons. In the early 1950s acquisition stumbled when it came to the issue of first-use, in the mid-1960s nerve agent weapons failed to overcome political aversion and negative publicity, and in the 1970s acquisition was subsumed by the unrelenting shift towards disarmament. A key underlying feature of nerve agents in postwar Britain is thus that fears of the Soviet threat, as well as the perceived military necessity and demands for a British nerve agent deterrent, were repeatedly overridden.

For the acquisition of nerve agents, unlike with riot-control agents, commitments to the 1925 Geneva Protocol remained sacrosanct and unyielding. From 1952, defence officials keen on nerve agent weapons witnessed the strong resistance to any re-interpretation of first-use and any deviation from the internationally recognised 1925 Geneva Protocol. This resistance came from within the Government, with the Foreign Office and members of the Cabinet Defence Committee actively opposing any reinterpretation of the widely accepted Geneva Protocol. Even though the nerve agent weapons represented a significant advance in the CW field, and despite the Soviet threat and need for a CW deterrent, military requirements never justified a unilateral reinterpretation of first-use. The 1952 Global Strategy Paper, in part, represents a crucial success for the primacy of normative values over military utility and deterrence. Further exemplars of the influence of international treaties and political aversion to nerve agent weapons were also seen in 1955, when military attempts to re-interpret the Geneva Protocol were again rebuffed.

In the 1960s, advocates of nerve agent weapons could also do little to overcome political considerations and adverse publicity, with Wilson's 'masterly silence' isolating and marginalising CW policy.[4] In light of this aversion to nerve agent weapons, and to the publicity thought necessary for deterrence, defence officials keen on acquisition could do little but bide their time, hopeful of a change in fortunes. The 1960s thus represents both the strengths and weaknesses of political oversight. For while politicians side-lined and quashed requests for nerve agent weapons, they did not proactively and definitively remove it. The CoS were left simply waiting for a more opportune time. Rather than enact any meaningful change from the mid-1960s, nerve agent policy was in essence frozen in place.

The finale to this ongoing political-military struggle only came with the rise of chemical weapons disarmament. After 1972, British commitments to chemical weapons disarmament started to take hold of policy considerations. This upsurge of support for chemical weapons disarmament effectively shifted the game, with the FCO directly questioning military requirements and striving for comprehensive disarmament measures and concessions from defence officials. Building on the success of the BWC, they had significant momentum. In 1976, one of the most critical shifts in this ongoing clash came with commitments to chemical weapons disarmament ultimately overriding military requirements for nerve agent weapons. In 1976, détente, adverse publicity, and disarmament talks ended the aspirations of defence officials for a nerve agent capability.

The British nerve agent debate over acquisition, deterrence, publicity, and disarmament was a fluctuating and often contradictory one. Policy was regularly muddled and a compromise: secrecy clashed with cooperation; deterrence clashed with publicity; military utility clashed with

[4] TNA, DEFE 13/557, 'Chemical Warfare and Biological Warfare—The Future of MRE and CDEE, Porton', Sir Joseph Percival William Mallalieu, 22 August 1967.

political aversion and arms control treaties; and disarmament clashed with deterrence. While it was often changing course, even with the high importance attached to CW deterrence and the acquisition of nerve agent weapons, Britain never domestically mass-produced large stockpiles of nerve agent weapons nor did it officially change its approach to first-use. In the postwar period, the advocates for acquisition, first-use, and domestic production failed to fully overcome the obstacles in their path. Conversely, those who critiqued and resisted the attempts to produce, research and publicise nerve agent capabilities succeeded in exerting considerable pressure and influence on British nerve agent policy.

Bibliography

Archival Sources

Files from the National Archives (UK)

Records of the Prime Minister (PREM files): PREM 3/89; PREM 8/960; PREM 11/49; PREM 11/617; PREM 11/3099; PREM 13/3464; PREM 13/3465; PREM 13/3468.

Records of Cabinet Committees (CAB files): CAB 21/3912; CAB 21/4505; CAB 80/94; CAB 81/67; CAB 81/132; CAB 81/143; CAB 121/100; CAB 128/44; CAB 129/55; CAB 129/146; CAB 131/3; CAB 131/5; CAB 131/6; CAB 131/8; CAB 131/9; CAB 131/12; CAB 131/14; CAB 131/17; CAB 131/23; CAB 131/24; CAB 131/28; CAB 148/71; CAB 158/3; CAB 158/14; CAB 158/15; CAB 158/20; CAB 158/24; CAB 158/28; CAB 158/39; CAB 158/45; CAB 158/48; CAB 158/51; CAB 158/56; CAB 158/59; CAB 168/169.

Records of the Ministry of Defence (DEFE files): DEFE 4/35; DEFE 4/47; DEFE 4/49; DEFE 4/53; DEFE 4/59; DEFE 4/78; DEFE 4/110; DEFE 4/119; DEFE 4/129; DEFE 4/149; DEFE 4/222; DEFE 4/224; DEFE 4/272; DEFE 4/282; DEFE 5/37; DEFE 5/4; DEFE 5/19; DEFE 5/43; DEFE 5/44; DEFE 5/50; DEFE 5/53; DEFE 5/55; DEFE 5/58; DEFE 5/66; DEFE 5/84; DEFE 5/131; DEFE 5/142; DEFE 5/162; DEFE 5/176; DEFE 5/193; DEFE 6/2; DEFE 6/29; DEFE 7/700; DEFE

W. King, *Nerve Agents in Postwar Britain*, Britain and the World,
https://doi.org/10.1007/978-3-030-70474-2

7/1395; DEFE 7/2140; DEFE 10/18; DEFE 10/30; DEFE 10/32; DEFE 10/33; DEFE 10/34; DEFE 10/171; DEFE 10/174; DEFE 10/264; DEFE 10/265; DEFE 10/281; DEFE 10/355; DEFE 10/356; DEFE 10/357; DEFE 10/417; DEFE 10/420; DEFE 10/445; DEFE 10/446; DEFE 10/447; DEFE 10/448; DEFE 11/660; DEFE 11/672; DEFE 13/265; DEFE 13/557; DEFE 13/846; DEFE 13/997; DEFE 13/998; DEFE 13/999; DEFE 13/1000; DEFE 13/1055; DEFE 13/1056; DEFE 13/1818; DEFE 24/6; DEFE 24/31; DEFE 24/404; DEFE 24/1739; DEFE 25/24; DEFE 25/609; DEFE 32/2; DEFE 41/156; DEFE 41/157; DEFE 48/1; DEFE 48/34.

Other files (Home Office, War Office, Air Ministry and Foreign and Commonwealth Office): HO 228/11; WO 32/20126; WO 32/20166; WO 32/21379; WO 32/21760; WO 32/21761; WO 32/21950/2; WO 188/785; WO 188/802; WO 188/2530; WO 188/2772; AIR 2/17791; AIR 2/17792; AIR 8/1936; AIR 19/1097; AIR 20/9440; FCO 66/220; FCO 66/312; FCO 66/391; FCO 66/497.

Files from the United States National Archives (NARA II)

RG 218, Chairman's File: General Bradley 1949–1953, Box 07; RG 218, Central Decimal File 1961, Box 49; RG 218, Central Decimal File 1951–53, Box 152; RG 218, Central Decimal File 1951–53, Box 153; RG 218, Central Decimal File 1954–56, Box 126; RG 330, Entry 241, Box 486.

Files from the National Security Archives, George Washington University (unpublished)

Chemical and Biological Warfare, Box 2; Chemical and Biological Warfare, Box 7; Chemical and Biological Warfare, Box 8; Chemical and Biological Warfare, Box 12; Chemical and Biological Warfare, Box 13.

Files from the Harvard-Sussex Program Archive (Including Unpublished Works)

Categories: B2 Historical; B3 Utility of CW Weapons; B4 CBW Intelligence; D5 1945–60; D6 Since 1960; F1 New History.

Col. Dr Walter Hirsch, 'Soviet BW and CW Preparations and Capabilities', 19 May 1951.

'The Role of Chemical Defense in Chemical Warfare, Chemical Deterrence, and Chemical Disarmament', adapted text of a Keynote Address by Matthew Meselson, 13 March 1991.

'The use of CS in Vietnam', Matthew Meselson, July 1969.

'The United States and the Geneva Protocol of 1925', Matthew Meselson, September 1969.

Gerald R. Ford Library Papers, Melvin Laird, Box C3, Chemical Warfare Biological Research, Documents 29–35, National Security Decision Memorandum 35, 25 November 1969. Found at: The Harvard Sussex Program Archive, Harvard.

Files from the University of East Anglia Archives

Solly Zuckerman Collection, CBW/1; Solly Zuckerman Collection, Communications with Lord Todd; Solly Zuckerman Collection, General Communications: C, Chemical Defence Establishment; Solly Zuckerman Collection, MOS (3)/1; Solly Zuckerman Collection, MOS (3)/2; Solly Zuckerman Collection, MOS (3)/3.

Files from the Liddell Hart Centre for Military Archives

H. C. Pincher, Press Cuttings, Vol.5 1953–1955, Vol.10 1959–1960, Vol.11 1960–1961.

Published Sources

'Protocol for the Prohibition of the Use in War of Asphyxiating, Poisonous or Other Gases, and of Bacteriological Methods of Warfare', Geneva, 17 June 1925.

Digital National Security Archive, George Washington University. (Includes National Intelligence Estimates and CIA Reports).

Foreign Relations of the United States, Office of the Historian.

Hansard (Documents and Debates): House of Lords and House of Commons.

Newspapers: *Daily Mail*, *The Times*, *The New York Times*, *The Manchester Guardian*, *The Guardian*, *The Economist*, *The Daily Telegraph*, *The Sunday Telegraph*.

United Nations General Assembly. 1971. *Convention on the Prohibition of the Development, Production and Stockpiling of Bacteriological (Biological) and Toxin Weapons and on Their Destruction*. 26th Session, Resolution 2826, December 16.

United Nations. 1966. *Question of General and Complete Disarmament.* 1484th Plenary Meeting, United Nations Resolution 2162 B (XXI), December 5.

———. 1969. *Chemical and Bacteriological (Biological) Weapons and the Effects of Their Possible Use.* Report of the Secretary General. New York: United Nations.

United States Arms Control and Disarmament Agency. 1976. *Documents on Disarmament: 1974.* Washington, DC: United States Government Printing Office, May.

———. 1978. *Documents on Disarmament: 1976.* Washington, DC: United States Government Printing Office, May.

Secondary Sources

Books, Chapters, Journal Articles and Dissertations

Abdulkarim, Ekzayez, M. Daniel Flecknoe, Louis Lillywhite, Preeti Patel, Andreas Papamichail, and Hassan Elbahtimy. 2019. Chemical Weapons and Public Health: Assessing Impact and Responses. *Journal of Public Health* 42: 3.

Abe, Tatsuya. 2017. Challenge Inspections Under the Chemical Weapons Convention: Between Ideal and Reality. *The Nonproliferation Review* 24: 1–2.

Adams, Valerie. 1989. *Chemical Warfare, Chemical Disarmament: Beyond Gethsemane.* Basingstoke: Macmillan.

Agar, Jon, and Brian Balmer. 1998. British Scientists and the Cold War: The Defence Research Policy Committee and Information Networks, 1947–1963. *Historical Studies in the Physical and Biological Sciences* 28: 2.

Aldrich, Richard. 2001. *The Hidden Hand: Britain, America and Cold War Secret Intelligence.* London: John Murray.

Aldrich, Richard, and Michael Coleman. 1989. The Cold War, the JIC and British Signals Intelligence, 1948. *Intelligence and National Security* 4: 3.

Aldrich, Richard, and Rory Cormac. 2016. *The Black Door: Spies, Secret Intelligence and British Prime Ministers.* London: William Collins.

Allen, Steven. 2007. *An Analysis of Factors Leading to a U.S. Renunciation of Biological Weapons.* PhD Diss., George Mason University.

Andrew, Christopher. 2009. *The Defence of the Realm: The Authorized History of MI5.* London: Allen Lane.

Ball, S.J. 1995. Military Nuclear Relations Between the United States and Great Britain Under the Terms of the McMahon Act, 1946–1958. *The Historical Journal* 38: 2.

Balmer, Brian. 1997. The drift of biological weapons policy in the UK 1945–1965. *Journal of Strategic Studies* 20: 4.

———. 2001. *Britain and Biological Warfare: Expert Advice and Science Policy 1935–65*. Hampshire: Palgrave.
———. 2006. The UK Biological Weapons Programme. In *Deadly Cultures: Biological Weapons Since 1945*, ed. Mark Wheelis, Lajos Rózsa, and Malcolm Dando. London: Harvard University Press.
———. 2010. Keeping Nothing Secret: United Kingdom Chemical Warfare Policy in the 1960s. *Journal of Strategic Studies* 33: 6.
———. 2012. *Secrecy and Science: A Historical Sociology of Biological and Chemical Warfare*. Farnham, Surrey: Ashgate.
———. 2015. An Open Day for Secrets: Biological Warfare, Steganography, and Hiding Things in Plain Sight. In *Absence in Science, Security and Policy: From Research Agendas to Global Strategy*, ed. Brian Rappert and Brian Balmer. Hampshire: Palgrave Macmillan.
———. 2021. "Science Was Digging Its Own Grave": The Women's International League for Peace and Freedom and the Campaign Against Chemical and Biological Warfare. *The Nonproliferation Review*, [online first].
Balmer, Brian, Alex Spelling, and Caitríona McLeish. 2016. Preventing "A Virological Hiroshima": Cold War Press Coverage of Biological Weapons Disarmament. *Journal of War & Culture Studies* 9: 1.
———. 2018. Tear Gas Epistemology: The Himsworth Committee and Weapons as Drugs. In *In Chemical Bodies: The Techno-Politics of Control*, ed. Alex Mankoo and Brian Rappert. London: Rowman & Littlefield International.
Bartlett, C.J. 1984. *The Global Conflict 1880–1970: The International Rivalry of the Great Powers*. London: Longman.
———. 1992. *'The Special Relationship': A Political History of Anglo-American Relations Since 1945*. London: Longman.
Baumslag, Naomi. 2005. *Murderous Medicine: Nazi Doctors, Human Experimentation, and Typhus*. London: Praeger.
Baylis, John. 1983. Introduction. In *In Alternative Approaches to British Defence Policy*, ed. John Baylis. London: Macmillan Press.
———. 1986. "Greenwoodery" and British Defence Policy. *International Affairs* 62: 3.
———. 1989. *British Defence Policy: Striking the Right Balance*. New York: Palgrave Macmillan.
———. 1995. *Ambiguity and Deterrence: British Nuclear Strategy 1945–1964*. Oxford: Clarendon Press.
———. 1997. *Anglo-American Relations since 1939: The Enduring Alliance*. Manchester: Manchester University Press.
Baylis, John, and Alan Macmillan. 1993. The British Global Strategy Paper of 1952. *Journal of Strategic Studies* 16: 2.
Baylis, John, and Kristan Stoddart. 2015. *The British Nuclear Experience: The Role of Beliefs, Culture, and Identity*. Oxford: Oxford University Press.

Békés, Csaba. 2010. East Central Europe, 1953–1956. In *The Cold War, Volume I: Origins*, ed. Melvyn Leffler Odd and Arne Westad. Cambridge: Cambridge University Press.

Bentley, Michelle. 2014. Strategic Taboos: Chemical Weapons and US Foreign Policy. *International Affairs* 90: 5.

Bolton, Talitha. 2008. *Consent and the Construction of the Volunteer: Institutional Settings of Experimental Research on Human Beings in Britain During the Cold War*. PhD Diss., University of Kent.

Borkin, Joseph. 1978. *The Crime and Punishment of I.G. Farben*. London: Free Press.

Boyle, Peter. 2005. The Hungarian Revolution and the Suez Crisis. *History* 90: 4.

Brown, Frederic. 1968. *Chemical Warfare: A Study in Restraints*. Princeton: Princeton University Press.

Bryden, John. 1989. *Deadly Allies: Canada's Secret War 1937–1947*. Toronto: McClelland & Stewart.

Bullock, Alan. 1983. *Ernest Bevin: Foreign Secretary*. Oxford: Oxford University Press.

Burr, William, and David Rosenberg. 2010. Nuclear Competition in an Era of Stalemate, 1963–75. In *In The Cambridge History of the Cold War, Volume 3: Endings*, ed. Melvyn Leffler and Odd Arne Westad. Cambridge: Cambridge University Press.

Calder, Nigel, ed. 1968. *Unless Peace Comes: A Scientific Forecast of New Weapons*. New York: Viking Press.

Callaghan, John. 2007. *The Labour Party and Foreign Policy: A History*. London: Routledge.

Carrington, Lord. 1988. *Reflect on Things Past: The Memoirs of Lord Carrington*. London: Collins.

Carter, Gradon B. 2000. *Chemical and Biological Defence at Porton Down 1916–2000*. London: The Stationery Office.

Carter, Gradon, and Brian Balmer. 1999. Chemical and Biological Warfare and Defence, 1945–90. In *Cold War Hot Science: Applied Research in Britain's Defence Laboratories 1945–1990*, ed. Robert Bud and Philip Gummett. Amsterdam: Harwood.

Carter, Gradon, and Graham S. Pearson. 1996. North Atlantic Chemical and Biological Research Collaboration: 1916–1995. *Journal of Strategic Studies* 19: 1.

Castaldi, Carolina, and Alessandro Nuvolari. 2009. Chariots of Fire: The Evolution of Tank Technology, 1915–1945. *Journal of Evolutionary Economics* 19: 4.

Chamberlin, Paul Thomas. 2018. *The Cold War's Killing Fields: Rethinking the Long Peace*. New York: HarperCollins.

Chapman, Geoffrey, Hassan Elbahtimy, and Susan B. Martin. 2018. The Future of Chemical Weapons: Implications from the Syrian Civil War. *Security Studies* 27: 4.

Clark, Ian, and Nicholas Wheeler. 1989. *The British Origins of Nuclear Strategy 1945–1955*. Oxford: Clarendon Press.

Clarke, Peter. 2004. *Hope and Glory: Britain 1900–2000*. London: Penguin Books.

Coffey, Patrick. 2014. *American Arsenal: A Century of Waging War*. Oxford: Oxford University Press.

Cole, Leonard. 1998. The Poison Weapons Taboo: Biology, Culture, and Policy. *Politics and the Life Sciences* 17: 2.

Cookson, John, and Judith Nottingham. 1969. *A Survey of Chemical and Biological Warfare*. London: Monthly Review Press Classics.

Costanzi, Stefano, and Gregory D. Koblentz. 2019. Controlling Novichoks After Salisbury: Revising the Chemical Weapons Convention Schedules. *The Nonproliferation Review* 26: 5–6.

———. 2020. Updating the CWC: How We Got Here and What Is Next, *Arms Control Today*, April.

Croddy, Eric, Clarisa Perez-Armendariz, and John Hart. 2002. *Chemical and Biological Warfare: A Comprehensive Survey for the Concerned Citizen*. New York: Copernicus Books.

Davy, Richard. 2009. Helsinki Myths: Setting the Record Straight on the Final Act of the CSCE, 1975. *Cold War History* 9: 1.

Deighton, Anne. 1992. "Arming the Key Battleground": German Rearmament, 1950–55. *Journal of Diplomacy & Statecraft* 3: 2.

———. 2010. Britain and the Cold War. In *The Cold War, Volume I: Origins*, ed. Melvyn Leffler and Odd Arne Westad. Cambridge: Cambridge University Press.

Dockrill, Saki. 1992. Britain's Strategy for Europe: Must West Germany Be Rearmed? 1949–51. In *British Intelligence, Strategy and the Cold War, 1945–51*, ed. Richard Aldrich. Oxon: Routledge.

Dumbrell, John. 2006. *A Special Relationship: Anglo-American Relations from the Cold War to Iraq*. Basingstoke: Palgrave Macmillan.

Dunn, Michael, and Frederick Sidell. 1989. Progress in Medical Defense Against Nerve Agents. *Journal of the American Medical Association* 262: 5.

Earl, Hilary. 2013. Prosecuting genocide before the Genocide Convention: Raphael Lemkin and the Nuremberg Trials, 1945–1949. *Journal of Genocide Research* 15: 3.

Eden, Anthony. 1960. *The Memoirs of Sir Anthony Eden: Full Circle*. London: Cassell & Company.

———. 1976. *Another World, 1897–1917*. London: Allen Lane.

Ellison, D. Hank. 2011. *Chemical Warfare During the Vietnam War: Riot Control Agents in Combat*. London: Routledge.

Evans, Rob. 2000. *Gassed: British Chemical Warfare Experimentation Humans at Porton Down*. London: House of Stratus.

Fredericks, William Curtis. 1988. *The Evolution of Post-World War II United States Chemical Warfare Policy*. PhD Diss., Oxford University.

Freedman, Lawrence. 1977. *U.S. Intelligence and the Soviet Strategic Threat*. London: The Macmillan Press.

French, David. 2011. *The British way in Counter-Insurgency, 1945–1967*. Oxford: Oxford University Press.

———. 2012. *Army, Empire, and Cold War: The British Army and Military Policy, 1945–1971*. Oxford: Oxford University Press.

———. 2013. Duncan Sandys and the Projection of British Power After Suez. *Diplomacy & Statecraft* 24: 1.

Fry, Geoffrey K. 2005. *The Politics of Decline: An Interpretation of British Politics from the 1940s to the 1970s*. Hampshire: Palgrave Macmillan.

Gaddis, John Lewis. 2005. *The Cold War*. London: Allen Lane.

Galbraith, Russell. 2000. *Inside Outside: The Man They Can't Gag*. Edinburgh: Mainstream Publishing.

Gellermann, Günther W. 1986. *Der Krieg, der nicht stattfand: Möglichkeiten, Überlegungen und Entscheidungen der deutschen Obersten Führung zur Verwendung chemischer Kampfstoffe im Zweiten Weltkrieg*. Koblenz: Bernard & Graefe Verlag.

Gilbert, Martin. 1986. *Winston S. Churchill, Vol. 7: Road to Victory 1941–1945*. London: Heinemann.

Goldblat, J. 1975. Chemical Disarmament Negotiations in 1974. In *Chemical Disarmament: New Weapons for Old*, ed. A SIPRI Monograph. Stockholm: Almqvist & Wiksell.

Goldman, David. 2009. The Generals and the Germs: The Army Leadership's Response to Nixon's Review of Chemical and Biological Warfare Policies in 1969. *The Journal of Military History* 73: 2.

Goodman, Michael. 2003. British Intelligence and the Soviet Atomic Bomb, 1945–1950. *Journal of Strategic Studies* 26: 2.

Goodman, Michael S. 2007. *Spying on the Nuclear Bear: Anglo-American Intelligence and the Soviet Bomb*. Stanford, CA: Stanford University Press.

Gordon, S. 2009. *The Great Cold War: A Journey Through the Hall of Mirrors*. Stanford, CA: Stanford Security Studies.

Gott, Richard. 1963. The Evolution of the Independent British Deterrent. *Royal Institute of International Affairs* 39: 2.

Gowing, Margaret. 1974. *Independence and Deterrence: Britain and Atomic Energy 1945–52, Volume 1: Policy Making*. London: Palgrave Macmillan.

Gregory, Jane, and Steve Miller. 1998. *Science in Public: Communication, Culture, and Credibility*. New York: Plenum Press.

Guillemin, Jeanne. 2017. *Hidden Atrocities: Japanese Germ Warfare and American Obstruction of Justice at the Tokyo Trial*. New York: Columbia University Press.

Guillemin, Jeanne, Matthew Meselson, Julian Perry Robinson, and Nicholas Sims. 2017. Witness Seminar: Origins of the Biological Warfare Convention. In *Biological Threats in the 21st Century: The Politics, People, Science and Historical Roots*, ed. Filippa Lentzos. London: Imperial College Press.

Hall, Charlie. 2017. *British Exploitation of German Science and Technology from War to Post-War, 1943–1948*. PhD Diss., University of Kent.

———. 2019. *British Exploitation of German Science and Technology, 1943–1949*. Oxon: Routledge.

Hammond, Peter, and Gradon Carter. 2002. *From Biological Warfare to Healthcare: Porton Down, 1940–2000*. Hampshire: Palgrave.

Hanhimäki, Jussi. 2003. "They Can Write it in Swahili": Kissinger, the Soviets, and the Helsinki Accords, 1973–75. *Journal of Transatlantic Studies* 1: 1.

Harris, Robert, and Jeremy Paxman. 1982. *A Higher Form of Killing: The Secret of Gas and German Warfare*. London: Chatto & Windus.

Haslam, Jonathan. 2015. *Near and Distant Neighbours: A New History of Soviet Intelligence*. Oxford: Oxford University Press.

Hathaway, Robert M. 1991. *Ambiguous Partnership: Britain and America, 1944–1947*. New York: Columbia University Press.

Healey, Denis. 1989. *The Time of My Life*. London: Michael Joseph.

Hennessy, Peter. 2007. *Cabinets and the Bomb*. Oxford: Oxford University Press.

Henry, Chris, and Mike Fuller. 2002. *The 25-pounder Field Gun 1939–72*. Oxford: Osprey Publishing.

Hershberg, James. 2010. The Cuban Missile Crisis. In *In The Cold War, Volume 2: Crises and Détente*, ed. Melvyn Leffler and Odd Arne Westad. Cambridge: Cambridge University Press.

Hill, Christopher. 2018. *Peace and Power in Cold War Britain: Media, Movements and Democracy, c.1945–1968*. London: Bloomsbury.

Howlett, Darryl. 1990. UK Arms Control and Disarmament Policy on Chemical and Biological Weapons. In *UK Arms Control in the 1990s*, ed. Mark Hoffman. Manchester: Manchester University Press.

Hoyt, Timothy. 2016. The United States and the Cold War Arms Race. In *Arms Races in International Politics*, ed. Thomas Mahnken, Joseph Maiolo, and David Stevenson. Oxford: Oxford University Press.

Humphrey, J.H., and Matthew Meselson. 1969. Preventing CBW. In *CBW: Chemical and Biological Warfare*, ed. Steven Rose. Boston, MA: Beacon Press.

Hunt, Richard. 2015. *Melvin Laird and the Foundation of the Post-Vietnam Military, 1969–1973, Vol. 7. Secretaries of Defense Historical Studies*. Washington, DC: Office of the Secretary of Defense.

Jefferson, Catherine. 2014. Origins of the Norm Against Chemical Weapons. *International Affairs* 90: 3.

Jenks, John. 2006. *British Propaganda and News Media in the Cold War*. Edinburgh: Edinburgh University Press.

Johnston, Andrew. 1998. Mr Slessor Goes to Washington: The Influence of the British Global Strategy Paper on the Eisenhower New Look. *Diplomatic History* 22: 3.

Jones, Matthew. 2017a. *The Official History of the UK Strategic Nuclear Deterrent: Volume I: From the V-Bomber Era to the Arrival of Polaris, 1945–1964*. Oxon: Routledge.

———. 2017b. *The Official History of the UK Strategic Nuclear Deterrence, Volume II: The Labour Government and the Polaris Programme, 1964–1970*. London: Routledge.

Jun, Niu. 2010. The birth of the People's Republic of China and the Korean War. In *In The Cold War, Volume I: Origins*, ed. Melvyn Leffler and Odd Arne Westad. Cambridge: Cambridge University Press.

Kaszeta, Dan. 2020. *Toxic: A History of Nerve Agents. From Nazi Germany to Putin's Russia*. London: C. Hurst & Co.

Kemp-Welch, A. 2008. *Poland under Communism: A Cold War History*. Cambridge: Cambridge University Press.

Kent, John, and John Young. 1992. The "Western Union" Concept and British Defence Policy, 1947–8. In *British Intelligence, Strategy and the Cold War, 1945–51*, ed. Richard Aldrich. Oxon: Routledge.

King, William. 2021. A Weapon Too Far: The British Radiological Warfare Experience, 1940–1955. *War in History*, [online first].

King, William, and Jeanne Guillemin. 2019. The Price of Alliance: Anglo-American Intelligence Cooperation and Imperial Japan's Criminal Biological Warfare Programme, 1944–1947. *Intelligence and National Security* 34: 2.

Kirby, Reid. 2006. Nerve Gas: America's Fifteen-Year Struggle for Modern Chemical Weapons. *CML Army Chemical Review*, January–June.

Kovačević, Damir, Afrimadona, and Martin Claar. 2019. Gas, Power, and Norms: Competing Logics for the Declining Use of Chemical Warfare. *The Nonproliferation Review* 26: 3–4.

Kraft, Alison. 2018. Dissenting Scientists in Early Cold War Britain: The "Fallout" Controversy and the Origins of Pugwash, 1954–1957. *Journal of Cold War Studies* 20: 1.

Langer, Elinor. 1967. Chemical and Biological Warfare(II): The Weapons and the Policies. *Science* 155: 3760.

Langford, R. Everett. 2004. *Introduction to Weapons of Mass Destruction: Radiological, Chemical and Biological*. Hoboken, NJ: John Wiley & Sons.

Leffler, Melvyn. 2010. The Emergence of an American Grand Strategy, 1945–1952. In *The Cold War, Volume I: Origins*, ed. Melvyn Leffler and Odd Arne Westad. Cambridge: Cambridge University Press.

Lego, Jeffrey W. 1995. *Cooperation under Fire: Anglo-German Restraint During World War II*. London: Cornell University Press.

Levy, David. 2015. The Traitor Hunter. *International Journal of Intelligence and Counter Intelligence* 28: 2.

Lewis, Julian. 2003. *Changing Direction: British Military Planning for Post-War Strategic Defence, 1942–1947*. London: Frank Cass.

Logevall, Fredrik. 2010. The Indochina Wars and the Cold War, 1945–1975. In *The Cold War, Volume 2: Crises and Détente*, ed. Melvyn Leffler and Odd Arne Westad. Cambridge: Cambridge University Press.

Lowe, Peter. 1989. The Significance of the Korean War in Anglo-American Relations, 1950–1953. In *British Foreign Policy 1945–1956*, ed. Michael L. Dockrill and John W. Young. London: Macmillan.

Maddrell, Paul. 1998. *Britain's Exploitation of Occupied Germany for Scientific and Technical Intelligence on the Soviet Union*. PhD Diss., University of Cambridge.

———. 2006. *Spying on Science: Western Intelligence in Divided Germany 1945–1961*. Oxford: Oxford University Press.

Mahan, Erin. 2002. *Kennedy, De Gaulle and Western Europe*. New York: Palgrave Macmillan.

Maiolo, Joe. 2010. *Cry Havoc: The Arms Race and the Second World War 1931–1941*. London: John Murray.

Mankoo, Alexander Chanan. 2019. *A Historical Sociology of Teargas in Britain and the Empire, 1925–1965*. PhD Thesis, University College London.

Mankoo, Alex, and Brian Rappert. 2018. *Chemical Bodies: The Techno-Politics of Control*. London: Rowman & Littlefield International.

Martin, Susan. 2016. Norms, Military Utility, and the Use/Non-Use of Weapons: The Case of Anti-Plant and Irritant Agents in the Vietnam War. *Journal of Strategic Studies* 39: 3.

Mastny, Vojtech. 2010. Soviet Foreign Policy, 1953–1962. In *The Cold War, Volume I: Origins*, ed. Melvyn Leffler and Odd Arne Westad. Cambridge: Cambridge University Press.

McCamley, N.J. 2006. *Secret History of Chemical Warfare*. Barnsley: Pen & Sword Military.

McLeish, Caitríona, and Brian Balmer. 2012. Development of the V-Series Nerve Agents. In *Innovation, Dual Use, and Security: Managing the Risks of Emerging Biological and Chemical Technologies*, ed. Jonathan Tucker. Cambridge, MA: The MIT Press.

Meselson, Matthew. 1969a. The Yemen. In *CBW: Chemical and Biological Warfare*, ed. Steven Rose, 99–102. Boston, MA: Beacon Press.

———. 1969b. Why Not Poison? *Science*, April 25.

———. 1970a. Behind the Nixon Policy for Chemical and Biological Warfare. *Bulletin of the Atomic Scientists*, January.

———. 1970b. CB Weapons: The Facts. In *The Supreme Folly: Chemical & Biological Weapons*, ed. Frank Barnaby. London: National Council of Labour Colleges Publishing Society for the Women's International League for Peace and Freedom.

———. 1972. Gas Warfare and the Geneva Protocol of 1925. *Bulletin of the Atomic Scientists*, February.

———. 1975. What Policy for Nerve Gas? *Arms Control Today* 5: 4.

———. 1991. The Myth of Chemical Superweapons. *The Bulletin of the Atomic Scientists* 47: 3.

Meselson, Matthew, and Robinson, Julian Perry. 1969. Escalation of Chemical Warfare. *New Scientist*, August.

Meselson, Matthew, and Julian Perry Robinson. 1980. Chemical Warfare and Chemical Disarmament. *Scientific American* 242 (4): 38–39.

Mikulak, Robert. 1978. Preventing Chemical Warfare. In *Chemical Weapons and Chemical Arms Control*, ed. Matthew Meselson. New York: Carnegie Endowment for International Peace.

Mirzayanov, Vil. 2009. *State Secrets: An Insider's Chronicle of the Russian Chemical Weapons Program*. Denver, CO: Outskirts Press.

Murphy, Seán, Alastair Hay, and Steven Rose. 1984. *No Fire, No Thunder: The Threat of Chemical and Biological Weapons*. London: Pluto Press.

Nunnerley, David. 1972. *President Kennedy & Britain*. London: The Bodley Head.

Park, Myunglim. 1998. The '"American Boundary", Provocation, and the Outbreak of the Korean War. *Social Science Japan Journal* 1: 1.

Paterson, Thomas. 1989. *Kennedy's Quest for Victory: American Foreign Policy, 1961–1963*. Oxford: Oxford University Press.

Peden, G.C. 2009. *Arms, Economics and British Strategy: From Dreadnoughts to Hydrogen Bombs*. Cambridge: Cambridge University Press.

Penkovsky, Oleg. 1965. *The Penkovsky Papers*. London: Collins.

Peyton, John. 2001. *Solly Zuckerman: A Scientist Out of the Ordinary*. London: John Murray.

Pierre, Andrew. 1972. *Nuclear Politics: The British Experience with an Independent Strategic Force, 1939–1970*. Oxford: Oxford University Press.

Pincher, Chapman. 1978. *Inside Story: A Documentary of the Pursuit of Power*. London: Sidwick & Jackson.

Pine, Melissa. 2009. Transatlantic Nuclear Cooperation: The British Perspective, 1945–1991. In *The British Way in Cold Warfare: Intelligence, Diplomacy and the Bomb, 1945–1975*, ed. Matthew Grant. London: Continuum Books.

Price, Richard. 2007. *The Chemical Weapons Taboo*. London: Cornell University Press.

———. 2019. Syria and the Chemical Weapons Taboo. *Journal of Global Security Studies* 4: 1.

Rees, Wyn. 1989. The 1957 Sandys White Paper: New Priorities in British Defence Policy? *Journal of Strategic Studies* 12: 2.

Reynolds, David. 2000a. *Britannia Overruled: British policy and world power in the twentieth century*. Harlow: Pearson Education.

———. 2000b. *One World Divisible: A Global History Since 1945*. London: Penguin Books.

Robb, Thomas. 2014. *A Strained Partnership? United States-UK Relations in the Era of Détente, 1969–77*. Manchester: Manchester University Press.

Robinson, Julian Perry. 1971. *The Rise of CB Weapons*. Stockholm: Almqvist & Wiksell.

———. 1981. Chemical Arms Control and the Assimilation of Chemical Weapons. *International Journal (Toronto)* 38: 3.

———. 1993. Origins of the Chemical Weapons Convention. In *Shadows and Substance: The Chemical Weapons Convention*, ed. Benoit Morel and Kyle Olson. Boulder, CO: Westview Press.

Rose, Steven, ed. 1968. *CBW: Conference on Chemical and Biological Warfare*. London: Harrap Publishing.

Rose, Steven, and Filippa Lentzos. 2016. Interview: Unconventional Weapons and Activist Scientists. In *Biological Threats in the 21st Century: The Politics, People, Science and Historical Roots*, ed. Filippa Lentzos. London: Imperial College Press).

Rosecrance, R.N. 1968. *Defence of the Realm: British Strategy in the Nuclear Epoch*. New York: Columbia University Press.

Sandford, Christopher. 2018. *Union Jack: John F. Kennedy's Special Relationship with Great Britain*. Cheltenham: The History Press.

Savranskaya, Svetlana, and William Taubman. 2010. Soviet Foreign Policy, 1962–1975. In *The Cold War, Volume 2: Crises and Détente*, ed. Melvyn Leffler and Odd Arne Westad. Cambridge: Cambridge University Press.

Schmidt, Ulf. 2004. *Justice at Nuremberg: Leo Alexander and the Nazi Doctors' Trial*. Hampshire: Palgrave Macmillan.

———. 2007. *Karl Brandt: The Nazi Doctor, Medicine and Power in the Third Reich*. Cornwall: Hambledon Continuum.

———. 2015. *Secret Science: A Century of Poison Warfare and Human Experiments*. Oxford: Oxford University Press.

Schulzinger, Robert. 2010. Détente in the Nixon-Ford Years, 1969–1976. In *The Cold War, Volume 2: Crises and Détente*, ed. Melvyn Leffler and Odd Arne Westad. Cambridge: Cambridge University Press.

Scott, Len. 1999. Espionage and the Cold War: Oleg Penkovsky and the Cuban missile crisis. *Intelligence and National Security* 14: 3.

Self, Robert C. 2010. *British Foreign and Defence Policy Since 1945: Challenges and Dilemmas in a Changing World*. Basingstoke: Palgrave Macmillan.

Shaw, Tony. 2001. *British Cinema and the Cold War: The State, Propaganda and Consensus*. London: I. B. Tauris.
Shoham, Dany. 1998. Chemical and Biological Weapons in Egypt. *The Nonproliferation Review* 5: 3.
Shpiro, Shlomo. 2009. Poisoned Chalice: Intelligence Use of Chemical and Biological Weapons. *International Journal of Intelligence and CounterIntelligence* 22: 1.
Sidel, Victor, and Robert Goldwyn. 1967. Chemical Weapons: What They Are and What They Do. *Scientist and Citizen* 9: 7.
Sidell, Frederick, and Jonathan Borak. 1992. Chemical Warfare Agents: II. Nerve Agents. *Annals of Emergency Medicine* 21: 7.
Sidell, Frederick, et al. 2008. Nerve Agents. In *Medical Aspects of Chemical Warfare*, ed. Martha Lenhart. Falls Church, VA: Office of the Surgeon General.
Sigmund, Elizabeth. 1980. *Rage Against the Dying: Campaign Against Chemical and Biological Warfare*. London: Pluto Press.
Sigmund, Elizabeth, and Gail Crowther. 2014. *Sylvia Plath in Devon: A Year's Turning*. Stroud: Fonthill Media.
Sims, Nicholas. 1987. *International Organization for Chemical Disarmament (SIPRI)*. Stockholm: Almqvist & Wiksell.
Sloan, Roy. 1998. *The Tale of Tabun: Nazi Chemical Weapons in North Wales*. Llanrwst, Wales: Gwasg Carreg Gwalch.
Snyder, W.P. 1984. Walter Bedell Smith: Eisenhower's Chief of Staff. *Journal of Military Affairs* 48: 1.
Speller, Ian. 2008. Corbett, Liddell Hart and the "British Way in Warfare" in the 1960s. *Defence Studies* 8: 2.
Spelling, Alex. 2016. "Driven to Tears": Britain, CS Tear Gas, and the Geneva Protocol, 1969–1975. *Diplomacy & Statecraft* 27: 4.
Spiers, Edward M. 1984. Bargaining with Binaries. *NATO Review* 32: 5.
———. 1986. *Chemical Warfare*. London: Macmillan.
———. 1989. *Chemical Weaponry: A Continuing Challenge*. London: Macmillan.
———. 2006. Gas Disarmament in the 1920s: Hopes Confounded. *Journal of Strategic Studies* 29: 2.
———. 2010. *A History of Chemical and Biological Weapons*. London: Reaktion Books.
Spohr, Kristina, and David Reynolds. 2016. *Transcending the Cold War: Summits, Statecraft, and the Dissolution of Bipolarity in Europe, 1970–1990*. Oxford: Oxford University Press.
Steininger, Rolf. 1997. "The Americans Are in a Hopeless Position": Great Britain and the War in Vietnam. *Diplomacy & Statecraft* 8: 3.
Synge, A. 1969. Chemical and Biological Warfare. *New Blackfriars* 50: 585.
Szinicz, L. 2005. History of Chemical and Biological Warfare Agents. *Toxicology* 214 (3): 167–181.

Taubman, William, Sergei Khrushchev, and Abbott Gleason. Introduction. In *Nikita Khrushchev*, ed. William Taubman, Sergei Khrushchev, and Abbott Gleason. London: Yale University Press.

Taylor, Richard. 1988. *Against the Bomb: The British Peace Movement 1958–1965.* Oxford: Clarendon Press.

Terrill, W. Andrew. 1991. The Chemical Warfare Legacy of the Yemen War. *Comparative Strategy* 10: 2.

Tezcür, Güneş Murat, and Doreen Horschig. 2020. A Conditional Norm: Chemical Warfare from Colonialism to Contemporary Civil Wars. *Third World Quarterly* 42: 2.

Thomas, Donald. 2002. *Cochrane: Britannia's Sea Wolf.* London: Cassell Military Paperbacks.

Troyanovsky, Oleg. 2000. The Making of Soviet Foreign Policy. In *Nikita Khrushchev*, ed. William Taubman, Sergei Khrushchev, and Abbott Gleason. London: Yale University Press.

Tucker, Jonathan B. 2006. *War of Nerves: Chemical Warfare from World War I to Al-Qaeda.* New York, NY: Pantheon Books.

Twigge, Stephen, and Len Scott. 2000. *Planning Armageddon: Britain, the United States and the Command of Western Forces 1945–1964.* Amsterdam: OPA.

Ucko, David H. 2019. Counterinsurgency as Armed Reform: The Political History of the Malayan Emergency. *Journal of Strategic Studies* 42: 3–4.

Utgoff, Victor A. 1990. *The Challenges of Chemical Weapons: An American Perspective.* London: Macmillan.

Vagts, Detlev F., and Theodor Meron. 2013. Legal Aspects of the Geneva Protocol of 1925. In *Humanizing the Laws of War: Selected Writings of Richard Baxter*, ed. Richard Baxter, Detlev F. Vagts, Theodor Meron, Stephen Schwebel, and Charles Keever. Oxford: Oxford University Press.

Varsori, Antonio. 2000. Reflections on the Origins of the Cold War. In *Reviewing the Cold War: Approaches, Interpretations, Theory*, ed. Odd Arne Westad. Frank Cass: London.

Vickers, Rhiannon. 2016. Foreign and Defence Policy. In *In Harold Wilson: The Unprincipled Prime Minister? Re-appraising Harold Wilson*, ed. Kevin Hickson and Andrew Crines. London: Biteback Publishing.

Walker, John R. 2012. *Britain and Disarmament: The UK and Nuclear, Biological and Chemical Weapons Arms Control and Programmes 1956–1975.* Farnham, Surrey: Ashgate.

Westad, Odd Arne. 2017. *The Cold War: A World History.* London: Allen Lane.

Wilkinson, Nicholas. 2009. *Secrecy and the Media: The Official History of the United Kingdom's D-Notice System.* Oxon: Routledge.

Wilkinson, Mark. 2018. *Before Intelligence Failed: British Secret Intelligence on Chemical and Biological Weapons in the Soviet Union, South Africa and Libya.* London: C. Hurst & Co.

Wright, Peter. 1987. *Spycatcher: The Candid Autobiography of a Senior Intelligence Officer*. London: Viking.
Young, John W. 1991. *Cold War Europe 1945–89: A Political History*. London: Edward Arnold.
———. 1997. *Britain and the World in the Twentieth Century*. London: Arnold.
———. 2003. *The Labour Governments 1964–70, Volume 2: International Policy*. Manchester: Manchester University Press.
Zubok, Vladislav, and Konstantin Pleshakov. 1996. *Inside the Kremlin's Cold War: From Stalin to Khrushchev*. Cambridge, MA: Harvard University Press.
Zuckerman, Solly. 1988. *Monkeys, Men, and Missiles: An Autobiography 1946–88*. London: W.W. Norton & Company.

Index[1]

A
Abyssinia, 184
Air Ministry, 10n22, 36, 41–45, 50, 52, 79–81, 79n103, 79n106, 79n108, 80n111, 85, 85n135, 114, 114n87, 117, 138, 162, 175, 176, 260
Albu, Austen, 101
Allaun, Frank, 151, 155n104
Amiton, 115, 150n86
Angolan Civil War, 214
Anti-Chemical and Biological Warfare Group, 198
Anti-Tank Defence Measures Sub-Committee, 60, 62, 69
Arab-Israeli War of 1973, 214
Arms Control and Disarmament Agency (US), 228, 234, 238n124, 239n129, 245n158, 246n161, 246n162
Attlee, Clement, 21, 22, 22n54, 22n56, 28, 49, 51–53, 56, 69, 70, 78, 79n105, 88, 101, 119, 147, 238, 251, 252, 254

B
Baltic Sea, 153
Barbara Castle, 151
BBC, 2n6, 89, 100, 144, 196, 196n123, 199–201, 201n158, 202n166, 204n176, 207
Berlin, 1, 14, 38, 133, 158, 173
Bernal Peace Library, 200
Bessell, Peter, 209
Bilateral disarmament discussions (US and the Soviet Union), 247

[1] Note: Page numbers followed by 'n' refer to notes.

W. King, *Nerve Agents in Postwar Britain*, Britain and the World,
https://doi.org/10.1007/978-3-030-70474-2

Biological Warfare, 5, 7, 7n19, 11n24, 24n66, 27n80, 28n84, 31, 33n111, 40, 40n146, 41n150, 46n175, 59n24, 62n37, 66n57, 68n62, 68n64, 75, 76, 79n107, 82n124, 83, 83n127, 85n133, 85n134, 86n138, 86n139, 88n149, 110n72, 127n152, 128n154, 137, 145, 146, 151n91, 152n93, 152n95, 158n120, 159n128, 162n142, 163n147, 164n155, 164n157, 164n159, 166n164, 166n166, 168n174, 171, 176n16, 177n24, 179n32, 179n34, 180n38, 181n41, 181n42, 185n56, 186n64, 186n65, 187n67, 189n80, 189n81, 189n83, 189n85, 190n90, 191, 191n91, 191n92, 191n93, 191n94, 191n95, 192n98, 193n106, 193n107, 194n112, 195n114, 195n115, 195n118, 196n121, 196n122, 199n139, 199n140, 200n151, 200n152, 202n162, 203, 209n207, 215, 223n54, 223n56, 225n65, 226, 226n69, 229, 229n84, 240n137, 243, 260, 261, 263n4

Biological weapons, 2, 16n15, 19, 28, 33, 40, 41, 41n150, 71, 75–77, 83, 106, 138n27, 146, 147, 155, 194n113, 196, 196n120, 208, 210n210, 214, 218, 219, 222–229, 225n66, 260

Biological Weapons Convention, 215, 225, 225n67, 229, 240n133

Blue Streak, 134

Brandt, Willy, 213

Brezhnev, Leonid, 213, 239

British Council of Churches, 219

Brook, Norman, 27n80, 98, 98n17, 98n18, 99n23, 122, 122n124, 122n127, 124, 124n136, 155, 155n103

Brundrett, Frederick, 122, 122n126, 123, 123n134, 126, 126n146, 127n149, 127n152, 128n153

Butler, Rab, 99, 99n25

BZ, 3, 193n106, 200

C

Cabinet Defence Committee, 5, 9, 16, 22n55, 28, 28n84, 32, 38n135, 42, 48–52, 48n187, 49n191, 49n194, 49n195, 50n202, 51n203, 51n204, 51n207, 51n208, 52n209, 52n210, 58, 58n17, 75–78, 75n92, 76n94, 76n95, 77n100, 93, 98, 99n20, 99n24, 103, 119, 121, 123n129, 123n131, 126–128, 130, 135, 148, 148n73, 149, 149n76, 151, 154, 166, 166n167, 167n169, 167n171, 168–171, 168n173, 168n175, 170n181, 171n186, 171n187, 175, 176, 181, 186, 205, 211, 221, 254, 262

Cabinet Office, 186, 192

Calder, Nigel, 200, 200n152

Campaign for Nuclear Disarmament, 144, 197–199, 199n143, 207, 207n194

Canada, 8, 36, 36n126, 36n127, 37, 37n131, 87, 88n147, 103, 128, 130, 169, 239, 246

Carrington, Peter (Lord Carrington), vii, 175n8, 214, 220n43, 234, 235, 235n110

Cawood, Walter, 164, 165

Central Intelligence Agency, 159, 160

Chalfont, Alun (Lord Chalfont), 229, 230

Chalfont, Lord, 220n41, 220n43, 240n137

Chemical and Biological Warfare, viii, 7, 8, 11n24, 12, 16, 19, 36, 47, 66n57, 70, 71, 71n76, 73, 75, 76, 105–107, 135, 142, 144–146, 151, 155, 158, 160, 162–164, 164n158, 175, 184, 186, 187, 187n71, 193n107, 194–204, 195n114, 195n115, 196n121, 197n128, 198n133, 200n152, 201n155, 203n170, 207, 208n203, 209, 215, 216, 219, 221–225, 223n54, 223n59, 224n61, 224n63, 225n65, 229n84, 233n102, 241, 243, 252
Chemical Corps (US), 29, 59n24, 150, 155, 156
Chemical Defence Advisory Board, 110n72, 162n143
Chemical warfare deterrence, 2, 5, 7, 8, 12, 20–23, 34, 39, 42, 44, 48, 48n186, 49, 63, 101, 102, 112, 118–120, 122–124, 132, 135, 139–141, 147, 154, 156, 182, 185–187, 189, 227, 232, 235–238, 237n122, 242, 254–256, 258–260, 264
Chemical Warfare Sub-Committee (CWSC), 20n48, 29n93, 36n126, 36n127, 39n139, 39n141, 39n142, 40n145, 40n146, 42n156, 43n159, 43n160, 43n161, 43n163, 44n164, 44n165, 45, 45n169, 45n170, 45n172, 46n173, 47n180, 51n206, 53, 60, 62–65, 62n38, 63n39, 63n42, 63n43, 64n45, 64n46, 65n49, 65n52, 66n53, 67, 67n60, 69, 69n67, 79n107, 85, 85n133, 85n135, 86n139, 87n142, 87n143, 102n41, 103n42, 103n45, 103n46, 119n111, 122, 141
Chemical Weapons Convention, 10n23, 215, 220n43, 222, 222n51, 225, 229, 237, 241n140, 244n152, 245–248, 249n176, 253, 254n1
Chemical weapons disarmament, 8, 10n23, 154, 215, 222, 222n51, 223, 226–231, 227–228n80, 233–235, 238–241, 243–249, 249n175, 251, 253, 255, 263
Chemical weapons (mass-)production, 6, 19, 23, 25, 27, 27n76, 29, 32, 33, 43, 45–51, 48n187, 49n193, 53, 58–60, 62, 64, 67, 69, 69n67, 70, 78–82, 86–88, 90, 90n166, 93, 98, 99, 104, 105, 107, 108, 111, 114, 117–121, 123, 127–129, 130n167, 132, 135–137, 136–137n17, 139, 147, 149, 151–154, 155n104, 157n116, 163, 169–171, 180, 181, 183, 184, 224, 225n65, 229–232, 239, 245, 246, 251, 252, 255, 262
Chiefs of Staff Committee, xiii, 14, 27–29, 32–34, 45–52, 45n169, 57, 58, 60, 62, 67, 69–76, 71n76, 78, 80, 82, 83, 83n129, 85, 87, 88, 90, 93, 99, 103–105, 108–113, 113n84, 122, 126, 138, 139, 156–158, 157n111, 161, 163–166, 164n157, 179–182, 186–188, 190–193, 191n92, 193n107, 205, 210, 211, 211n217, 216, 221, 222, 226, 228, 230–235, 233n103, 237, 238, 240–247, 249, 252–254, 256, 260, 263
Childs, A. E., 78, 80
China, 2, 15, 15n9

Churchill, Winston, 14, 20–22, 20n45, 20n46, 20n47, 26, 27n80, 56, 57, 58n16, 68, 70, 71n75, 72, 72n79, 73n82, 74n86, 75–78, 75n87, 76n93, 80n111, 88–92, 89n156, 89n158, 90n165, 91n166, 91n168, 91n169, 91n171, 92n184, 96, 98, 106, 118, 119n111, 149, 255
Civil defence, 51, 64–67, 69, 100, 101, 114, 147, 159
CN, 3
Cockayne, William, 207, 207n195, 207n197, 208, 208n200
Committee of 100, 144, 195
Continental defence, 38, 39, 46, 47, 179
Cooper, Frank, 183, 184
Crawford, Kenneth, 68n63, 83–86, 83n128, 83n129, 84n130, 141
CSCS gas, 3, 3n11, 194, 194n113, 209, 213–249, 216n11, 216n12, 219n36, 220n40, 220n43
Cuban Missile Crisis, 12, 134, 165–167, 175

D

The Daily Mail, 20n47, 66, 66n55, 67, 67n58, 92, 92n181, 100n31, 148, 148n74, 167n168
Dalyell, Tam, 204–207, 204n174, 204n176, 206n189, 207n191, 209, 210, 240n137, 253
Davies, Rhys, 47
Defence Research Committee, 186, 187
Defence Research Policy Sub-Committee, 32–34, 33n108, 39–43, 45–47, 53, 60, 62, 67–70, 80, 82, 82n121, 102–105, 108, 109, 109n72, 111, 113–117, 119, 122, 126, 138, 141, 142, 156, 157, 157n116, 163, 166, 186, 254
Defence White Paper (1957), 97
de Freitas, Geoffrey, 100, 101
Department of Defense (US), 87, 180
Dickens, James, 207
Dickson, William, 81, 83
Disposal (of CW agents), 31n101, 44, 58, 59, 117, 120, 124, 127, 152, 153
D-Notice system, 30, 30n95, 31, 59
Dyhernfurth, 23, 25

E

Eden, Anthony, 11, 58n16, 76, 76n94, 96, 98, 118, 118n108, 119, 119n112, 121–127, 121n122, 121n123, 122n124, 122n127, 123n130, 123n131, 123n132, 123n134, 125n145, 126n148, 129, 132, 135, 138, 139, 147, 149, 238, 252, 255, 259
Egypt, 96, 118, 125, 167, 196, 196n120
Eighteen-Nation Disarmament Conference, 223, 226
Eisenhower, Dwight, 57n15, 89, 89n155, 96, 124–126, 132, 149
Exercise Tureen, 175, 176n12, 176n13

F

First-use, 2, 7, 9, 11, 19, 20, 34, 35, 38, 44, 45, 47, 52, 61–63, 68, 71–76, 71n76, 78, 79, 84, 93, 99, 103, 104, 106, 108–114, 116, 119, 120, 123, 126, 127, 132, 154, 161, 163, 182,

188n78, 222, 225, 238, 251, 252, 255, 259, 262–264
First World War, 16, 18, 19, 77, 119, 197, 220, 259
Foreign and Commonwealth Office, 10, 10n23, 205n179, 216–219, 226–230, 228n81, 234, 234n107, 235, 235n111, 235n113, 237, 237n118, 238, 238n123, 240–243, 245, 246, 248, 249, 260, 263
Foreign Office, 8, 10, 14, 75, 77, 78, 93, 99, 104, 110, 111, 113, 139, 149, 149n81, 205, 205n179, 215, 252, 262
France, 2, 19, 137n19, 139n28, 170, 227, 239

G
Garner, Harry, 67
Geneva Protocol, 3n11, 9, 19, 26, 61, 61n34, 62n35, 68, 70–75, 77, 78, 82, 93, 104–106, 109–113, 120, 146, 151, 151n92, 158, 184, 212, 215–220, 216n12, 216n16, 217n18, 217n20, 217n21, 218n24, 218n25, 218n28, 219n34, 219n36, 220n43, 222, 223n54, 225, 225n65, 226, 227n76, 240, 240n133, 262
Global Strategy Paper (1952), 11, 46, 53, 55–93, 57n15, 75n91, 97, 98, 103, 251, 263
Grapple test series, 97
Grass-roots movements, 194
The Guardian, 1n1, 2n6, 11n24, 105n50, 145n56, 145n58, 153n100, 199n147, 200n149, 203, 222n52, 223n53, 226n72, 243n147, 246n164, 247n169

H
Hamilton, Michael, 189, 243
Harding, John, 81, 83, 85
Healey, Denis, 174, 175, 175n8, 181–183, 181n43, 182n45, 185, 186, 188–190, 192, 193, 199n141, 202, 205n183, 206n185, 208, 211, 212, 214, 216–219, 220n43, 227, 228, 232, 234, 235, 260
Heath, Edward, 207, 214
Helsinki accords, 214, 214n4
Henderson, Arthur, 151
Himsworth Committee, 216, 216n13
Hitler, Adolf, 20n45, 20n47, 31, 31n104, 76n96
Hogg, Ian, 210, 211
Hooley, Frank, 220
House of Commons, 20n48, 21n53, 22, 22n56, 23n57, 37n133, 45n168, 47n181, 59, 59n20, 99, 100, 100n32, 101n34, 101n35, 105, 105n50, 105n51, 106n52, 106n53, 106n55, 145n56, 145n59, 145n60, 146n63, 146n65, 151n89, 151n90, 151n91, 152n93, 152n94, 152n95, 152n96, 153n100, 155n104, 196n121, 204n178, 206n186, 206n187, 206n189, 207n191, 209n204, 209n207, 219n35, 220n39, 230n88, 240n137, 243n146, 243n148, 247n169, 248n170, 248n171
House of Lords, 143, 143n43, 224n64, 240n137
Hudleston, Edmund, 137, 138, 140
Hughes, Emrys, 145, 147, 207

I

Imperial Chemical Industries, 115, 131, 150, 150n86
Incapacitating agents, 3, 136, 158, 166n165, 217–219, 227, 232
Inch, T., 199
Indo-Pakistani War of 1971, 214
Intelligence assessments/reports, xiii, 9, 16n15, 21n49, 24, 24n66, 25, 35, 35n121, 36, 64, 64n45, 65, 67, 88n149, 105, 107, 108, 108n66, 113, 130–132, 131n169, 148, 156, 158–162, 176, 178, 185, 187, 188, 227, 260, 261

J

Japan, 16n15, 26, 26n71, 55n1, 238, 244, 246
JIC, 25, 35, 35n119, 64, 107, 130n167, 131, 131n169, 156, 159, 160, 176, 231
Johnson, Eric, 145, 243n147
Johnson, Lyndon B., 174, 194, 195
Joint Planning Staff, 34n116, 111–113, 138n27
Jones, Elwyn, 218

K

Khrushchev, Nikita, 96, 96n4, 133, 133n2, 134, 134n4, 146, 161
Korean War, 13, 15, 15n9, 48, 49, 49n193, 51, 55, 55n1, 56n6, 61, 66n57, 68, 74, 86, 95, 251, 262

L

Langer, Elinor, 195, 195n115, 224n61
Lipton, Marcus, 151
Lucky Dragon incident, 100
Lyttelton, Oliver, 76

M

Macmillan, Harold, 4n15, 31n103, 70, 70n74, 75n91, 96–98, 106, 117, 117n103, 117n105, 127, 134, 134n3, 134n6, 135, 146–149, 148n72, 155, 167n168, 168–171, 181, 183, 199n139, 215n9, 240n134, 252, 255, 255n2
Malaya, 61, 90
Mallalieu, Joseph Percival William, 189, 189n83, 189n85, 190, 263n4
Malone, Adrian, 201
Mao Zedong, 15
Mason, Roy, 189–193, 189n81, 189n82, 190n90, 191n91, 191n93, 191n95, 192n96, 192n97, 192n100, 198, 202n162, 208n201, 209n206, 210–212, 221, 228
Ministry of Supply, 52, 67, 85n135, 86, 86n136, 86n137, 86n141, 87n143, 93n185, 99, 110n72, 115, 116, 116n99, 116n100, 128n153, 136n13, 162n143
Monckton, Walter, 117–125, 117n103, 117n105, 123n130, 123n131, 123n132, 123n134, 124n136, 125n142, 128
Morris, John, 201–203, 206, 208, 209, 240
Mulley, Fred, 205, 205n183, 206n185, 223, 223n54, 223n56, 226, 227n76
Mustard gas, 17
Mutual Weapons Development Program, 129

N

Nancekuke, 46n175, 47–49, 48n187, 58, 67, 69, 70, 80, 82, 90, 93, 99, 129, 136, 136n14, 192, 199, 241, 243, 247
National Intelligence Estimate (United States), 108n64, 130n167, 131n169, 132n174, 160n136
NATO, 5, 8, 38, 56, 64, 73, 97, 102, 105, 109, 112, 123, 128n156, 166–170, 178, 180, 188, 188n78, 191, 191n92, 227n80, 237, 239, 245, 254
Navalny, Alexei, 1, 1n1, 1n2
Nerve agent acquisition, 3, 6, 8, 9, 17, 19, 23, 25–28, 30, 32, 33, 42, 43, 50, 52, 62, 63, 68, 80–82, 88, 105, 108, 112, 114, 115n91, 119, 151n92, 159, 164, 166, 168, 171, 175, 176, 179–181, 183–186, 188, 192, 205, 211, 215, 220, 228, 229, 232–234, 233n103, 238, 241, 246, 251–255, 258, 259, 261–264
Nerve agent deal (Britain-United States), 80, 86–88, 90, 90n166, 91, 93, 98, 99, 180, 262
Nerve agent myth, 236, 237
Nixon, Richard, 174, 210n210, 213, 214n5, 224, 225n65, 239, 240
Non-Aligned Countries, 244, 246
Non-Aligned Movement, 228, 238, 239, 254
Northern Ireland, 189, 216, 219
North Korea, 2, 2n6
Novichok agents, 1, 2
Nuclear weapons, 2, 7, 8, 16, 23, 28, 30, 33–35, 40, 41, 57, 58, 63, 72, 73, 81–84, 97, 98, 100–102, 106, 109, 114, 121, 122, 124, 134, 135, 143, 144, 147, 157, 158, 165, 173, 184, 189, 190, 191n92, 223, 224, 237, 238, 259, 260

O

O'Brien, T. H., 114
The Observer, 197, 204, 207
Operation Dismal, 30
Operation Fearless, 208
Owen, Wilfred, 197

P

Palmer, Arthur, 204, 204n172
Paris Agreements (1954), 105, 106
Parliamentary Select Committee on Science and Technology, 203, 204
Penkovsky, Oleg, 159–161, 159n130, 160n133, 161n139, 176
Peters, John, 201, 201n158, 202n166
Phosgene, 17, 27, 28, 43, 49, 53, 79
Pincher, Henry Chapman, 92n179, 148, 148n70, 148n72, 150–152, 150n84, 150n86, 150n87, 151n89, 154, 155, 157, 157n114, 252
Plant Protection Limited (PPL), 115
Porton Down, 17, 18, 34, 37n130, 46, 78, 78n102, 90–92, 92n182, 107, 115, 129, 137, 145, 146n65, 189, 189n81, 189n83, 189n85, 190n90, 191n91, 191n93, 191n95, 194n108, 195, 197–199, 198n135, 201–204, 201n158, 201n159, 202n161, 202n162, 202n163, 202n165, 202n166, 202n167, 203n169, 203n170, 205n180, 205n181, 207, 240, 241, 247n169, 263n4
Prior-Palmer, Otho, 101, 102
Private Eye, 207
Psychological effects, 61, 236

Q

The Quakers, 219
Quinlan, Michael, 185, 185n58, 187

R

R-33, 131, 131n173
Reading, Eric, 143, 143n44
Respirators, 51, 108, 108n66
Richard, Ivor, 209
Robinson, Robert, 201
Rose, Steven, 17n24, 200, 200n152, 209, 209n205
Rotblat, Joseph, 100, 146
Royal Air Force, 29, 63n43, 88, 177, 233n103
Royal Navy, 175–177, 189
Russell, Bertrand, 100, 203n171

S

SALT, 213
Sandys, Duncan, 97, 98, 127, 127n151, 135n11
Sanity (CND magazine), 199, 207, 207n194
Sarin, 2, 4n12, 17, 25–27, 27n76, 29, 29n93, 31, 32, 41, 45, 46n173, 47–50, 53, 59, 67, 69, 80, 85–93, 99, 115, 116, 120, 121, 129, 136, 150, 167, 170, 177, 178, 209n207, 251, 255
Second World War, 2, 13, 15, 16n18, 18, 21–23, 27, 27n81, 34, 39, 39n143, 42, 48, 60, 61, 63, 75, 89, 101, 102, 105, 112, 114, 118, 143, 146, 153, 162, 175n8, 189, 238, 251, 255, 259
Shinwell, Emanuel, 37n133, 47, 48n187, 49, 145
Shinwell, Manny, 47
Sigmund, Elizabeth, viii, 7, 11n24, 196–198, 196n122, 196n124, 197n125, 197n126, 197n128, 197n129, 198n133, 199n148, 200n154, 207, 221, 253
Signals intelligence, 35
Silverman, Sydney, 146, 151, 152
Sino-Soviet split, 173
Soames, Christopher, 152, 153, 155
Soman, 4n12, 17, 25, 26, 27n76, 29, 32, 45
Soviet CW capabilities, 24, 130, 131, 159, 261
Soviet T.34 (tank), 102
Soviet Union, 2, 9, 13, 14, 20n47, 23, 23n60, 25, 26, 34–36, 39, 40, 42, 49–51, 53, 55, 57, 60, 60n27, 62–65, 69, 72, 79, 88n149, 97, 103, 107, 111, 121, 130, 146n63, 148, 156, 159, 166, 176, 178, 181, 227, 231, 237, 239, 243, 244, 246, 247, 261
Special Branch, 145, 196
Speer, Albert, 31
Stalin, Joseph, 14, 56, 96, 96n4, 108n63
State Department (US), 180
Stewart, Michael, 219, 219n36, 220, 220n43
Stockholm International Peace Research Institute, 200
Suez crisis, 11, 96, 118, 125, 126, 127n151, 174, 183, 193
Sulfur mustard, 3, 27, 49, 53, 63, 63n43, 65, 79, 80, 117, 118, 136, 170
Syria, 2, 2n6

T

Tabun, 4n12, 16n17, 17, 18, 23, 25–32, 27n76, 30n94, 31n101, 34, 41, 44, 45, 49, 50, 53, 66, 79, 115, 117, 153

The Telegraph, 2n6, 11n24, 31n101, 31n104, 76n96, 83n128, 114, 114n88, 143, 143n44, 143n45, 157n114, 167n168, 189n82, 203, 210n209, 220n40, 243n145, 246n164, 248n171
Third World, 2n5, 9, 12, 165, 166, 172, 176, 179–181, 180n37, 184, 188, 190, 193, 214, 252, 253n1, 254, 261, 262
Thorneycroft, Peter, 165–168, 166n165, 190, 193
The Times, 11n24, 20n47, 31, 31n101, 31n103, 31n104, 32n105, 92n182, 152, 153n98, 155, 155n105, 189n84, 199n143, 203, 209, 209n205, 209n208
Tizard Report, 21
Todd, Alexander, 162–164, 162n143, 162n144, 164n158
Trend, Burke, 223
Tripartite cooperation, 5, 128, 129, 141, 149, 261
Tripartite Meetings, 129, 130
Truman, Harry (President), 14, 22, 56

U
UN Conference on Disarmament at Geneva, 230
UN Group of Consultant Experts on CBW, 223, 224
United Nations (UN), 5, 184, 205, 215, 217, 219, 220, 222–226, 223n54, 223n55, 223n56, 224n60, 238, 239, 245–247, 254
United States, viii, 2, 8, 10, 12–14, 15n14, 17, 23, 26, 29, 36–38, 36n124, 36n126, 36n127, 37n131, 45, 46n173, 49, 50, 53, 55–57, 57n13, 59, 59n24, 66n57, 69, 71n76, 72–75, 80–82, 84–91, 86n139, 88n147, 88n149, 93, 96–100, 98n17, 103, 103n44, 108, 109, 111, 115, 115n91, 121, 123–126, 128–131, 128n156, 130n165, 131n169, 133–143, 134n8, 136n17, 137n19, 139n28, 140n32, 142n42, 143n45, 146–152, 154–156, 155n105, 155n107, 165, 166, 167n168, 169–171, 170n182, 170n184, 174, 176, 176n18, 180, 187, 193n106, 194, 194n113, 195, 195n114, 195n115, 196n122, 199, 201, 208–210, 212, 213, 216, 217, 224, 225n65, 227, 228, 231, 234, 235, 238, 238n124, 239, 239n129, 242, 243, 245–247, 245n156, 245n158, 246n161, 246n162, 247n166, 247n167, 248n173, 253n1, 255, 257, 258, 261, 262

V
V-agents, 2, 115, 116, 120, 121, 128–131, 147–151, 147n67, 148n71, 149n79, 154, 156n108, 157
Verification (arms control), 226n71, 227, 231, 234, 237–239, 244–249, 248n172, 249n175
Vietnam, 9n20, 167, 174, 174n2, 194n111, 194n112, 194n113, 196, 209, 209n207, 212, 214, 216
Vietnam War, 194
VR-55, 176, 178
VX, 2, 2n6, 116, 116n97, 121, 129–131, 147, 149, 154, 170, 176–178, 180n37, 209, 225n65, 228

W
Wales, 16n17, 29, 30, 58
Warbey, William, 151

War Office, 10, 10n22, 39–42, 44, 45, 80, 85, 108n62, 109, 109n67, 109n69, 114, 114n87, 117, 128n156, 140n33, 140n36, 142n42, 152, 156, 157, 161, 163, 164, 175–177
Warsaw Pact, 176, 178, 227, 231, 248, 254, 261
Watkinson, Harold, 151
West Germany, 16, 38n137, 95, 106, 170, 177, 178, 188, 213, 227
Wilson, Harold, 174, 174n5, 181–183, 181n43, 182n45, 189, 192, 196, 199, 200, 200n150, 202, 203, 206, 207, 211, 212, 214–216, 217n20, 218, 218n25, 219n36, 220, 220n43, 222, 223, 223n54, 223n55, 223n56, 226, 226n73, 240, 252, 255, 263

Y

Yemen, 167, 167n168, 195n114, 196, 196n120, 196n121

Z

Zilliacus, Konni, 151
Zuckerman, Sir Solly, 12, 110n72, 135, 156–172, 157n111, 157n112, 157n116, 159n125, 162n143, 162n144, 175n9, 179–181, 188, 188n75, 201n155, 203, 204, 204n172, 217n20, 218, 218n25, 219, 223, 223n54, 223n55, 223n57, 223n59, 224, 224n60, 224n63, 240n134, 241, 252

The manufacturer's authorised representative in the EU is Springer Nature Customer Service Centre GmbH, Europaplatz 3, 69115 Heidelberg, Germany. If you have any concerns regarding our products, please contact ProductSafety@springernature.com

Printed and bound by CPI Group (UK) Ltd, Croydon, CR0 4YY

30/09/2026

02225451-0010